Certified Wireless Solutions Administrator (CWSA)

Study and Reference Guide

(CWSA-101)

Errata, when available, for this study guide, can be found at: www.cwnp.com/errata/

First printing: September 2019, version 1.0

ISBN: 9780997629057

In addition to the authors of this book, listed in the About the Authors section of the Introduction, CWNP would like to say a special thanks to all those involved in the development of materials for CWSA-101 from the Job Task Analysis (JTA) through to materials review and feedback. These individuals include Ryan Adzima, Robert Bartz, Ian Beyer, Tom Carpenter, Joshua Gochee, Bryan Harkins, Jason Hintersteiner, and Manon Lessard. If we have left out your name it is only because so many helped and not because you were not appreciated. Many thanks to all of you.

Table of Contents

Introduction

The Certified Wireless Solutions Administrator (CWSA) implements, administers and troubleshoots technologies that heavily rely upon, or directly integrate with, wireless systems in enterprise, government, and manufacturing environments. This individual is able to install, customize, and coordinate appropriate solutions to meet an organization's requirements and constraints.

The CWSA-101 exam consists of 60 multiple choice, single correct answer questions and is delivered through Pearson VUE. The candidate can register for the exam at the Pearson VUE website (https://home.pearsonvue.com/). The candidate will have 90 minutes to take the exam and must achieve a score of 70% or greater to earn the CWSA certification. If the candidate desires to become a Certified Wireless Network Trainer (CWNT) the passing score must be 80% or greater. A CWNT is authorized to teach official CWNP courses for certifications in which they hold the CWNT credential.

Book Features

The CWSA Certified Wireless Solutions Professional Study and Reference Guide includes the following features:

- Periodic Beyond the Exam excerpts to reinforce the learning.

- End of chapter review questions. The review questions at the end of each chapter, when appropriate, are intended to help you ensure proper reading of the chapter. They are not intended to simulate exam questions and should not be assumed to be questions that will be presented on the exam.

- Notes with special indicators. The notes throughout the book fall into one of three categories, as outlined in Table i.1.

- CWNP official glossary. A glossary of terms provided at the end of the book that helps you as a reference while reading.

- Complete coverage of the CWSA-101 objectives. Every objective is covered in the book, and each chapter lists the major objective category covered within.

Icon	Description
	Note: A general note related to the current topic.
	Defined Note: A note providing a concise definition of a term or concept.
	Exam Note: A note providing tips for exam preparation.

Table i.1: Book Note Icons

Authors

The following individuals wrote one or more chapters in this book:

- Ryan Adzima
- Tom Carpenter
- Joel Crane
- Darrell DeRosia
- Jason Hintersteiner
- Manon Lessard
- Ayman Mukaddam
- Fehmi Sakkal
- Jake Snyder

All of these authors are either CWNEs or experts in the field with more than ten years of experience working with various wireless technologies.

At CWNP, we would like to thank all of these authors for their hard work and dedication in making this book a reality. You have brought significant value to thousands.

The authors would like to dedicate this book to future CWSAs: you and your passion for wireless are what makes this all worthwhile.

CWSA-101 Objectives

The CWSA-101 exam tests your knowledge against five knowledge domains as documented in Table i.2. The CWSA candidate should understand these domains before taking the exam. The CWSA-101 objectives follow.

Knowledge Domain	Percentage
Wireless Technologies	15%
Radio Frequency Communications	15%
Planning Wireless Solutions	30%
Implementing Wireless Solutions	20%
Supporting Wireless Solutions	20%

Table i.2: CWSA-101 Exam Knowledge Domains with Percentage of Questions in Each Domain

1.0   Wireless Technologies (15%)

1.1   Maintain continued awareness of wireless technologies and applications of those technologies

    1.1.1   Understand research and lab testing skills to maintain technology awareness

    1.1.2   Understand the most common applications of wireless technologies, the frequencies used and communication protocols

1.2   Understand industry standard, certification and regulatory organizations and standards development processes

    1.2.1   Institute of Electrical and Electronics Engineers (IEEE)

    1.2.2   Internet Engineering Task Force (IETF)

    1.2.3   Wi-Fi Alliance

    1.2.4   International Telecommunication Union (ITU)

    1.2.5   Bluetooth Special Interest Group (SIG)

    1.2.6   3rd Generation Partnership Project (3GPP)

    1.2.7   Zigbee Alliance

    1.2.8   WLAN Association

    1.2.9   Explain the roles of regulatory agencies such as the FCC, IC, CE and others

2.2.10   Additional modulation methods (AM, FM, and CW)

2.3   Explain the basic capabilities of components used in RF communications

    2.3.1   Radios (receivers, transmitters, and transceivers)

    2.3.2   Antennas

    2.3.3   Intentional radiator and Equivalent Isotropically Radiated Power (EIRP) and Effective Radiated Power (ERP)

    2.3.4   RF cabling and connectors

    2.3.5   Link types including PTP, PTMP, mesh, ad-hoc and on-demand

2.4   Describe the basic use and capabilities of the RF bands and other wireless carriers (light and sound) used for communications

    2.4.1   Radio Frequency Bands

    2.4.2   Light-based communications (Li-Fi and others)

    2.4.3   Soundwave-based communications (underwater)

3.0   Planning Wireless Solutions (30%)

3.1   Identify and document the wireless system requirements

    3.1.1   Use cases and applications

    3.1.2   Capacity requirements

    3.1.3   Security and monitoring requirements

    3.1.4   Integration requirements (automation, data transfer/conversion, APIs, cross-platform integration)

    3.1.5   Stakeholder identification

3.2   Identify and document system constraints

    3.2.1   Budgetary constraints

    3.2.2   Security constraints

    3.2.3   Technical constraints

    3.2.4   Business policies and requirements

    3.2.5   Regulatory constraints

    3.2.6   System dependencies

    3.2.7   Evaluate existing network infrastructure and understand its limitations in the context of the new wireless system

3.3   Select appropriate wireless solutions based on requirements and constraints

5.0   Supporting Wireless Solutions (20%)

Each chapter of the CWSA Study and Reference Guide lists the major objectives covered in that chapter on the chapter title page.

Extended Table of Contents

Chapter 1: Introducing Wireless Technologies

Objectives Covered:

1.1   Maintain continued awareness of wireless technologies and applications of those technologies

1.2   Understand industry standard, certification and regulatory organizations and standards development processes

As an IT professional, you may already know that a career in IT is synonymous with continuous education. We understand that you have selected this book, in part, for this very reason. We all strive to better ourselves as network engineers because no two wireless environments or networks are the same. This reality is the number one reason why CWNP has felt the need to expand their wireless certification program to allow professionals whose responsibilities are more towards day to day operations and solution administration to further their knowledge of wireless systems and automatization. The Certified Wireless Solutions Administrator (CWSA) is the conductor, the person responsible for ensuring all these components coexist and work together. This is accomplished through understanding wireless fundamentals and knowledge of how each solution operates.

In this chapter, we will cover the fundamentals of wireless, an overview of wireless technologies, key terms and concepts you will need to understand as part of the CWSA certification. The intent is to provide you with a solid foundation so that you may be able to stay up to date on your knowledge of wireless technologies and understand the possible applications and implications of such technologies in the market. If you are responsible for a wireless system, you will be better equipped to support this solution as a result of this program. From testing wireless solutions in lab environments, staging or production having the foundational knowledge will help you understand and isolate problems if they occur, or just keep the solution working flawlessly. We will also cover the most common use cases, protocols, and frequencies used in wireless today.

The second objective of this chapter is to introduce you to industry standards, certification authorities and regulatory organizations so that you may be able to better target your research in maintaining your wireless knowledge. Also, we will review the standards development process used by most of these organizations. This will allow you to better understand the nomenclature used in the wireless industry.

Wireless technologies are a part of everyday life in most parts of the world. Whether we are talking of BLE, Cellular, Wi-Fi, Zigbee, or any other flavor of wireless networks first starts with radio signals. We expect these technologies to work at all times, wherever we go.

History of Wireless

Radio signals come from electromagnetic energy which is radiated in a specific direction. When Heinrich Hertz proved the existence of radio signals in the late 19th century, little did he know his discovery would have the importance we know them to have today.

Many scientists contributed to the initial invention of wireless networks: Tesla, Popov, Fessenden, and more commonly Marconi are recognized as pioneers who developed the idea of wireless transmissions. Using amplitude modulation, a technique which uses the strength of a radio wave to represent a symbol, such as a dash or a dot in Morse code, RF waves were tamed into a palatable language.

Through the years, the use of the radio evolved, from the wireless telegraph to full-blown radio casts. During the World Wars, armies started using radio technologies to help the war effort. However, the problem with a fixed frequency such as the one commonly used in amplitude modulation is that a specific, narrow band of RF frequencies are used. The enemy camp could overload the frequency with energy to disrupt the communications. This is what jamming is in RF terms. It prevents the possibility of being able to decode anything by saturating the RF environment.

Hedy Lamarr and George Antheil figured a way around this issue around the Second World War: frequency-hopping spread spectrum (FHSS). The unpredictable nature of FHSS would prevent any attempt at jamming transmissions. Spreading data over wider channels was the initial method used by many radio-guided systems and is still used by some wireless technologies today, especially Bluetooth. Since the initial modulation used by the first 802.11 standard was FHSS, Lamarr is often called the Mother of Wi-Fi.

In the 20 years since the 802.11 standard was developed, a lot has changed for Wi-Fi. We have seen maximum data rates go from 2 Mbps to over 3 Gbps on commonly available consumer devices. This has been achieved by improvements in the hardware to be more accurate and precise, new modulation rates (encoding schemes), more spectrum and better use of what is available. Many people solely rely on secure wireless communications for everything. Wi-Fi is not the only wireless network that has seen these types of improvements. Cellular carriers around the world, first commercialized data offerings on 2G Networks (CDMA or GSM). These evolved to the

LTE (4G), and the collection of standards 5G utilizes that are operated today. This evolution brought data speeds from 9.6 Kbps to over 1 Gbps today.

Understanding Radio Waves and Frequencies

Chapter 4 will get into more detail on radio waves themselves. What a wave is, how to know the frequency of the wave, etc. Let's touch on the core concepts at a high level though. The rate or number of cycles which the wave goes through per second is called the frequency. The measuring unit of radio frequencies is the Hertz, and one Hertz is one wave cycle. The term wavelength refers to the distance between two identical points on a wave. Stated differently, it is measured from one point on the wave to the recurrence of that point (see Figure 1.1). The power of a radio wave is called its amplitude. Phase, unlike frequency, wavelength, and amplitude, phase is not a characteristic of a single RF wave but is instead a comparison between two RF waves. All of these four characteristics of RF are important to know because they are the key to modulation and encoding and they will be covered in greater detail in Chapter 4. Modulation is the process of taking a radio wave and adding data to it. It can use the amplitude, frequency, phase or wavelength of a radio wave. Coding is the lexicon used to represent the data. To use a non-RF example, modulation the voice of a speaker giving a presentation. Coding is the language in use.

A radio band is a term used to define a continuous section of the RF spectrum. Some bands are licensed for specific uses, such as HAM radio and weather radars. Other bands are licensed to specific operators, and the carrier allocates a technology, such as LTE, 4G, 5G, for use in the spectrum. Others are unlicensed and available for everyone to use. The two well-known unlicensed bands are the 2.4 GHz Industrial Scientific and Medical (ISM) band and the 5 GHz Unlicensed National Information Infrastructure (UNII) band. Different rules apply to each frequency based on regulatory domains; however, the 2.4 GHz and 5 GHz ranges tend to be available in most parts of the world (or, at least, portions of them) and thus have been adopted by many wireless technologies. The consistency of available spectrum has made it a natural place for wireless hardware manufacturers to utilize while reducing costs. Wireless networks that operate within these frequencies include Bluetooth, LTE-U, LAA, Zigbee, LoRa, and Wi-Fi. Because this spectrum is unlicensed many other devices, such as microwave ovens, cordless phones and baby monitors also compete for airtime.

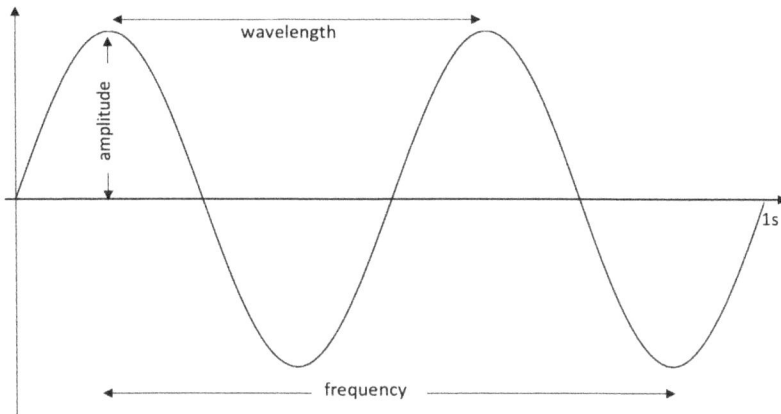

Figure 1.1: Amplitude, frequency and wavelength

Mobile devices such as cell phones use bands which are licensed for use by the providers. It is one of the reasons why they tend to be less prone to interference. Unlike BLE or Wi-Fi, mobile (cellular) networks have a wider coverage area. It is one reason why some IoT vendors have chosen to initially offer their IoT products over mobile networks and not local wireless networks.

Wireless Technologies and Related Components (An Overview)

Cellular and Wi-Fi for Internet access are the most common wireless technologies. Wi-Fi is a pervasive technology in most homes, retail, business, etc. The technology and protocols for Wi-Fi are very fault-tolerant, and many people think it is just magic, after all "How hard is it to not run wires?". Wi-Fi has even become the generic term for using the Internet to many. As such, understanding the components of the eco-system that provides Internet access is critical for a successful deployment and ongoing operations of the network. Wi-Fi is not the only wireless system though and understanding how these technologies fit into the Open Systems Interconnection (OSI) model is necessary for system administrators. As a solutions administrator you must be skilled at identifying issues with the system and where these issues fit in the OSI model to resolve them. If you have taken the CWNA, or another networking class, this section may be very familiar to you.

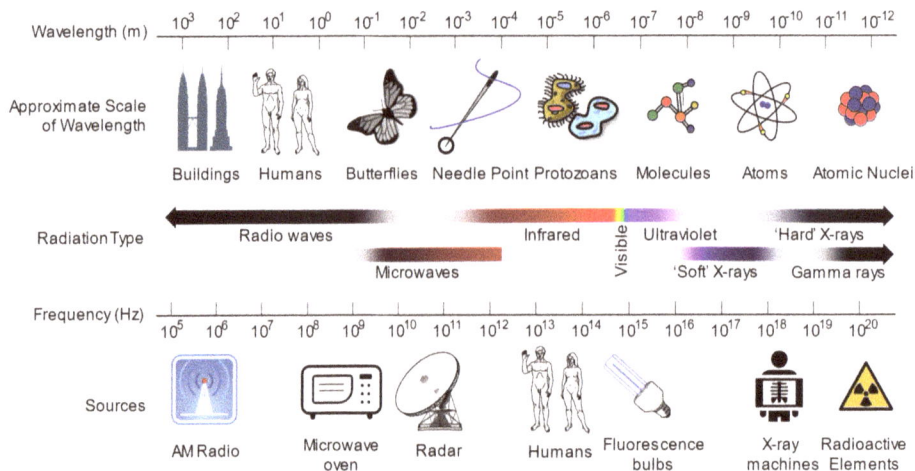

Figure 1.2: Electromagnetic Spectrum (Source: Industry Canada)

Physical Layer

The Physical Layer, sometimes called the PHY, is responsible for providing the mechanical, electrical, functional, or procedural means for establishing physical connections between data-link entities. The connections between all other layers are logical, as the only real physical connection that results in true transfer of data is at Layer 1 — the Physical layer. For example, we say that the Layer 7 HTTP protocol on a client creates a connection with the Layer 7 HTTP protocol on a web server when a user browses an Internet website; in reality, this connection is logical, and the real connections happen at the Physical layer within a segment of the network, which is connected to another segment, which is connected to another segment, and so on, until the destination is reached.

It is really amazing to think that my computer — the one I'm using to type these words — is connected to a Wireless Access Point (AP) in my office, which is connected to my local network, that is in turn connected to the Internet. Through connections — possibly both wired and wireless — I can send signals (that's what happens at Layer 1) to a device on the other side of the globe. To think that there is a potential electrical-connection path between these devices and millions of others is really quite amazing.

It is Layer 1 that is responsible for taking the data frames from Layer 2 and transmitting them on the communications medium as binary bits (ones and zeros). This medium may be wired or wireless. It may use electrical signals or light pulses (both actually being electromagnetic in nature). Whatever you've chosen to use at Layer 1, the upper layers can communicate across it as long as the hardware and drivers abstract that layer so that it provides the services demanded of the upper layer protocols.

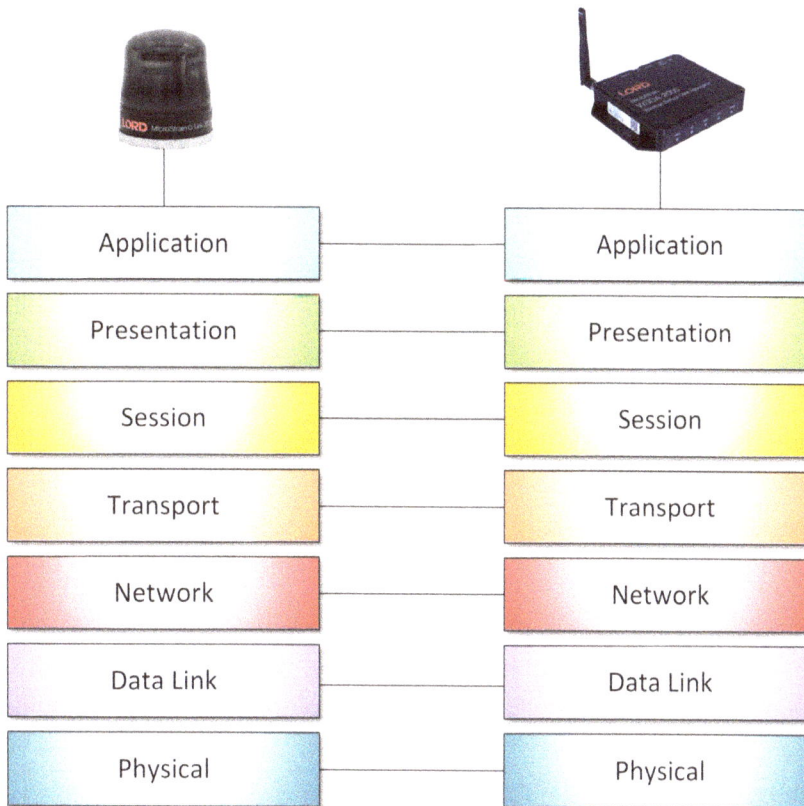

Figure 1.3: ISO OSI Model for Network Communications

Examples of Physical layer protocols and functions include Ethernet, Wi-Fi, and DSL. You probably noticed that Ethernet was mentioned as an example of a Data Link layer protocol. This is because Ethernet defines both the MAC sub-layer functionality within Layer 2 and the PHY for Layer 1. Wi-Fi technologies (802.11) are similar in that both the MAC and PHY are specified in the standard. Therefore, the Data Link and Physical layers are often defined in standards together. You could say that Layer 2 acts as an intermediary between Layers 3 through 7 so that you can run IPX/SPX (though hardly anyone uses this protocol today) or TCP/IP across a multitude of network types (network types being understood as different MAC and PHY specifications).

The Physical layer for wireless technologies is air and the required subsystems to enable these transmissions. While wireless devices operate on several layers of the OSI, having the device, powering/connecting this device, and the radio transmitter all fall into the physical layer. Chapters 4 and 5 will go into more depth on how the transmissions happen, encoding and decoding of signals. For our purposes in this chapter, we need to understand that wireless radios are half-duplex and need clear airwaves to transmit. If two wireless systems are in the same frequency band, say a baby monitor and Wi-Fi Access point, you will need to ensure they are using separate channels. If they are not on separate channels you can expect some issues from time to time when they are both contending for airtime. Key components to remember of the Physical layer are:

- Operating Frequency of the wireless device
- PHY rates
- Physical (Uplink) Connection
- Power

Data Link Layer

The Data Link Layer is defined as providing communications between connectionless-mode or connection-mode network entities. This may include the establishment, maintenance, and release of connections for connection-mode network entities. The Data Link layer is also responsible for detecting errors that may occur in the Physical layer. Therefore, the Data Link layer provides services to Layer 3 and Layer 1. The Data Link layer, or Layer 2, may also correct errors that are detected in the Physical layer automatically.

The IEEE has divided the Data Link layer into two sublayers, the Logical Link Control (LLC) sublayer, and the Medium Access Control (MAC) sublayer. The LLC sublayer is not actually used by many transport protocols, such as TCP. The varied IEEE standards identify the behavior of the MAC sublayer within the Data Link layer and the PHY layer as well.

The results of the processing in Layer 2 are that the packet becomes a frame that is ready to be transmitted by the Physical layer or Layer 1. So, the segments became packets in Layer 3, and now the packets have become frames. Remember, this is just the collection of terms that we use; the data is a collection of ones and zeros all the way down through the OSI layers. Each layer is simply manipulating or adding to these ones and zeros to perform that layer's service. Like the other layers before it, the services and processes within the Data Link layer are named after the layer and are called data-link entities. Key components to remember of the Data Link layer for wireless technologies are:

- Medium Access Control (MAC)
- Logical Link Control (LLC)

Network Layer

The Network Layer is defined as providing the functional and procedural means for connectionless-mode (UDP) or connection-mode (TCP) transmission among transport entities and, therefore, provides to the transport entities independence of routing and relay considerations. In other words, the Network layer says to the Transport layer, "You just give me the segments you want to be transferred and tell me where you want them to go. I'll take care of the rest." This is why routers do not usually have to expand data beyond Layer 3 to route the data correctly. For example, an IP router does not care if it's routing an email message or voice conversation. It only needs to know the IP address for which the packet is destined and any relevant QoS parameters to move the packet along.

Examples of Network layer protocols and functions include IP, ICMP, and IPSec. The Internet Protocol (IP) is used for addressing and routing of data packets in order to allow them to reach their destination. That destination can be on the local network or a remote network. The local machine is never concerned with this, with the exception of the required knowledge of an exit point, or default gateway, from the local machine's network. The Internet Control Message Protocol (ICMP) is used for testing the TCP/IP communications and for error message handling within Layer 3. Finally, IP Security (IPSec) is a solution for securing IP communications using authentication and/or encryption for each IP packet. While security protocols such as SSL, TLS, and SSH operate at Layers 4 through 7 of the OSI model, IPSec sits solidly at Layer 3. The benefit is that, since IPSec sits below Layer 4, any protocols running at or above Layer 4 can take advantage of this secure foundation. For this reason, IPSec has become more and more popular since it was first defined in 1995.

The services and processing operating in the Network layer are known as network entities. These network entities depend on the services provided by the Data Link layer. At the Network layer, Transport layer segments become packets. These packets will be processed by the Data Link layer.

For the purposes of this certification, you will need to understand which protocols are commonly used and how to troubleshoot these. The most common Network Layer protocols to run over wireless technologies are:

- IP
- ICMP
- IPSec

Transport Layer

Layer 4, the Transport Layer is defined as providing transparent transfer of data between session entities and relieving them from any concern with the detailed way in which reliable and cost-effective transfer of data is achieved. This simply means that the Transport layer, as its name implies, is the layer where the data is segmented for effective transport in compliance with Quality of Service (QoS) requirements and shared medium access.

Examples of Transport layer protocols and functions include TCP and UDP. The Transmission Control Protocol (TCP) is the primary protocol used for the transmission of connection-oriented data in the TCP/IP suite. HTTP, SMTP, FTP, and other important Layer 7 protocols depend on TCP for reliable delivery and receipt of data. The User Datagram Protocol (UDP) is used for connectionless data communications. For example, when speed of communications is more important than reliability, UDP is frequently used. Because voice packets either have to arrive or not arrive (as opposed to arriving late), UDP is frequently used for the transfer of voice and video data.

TCP and UDP are examples of transport entities at Layer 4. These transport entities will be served by the Network layer. At the Transport layer, the data is broken into segments if necessary. If the data will fit in one segment, then the data becomes a single segment. Otherwise, the data is segmented into multiple segments for transmission.

For the purposes of this certification, you will need to understand which protocols are commonly used and how to troubleshoot these. The most common Transport Layer to run over wireless technologies are:

- TCP
- UDP
- IPSec

Session Layer

The Session layer is defined in sub-clause 7.3 of the OSI Reference Model as providing the means necessary for cooperating presentation entities to organize and to synchronize their dialog and to manage their data exchange. This is accomplished by establishing a connection between two communicating presentation entities. The result is simple mechanisms for orderly data exchange and session termination.

A session includes the agreement to communicate and the rules by which the communications will transpire. Sessions are created, communications occur, and sessions are destroyed or ended. Layer 5 is responsible for establishing the session, managing the dialogs between the endpoints, and the proper closing of the session.

Examples of Session layer protocols and functions include the iSCSI protocol, RPC and NFS. iSCSI is a protocol that provides access to SCSI devices on remote computers or servers. The protocol allows SCSI commands to be sent to the remote device. The Remote Procedure Call (RPC) protocol allows subroutines to be executed on remote computers. A programmer can develop an application that calls the subroutine in the same way as a local subroutine. RPC abstracts the network layer and allows the application running above Layer 7 to execute the subroutine without knowledge of the fact that it is running on a remote computer. The Network File System (NFS) protocol is used to provide access to files on remote computers as if they were on the local computer. NFS actually functions using an implementation of RPC known as Open Network Computing RPC (ONC RPC) that was developed by Sun Microsystems for use with NFS; however, ONC RPC has also been used by other systems since that time. Remember that these protocols are provided only as examples of the protocols available at Layer 5 (as were the other protocols mentioned for Layers 6 and 7). By learning the functionality of protocols that operate at each layer, you can better understand the intention of each layer.

The services and processes running in Layer 5 are known as session entities. Therefore, RPC and NFS would be session entities. These session entities will be served by the Transport layer.

For the purposes of this certification, you will need to understand which protocols are commonly used and how to troubleshoot these. The most common Session Layer to run over wireless technologies are:

- RPC
- NFS
- IPSec

Presentation Layer

The Presentation layer is defined in sub-clause 7.2 of the OSI Reference Model as the sixth layer of the OSI model and it provides services to the Application layer above it and the Session layer below it. The Presentation layer, or Layer 6, provides for the representation of the information communicated by or referenced by application entities. The Presentation layer is not used in all network communications and it, as well as the Application layer and Session layer, is similar to the single Application layer of the TCP/IP model. The Presentation layer provides for syntax management and conversion, as well as encryption services. Syntax management refers to the process of ensuring that the sending and receiving hosts communicate with a shared syntax or language. When you realize this, you will realize why encryption is often handled at this layer. After all, encryption is really a modification of the data in such a way that must be reversed on the receiving end. Therefore, both the sender and receiver must understand the encryption algorithm in order to provide the proper data to the program that is sending or receiving on the network.

Don't be alarmed to discover that the TCP/IP model has its own Application layer that differs from the OSI model's Application layer. The TCP/IP protocol existed before the OSI model was released. For this reason, we relate the TCP/IP protocol suite to the OSI model, but we cannot say that it complies with the model directly. It's also useful to keep in mind the reality that the TCP/IP protocol is an implemented model and the OSI model is only a "reference" model.

Examples of Presentation layer protocols and functions include any number of data representation and encryption protocols. For example, if you choose to use HTTPS instead of HTTP, you are indicating that you want to use Secure Sockets Layer (SSL) encryption. SSL encryption is related to the Presentation layer or Layer 6 of the OSI model. SSL, the Netscape solution, and TLS, the IETF solution, both operate at Layer 6 of the OSI model.

Ultimately the Layer 6 is responsible, at least in part, for three major processes: data representation, data security, and data compression. Data representation is the process of ensuring that data is presented to Layer 7 in a useful way and that it is passed to Layer 5 in a way that can be processed by the lower layers. Data security usually includes authentication, authorization, and encryption. Authentication is used to verify the identity of the sender and receiver. With solid authentication, we gain a benefit known as non-repudiation. Non-repudiation simply means that the sender cannot deny the sending of data. This is often used for auditing and incident-handling purposes. Authorization ensures that only valid users can access the data and encryption ensures the privacy and integrity of the data as it is being transferred.

The processes running at Layer 6 are known as presentation entities in the OSI model documentation. Therefore, an application entity is said to depend on the services of a presentation entity and the presentation entity is said to serve the application entity.

Application Layer

The seven layers of the OSI model that we have and are discussing are defined in clause 7 of the document ISO/IEC 7498-1. The Application layer is defined in sub-clause 7.1 as the highest layer in the reference model and as the sole means of access to the OSIE (Open System Interconnection Environment). The Application Layer is the layer that provides access to the other OSI layers for applications, and to applications for the other OSI layers. Do not confuse the Application layer with the general word "application," which is used to reference programs like Microsoft Excel, Corel WordPerfect and so on. The Application layer is the OSI layer that these applications communicate with when they need to send or receive data across the network. You could say that the Application layer exposes the higher-level protocols that an application needs to talk to. For example, Microsoft Outlook may need to talk to the SMTP protocol in order to transfer email messages.

Examples of Application layer protocols and functions include HTTP, FTP, and SMTP. The Hypertext Transfer Protocol (HTTP) is used to transfer HTML, ASP, PHP and other types of documents from one machine to another. It is the most heavily used Application layer protocol on the Internet and, possibly, in the world. The File Transfer Protocol (FTP) is used to transfer binary and ASCII files between a server and a client. Both the HTTP and FTP protocols can transfer any file type. The Simple Mail Transport Protocol (SMTP) is used to move email messages from one server to another and usually works in conjunction with other protocols for mail storage.

Application layer processes fall into two general categories: user applications and system applications. Email (SMTP), file transfer (FTP), and web browsing (HTTP) functions fall into the user application category as they provide direct results to applications used by users such as Outlook (email), WS_FTP (file transfer), and Firefox (web browsing). Notice that the applications or programs used by the user actually take advantage of the application services in the Application layer or Layer 7. For example, Outlook takes advantage of SMTP. Outlook does not reside in Layer 7, but SMTP does. As examples of system applications, consider DHCP and DNS. The Dynamic Host Configuration Protocol (DHCP) provides for dynamic TCP/IP configuration and the Domain Name Service (DNS) protocol provides for name-to-IP-address resolution. Both of these are considered system-level applications because they are not usually directly accessed by the user (though this is open for debate since administrators are users too and they use command-line tools or programs to directly access these services quite frequently).

The processes operating in the Application layer are known as application-entities. An application entity is defined in the standard as an active element embodying a set of capabilities which is pertinent to OSI and which is defined for the Application layer. Application entities are the services that run in Layer 7 and communicate with lower layers while exposing entry points to the OSI model for applications running on the local computing device. SMTP is an application entity as is HTTP and other Layer 7 protocols.

For the purposes of this certification, you will need to understand which protocols are commonly used and how to troubleshoot these. The most common Application Layer to run over wireless technologies are:

- DNS

- DHCP
- HTTP/HTTPS

Wireless Technologies and the OSI Model

Wireless networks (802.11) are related to the OSI Model in three ways:

- They carry upper layer (Layers 3-7) data across the RF medium.
- They define Layer 2.
- They define Layer 1.

Common Components of Wireless Solutions

Now that we have looked at the OSI Model, let's take a look at several of the components involved in a wireless solution in a little more detail and where they reside within this model. Our areas of focus will be as follows:

Physical connectivity (Layer 1)

- Structured Cabling
- Powering the device
- Wireless PHY

LAN Networking Requirements

- DHCP, DNS, NTP (Layer 7)
- Network Security (Layer 3)

Hardware in use (Layer 1)

- Access Points
- Antennas
- Clients
- Controllers

Implementing Wireless Solutions

- Lab + Staging
- Documentation
- Security Updates

Structured Cabling

All wireless solutions need some wires at one point or another. Historically, cabling has evolved more slowly than any telecommunications technology but in the last few years, with the need for higher transmission speeds and the evolution of power over Ethernet, it is important that you learn the various capacities of today's cabling. Please refer to Table 1.1 to see the capabilities of each cable category.

Ethernet Standard	Maximum Length	Cabling Category
10BASE-T (10 Mbps)	100 meters	CAT-3
100BASE-TX (100 Mbps)	100 meters	CAT-5 or higher
1000BASE-T (1 Gbps)	100 meters	CAT-5 or higher
2.5GBASE-T (2.5 Gbps)	100 meters	CAT-8
5GBASE-T (5 Gbps)	100 meters	CAT-6
10GBASE-T (10 Gbps)	100 meters	CAT-6A
25GBASE-T (25 Gbps)	30 meters	CAT-8
40GBASE-T (40 Gbps)	30 meters	CAT-8

Table 1.1: Ethernet Standards and Cabling Categories

Power over Ethernet

Many wireless devices are installed in ceilings or under raised floors. These sites are referred to in building codes as plenum. It is usually forbidden to power devices installed in plenums through a standard electrical outlet. This is why power over Ethernet was invented. Although initially PoE was based on vendor-specific implementations, the Institute for Electrical and Electronics Engineers (IEEE) has ratified the following standards:

	802.3af-2003	802.3at-2009	802.3bt-2018 "PoE ++"	802.3bt-2018 "PoE ++"	Cisco UpoE
Type	Type 1	Type 2	Type 3	Type 4	Proprietary
Power per port: source	15.4 W	30 W	60 W	100 W	60
Power per device (Max)	12.95 W	25.5 W	51W	71W	51
Supported Cabling (Min)	Cat 3	Cat5	Cat 5	Cat5	Cat 5

Physical Layer Wireless PHYs

Each wireless standard has its own PHY rates. Increasingly, many of these are starting to look similar in nature. Within 802.11ax many concepts were carried over from cellular technologies (OFDMA, Resource Units, etc.) to make them more efficient. We go into more detail about PHY rates and support later in this book; however, some key things to know are that data rates are:

- Channel width in-use / available
- Signal Strength vs. Noise on the frequency
- Spatial Streams Available

- Supported Technologies

For Wi-Fi, the channel width and modulation used have a significant impact on the actual data rates available. This fact holds true for all wireless communication technologies. Each PHY supports specific data rates based on the combination of channel width, modulation, coding, and a few other features. Modulation and coding are addressed in detail later in this book. The data rates are not in some way arbitrarily variable (for example, going from 11 Mbps to 10.9 Mbps to 10.8 Mbps and so forth), but they are specific data rates supported based on combinations of hard and true factors (for example, going from 11 Mbps to 5.5 Mbps to 2 Mbps and so forth).

All wireless communication systems have varied, but fixed, data rates. The data rates are based on the channel bandwidth, the modulation used, and other elements that result in one or more fixed data rates available in the system. It is important to note that some systems, by design, have only a few limited data rates. For example, one vendor's wireless bridging solution will rate switch from 150 Mbps to 11 Mbps and those are the only rates supported.

It's not like we can slow down the RF waves to lower the data rate or speed them up. Yes, we can change the frequency, but we cannot change the rate at which they travel. So, when we lower or raise data rates, we are changing how we put information onto the waves and not changing the speed of the wave propagation.

Hardware Used

As mentioned above, multiple pieces of electronic components and antennas are required for wireless communications. These typically fall into one of 4 categories wireless antennas, wireless base stations, wireless clients, and wireless control systems.

Wireless Antennas

We will cover antennas in much more depth in Chapters 4 and 5. Antennas are the physical device that takes the signal from the radio and transmits it over the air. APs will use either internal or external antennas. The antenna has no manageable configuration on its own: it is a device which is made to meet the criteria required to be able to emit or receive energy in a specific band. There are many characteristics to antennas, but for the purpose of this chapter, you should remember that an antenna is a passive element: it can receive and transmit RF energy, may shape the way the RF energy is emitted due to its physical properties, but it cannot be configured beyond that.

Wireless Base Stations

The wireless base station is transmitting infrastructure that connects wireless clients to the larger network, be it via a repeater, mesh, or a wired connection. In Wi-Fi these are referred to as Access Points. In cellular you may hear terms like nano cell, picocell, etc. These and many others are referring to different types of wireless base stations.

It is a good idea to understand the types of wireless infrastructure link technologies you will hear about as well. Wireless links can be point-to-point, mesh, or point-to-multipoint networks. In many cases single cells are a single wireless base station to the network that service wireless clients.

A point-to-point or point-to-multipoint network creates a network of base stations to either extend a link wirelessly (ex: between two buildings) or share some form of resource. Whereas, a mesh network is the expanded version of a point-to-multipoint network: it extends wireless services to clients through a point-to-multipoint network which connects them to the larger network. Typically, this type of network consists of a root base station and a non-root or mesh wireless base station. The clients connect to the mesh base station and the base station device backhauls traffic wirelessly to the root. This type of network is becoming popular in homes where the coverage of a single AP is not enough, but it is also heavily used in the enterprise, industry, and other installations.

Wireless mesh topologies: this term refers to the layout of a network. Common layouts or architectures used in wireless technologies are full mesh networks and star networks.

Full and partial mesh: common in ZigBee and 802.11 deployments, a mesh topology uses the notion of a web of devices which are all linked together. In the last few years the increase in use for BLE IoT devices has also brought forward the need for a BLE mesh topology. In addition, in standards beyond Bluetooth 4.0, mesh topologies will also be supported on classic Bluetooth implementations.

Star: common in either wireless networks called Independent basic service set (IBSS) and BLE, the star topology's center often called a "smart hub" will allow devices to be connected and managed through a single point.

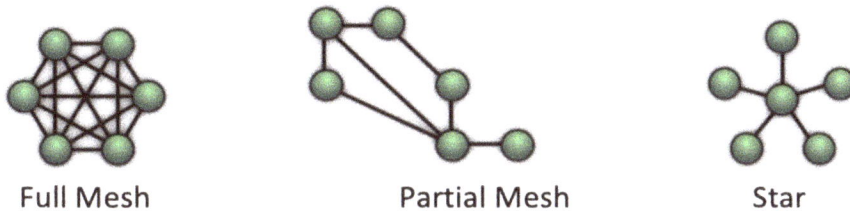

Figure 1.4: Topologies: Mesh and Star

Wireless technologies could be classified by their coverage area, from largest to smallest: A wireless wide area network is very large. Think of it as the coverage offered by a mobile phone provider. It can extend from a city to a region or state. Examples of such technologies are LTE, GSM, and WiMAX. The mobile phone industry is the primary operator of such networks which use licensed frequencies. Spectrum allocation is managed by local RF regulatory agencies (ex: the FCC, ETSI or CRTC). The capabilities of these networks will depend on the area and the technologies in use.

Wireless Clients

Wireless clients are devices with radios that are connecting to the wireless base stations. Some use Wi-Fi connections based on the 802.11 standard, and others are based on other technologies such as Bluetooth, BLE, Zigbee, LoRa, or Wi-Fi direct.

Wireless clients will be covered throughout this book, and they include any device that is not serving other devices. They can be laptops, desktops, robots, wearables or IoT. Regardless of the protocol which they are using to communicate, they share a few characteristics. Most are using drivers as well as algorithms to dictate their behavior which is proprietary. Due to the cost of constant recalibration, most of these devices are not using calibrated wireless adapters. Though you may think that client devices are dependent on the wireless infrastructure, in reality it is the opposite. Should you be interested in learning more about wireless adapters in use for Wi-Fi purposes, you may take a look at the compilation made by Mike Albano at clients.mikealbano.com

In the health vertical, BLE allows for wireless patient monitoring. This enhances patient mobility. It can also track assets, which are a huge expense in hospitals. Networks of BLE beacons, small, battery-operated transmitters, can provide useful information on the traffic patterns in warehouses and stores. BLE is also behind keyless entry and engine ignition in cars. It can power lights on or off, a classic IoT scenario which can both benefit the environment and save money to businesses. It can lock the door to your house. The possibilities for IoT devices are endless.

Should your network support or have to support any such technologies, you should be aware of the additional dependence this brings on the wireless network. Once the life of a patient depends on getting a medical device located on time, it becomes very important to maintain your network up to date, secure and in the best shape possible. This is what brings us to the topic of technology watches.

Wireless Control Systems

In enterprise Wi-Fi networks, wireless controllers are common. These devices control and manage the wireless base stations, keeping configurations and code up-to-date; they also control inter-base station communications. Not all wireless networks use separate control systems as this can run on the wireless base station in many cases. The two most common cases for wireless networks are a centralized or distributed control model.

A centralized wireless control system is a network where a piece of equipment called a controller (or group of controllers) is managing the configuration of the entire network. The clients' connections are established as tunnels to the controller through the AP. The APs get a configuration from the WLC and is then called a thin AP.

A distributed wireless control system features autonomous wireless base stations. In Wi-Fi terms, this is where you will find your "thick" AP, with full configuration and decision power, authentication, switching, etc. In larger distributed environments, a wireless network management system (WNMS) is used to administer the network as a whole from a single pane of glass.

Hybrids of the centralized topology exist with many vendors to accommodate for deployments with a lot of remote sites which have Internet access yet could use some form of data tunneling and disaster recovery in case the connection to the central WLC fails.

LAN Networking Requirements

Network services coming from the LAN essential for happy (functioning) wireless clients. The typical network connection involves a successful connection to the wireless network, and then the client will need the correct network credentials to do anything on the network. For devices to be able to access the network, it is very common for deployments to use Dynamic Host Configuration Protocol (DHCP) servers to automatically grant IP addresses to clients. You can use this service to provision domain name servers (DNS) which will be used to resolve names, such as the ones you may type in a browser URL, to IP addresses. Last but not least, you may require some form of Network Time Protocol server (NTP). These services are common across almost all IP.

Network Security

Most wireless systems make use of some form of authentication. Without going into too many details, there are three ways in which wireless users can be authenticated. Open authentication, which allows anyone to connect. The second is a pre-shared key that both systems have configured. The third is EAP based authentication. Most enterprise-grade Wi-Fi networks use EAP authentication using the SIM, certificates and/or usernames and passwords. Enterprise-grade Wi-Fi authentication often relies on the RADIUS standard, a non-proprietary standard which is available in solutions both sold through manufacturers or as Open Source code. These systems are often referred to as AAA servers. AAA stands for authentication, authorization, and accounting.

In addition to authentication, some wireless networks may make use of ACLs, DMZ, and VRFs. An ACL, or access control list, is a security measure to define what types of traffic is allowed or denied. The network traffic is matched, sequentially, to a list of protocols, IP addresses, mac addresses, ports, and other such criteria. When a match is found, the sequential process stops, and the traffic is either allowed or denied. A DMZ, or demilitarized zone, is a part of the network which is not protected. A VRF is a virtual instance of a network which is often used to isolate traffic which the network administrator wants to segregate, such as guest traffic. It is always preferable to use VRFs in a DMZ for those users which you wish to grant access to the Internet yet not allowed to use your internal services.

Implementing Wireless Solutions (Lab & Pilot Implementation)

Building a lab and testing the network is the best way to avoid problems in production environments. Depending on your use cases and business use of the network, your lab requirements may vary. If your business can afford to have devices which are identical to those which are found on your infrastructure, installing them in the lab and running the same software version and a basic configuration fitting for your installations is a time-saver. Not only will it allow you to quickly roll out spares which only need a quick scrub-configuration-reload in case of an outage, but it will also allow you to gain more experience with your infrastructure without impacting clients, or replicate problems in a non-production environment.

Should it be impossible for you to set up a lab with real gear, several virtual options exist. The most obvious virtual machines, VMWare, could allow you to spin an instance of just about any flavor of Windows, Linux or prepackaged VM platforms offered by manufacturers such as Cisco and Aruba. Systems such as RADIUS servers, DNS, DHCP, and others could be tested in this way. As for networking gear, GNS3 and EveNG are open source options one can look into, which would allow creating a lab environment simulating various devices from a plethora of manufacturers.

You should be aware that there are differences between network simulators and emulators. A simulator, such as Eve-NG, allows you to create a virtual topology comprised of several devices which you will have to configure. These devices will behave like real devices, but you will not be able to test traffic patterns, load, or any other such traffic conditions. If you would like to create a proof of concept of new infrastructure or modification of your network deployment, the first place to start would be either a real lab scenario or a network simulator scenario.

A network emulator will allow you to configure a lab for specific conditions related to your infrastructure. The emulation of specific types of traffic or network conditions such as a broadcast storm or DDOS becomes possible with an emulator, or through other traffic generating scenarios as detailed in Common Vulnerabilities and Exposures (CVE) alerts and their related papers. Spirent, Ixia, and NetSim by Tetco are examples of network emulators you may want to have in your lab.

Any lab test should begin by creating a plan, which identifies:

- The type of test

- The goal of the test
- The problem statement (if applicable)
- The expected result
- The topology required
- The steps required to test the scenario
- The expected results
- Any additional details such as expected behavior of applications, temporary service cuts, and impacts.

Using this plan, you can start setting up your lab environment and getting ready to execute the steps you have defined. Bear in mind that depending on the size of the enterprise where you are conducting those tests, you may not be the person who will do the change in the production environment and that person may not be as experienced as you are. If there are only a few maintenance windows available and your change falls short because your Padawan got scared by a command that "didn't work as expected," he or she could very well push the change until the information has been clarified with you. Concise and clear documentation will save you from such an experience.

Once your tests have been completed, gather the various results, and document them. Compare your results with what you were initially expecting. Should you face a situation where you were unable to recreate an issue or exploit, review the bug reports to ensure you've recreated the issue correctly. In lab, just like in life, it is often a question of balance: it is not necessary to spend weeks testing an unlikely scenario or documenting an issue for hours if it is not useful for you and your business or client.

A lab provides an environment in which you can test wireless equipment without concern for how it will impact your production network. You can develop practical skills and test various configurations to discover what will work best in your environment.

Staging

Depending on the size of the organization and scale of the network, many large-scale environments utilize a staging environment between the lab and production deployments. The purpose of a staging environment is to move a small segment of production traffic to this environment to monitoring closely for problems prior to the new configuration going into production. In a large enterprise, service providers, and carrier networks, these are required. For smaller entities with on-site support staff sometimes things are just rolled into production.

Nowadays, around a large percentage of telecommunications outages are attributable to human error. Many strategies are now used to avoid such risk. Among them, maintenance windows, testing procedures, and proper implementation practices are most typical. This is why you should always carefully plan any change, patch, or code update before applying this change to your production network, the lab to staging approach accomplishes this. It will help you in making your maintenance window more productive and bringing down the risk of causing issues which would cost your business and impact productivity.

Documentation

A systematic approach is recommended to implementing and supporting your wireless solutions. If you are a newcomer to the world of professional networking, you may not yet be aware of the impact "your layer" can have on every other aspect of running a business. Nowadays, with the number of technologies which are part of our lives, very few businesses can run without network services for a very long time. The phones, asset tracking, email, point of sale terminals (POS) and everything which lives in the cloud is dependent on the links networks provide. This is why adopting a systematic methodology to carry forward add moves and changes, roll out updates and perform system maintenance is very important.

The first step is to ensure you have proper documentation for all your infrastructures. This documentation should include:

- Physical and logical diagrams of all systems, interconnections, and interactions within these systems should exist. This should include network connections, interface types, equipment, VLANs, IP addresses and IP spaces in use
- Maps of all telecom closets, wireless base station, and other RF fixed systems, including intended coverage areas
- List of all service contracts, devices serial numbers and manufacturer's support contacts
- Code versions for all devices, including patch level

The means of documenting an entire deployment are up to you: some large environment will deem it worthy to have an online system such as a wiki or database. Some will even have an entire ecosystem of documentation, procedures, and a methodology to document thoroughly every single part of their ecosystem. The point though is to be able to access the information when you really need it: while conducting a weekly tech watch, getting notices from vendors about vulnerabilities or when a system (or more) is down. The last thing you would want to have to do when you have a serious issue affecting production and dollars are flying out the window by the minute is to have to fuss with Visio to create a network diagram for your vendor's technical assistance center. Be aware that the hardest part of documenting is not creating the documentation itself but to keep it up to date. It is up to you to be meticulous in maintaining its accuracy.

Security Updates

Getting in the habit of following security advisories and bugs is a practice that can save you valuable time and energy, especially in network administration roles. You should aim at spending some time every week on a technology watch which includes sites such as the following:

- CVE list (https://cve.mitre.org/cve/)
- NDS list (https://nvd.nist.gov)
- SANS Internet Storm Center: (Isc.sans.org)
- CISA (https://www.us-cert.gov)

Note that most manufacturers offer mailing lists or web pages which will alert you to bugs, vulnerabilities, and code updates. You should consult your resources regularly to ensure your infrastructure, whether it is BLE, 802.11 or Zigbee, is safe and secure. These few minutes a week can save you a couple of hours to a couple of days.

Industry Organizations

Three types of organizations guide the wireless industry. They are regulation, standardization and compatibility/certification. The Federal Communications Commission (FCC) and the European Telecommunications Standards Institute (ETSI) are examples of regulatory bodies that provide regulations in North America and Europe, respectively. The Institute of Electrical and Electronics Engineers (IEEE) and the Internet Engineering Task Force (IETF) are examples of standards development organizations. The Wi-Fi Alliance is a compatibility testing and certification group. Other wireless technologies have similar groups, and we will highlight these below.

It is essential to understand what these organizations do, and it is also essential to understand how they work together. For example, consider the interdependency between the FCC and the IEEE, or the relationship between the Wi-Fi Alliance and the IEEE. The FCC sets the legal boundaries within which the IEEE standards may function. The Wi-Fi Alliance tests equipment based on portions of IEEE standards and certifies it as being interoperable. These three organizations provide regulation, standardization and compatibility services for wireless Local Area Network (WLAN) technologies.

The benefits of these organizations to the consumer are clear and are depicted in Figure 1.5. When regulations are in place, such as power output limits, it is possible to implement local wireless networks with less interference from nearby networks. When standards are in place, like the IEEE 802.11 standard, it is possible to purchase devices that are compatible even though they come from different vendors. When certifications are in place to validate interoperability, consumers may buy products with confidence that those devices sharing the same certifications should be interoperable and fewer man-hours will be required for compatibility testing.

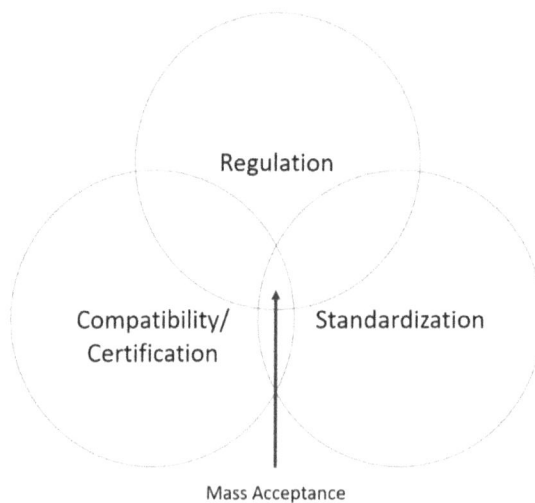

Figure 1.5: Requirements for Mass Acceptance of a Technology

Regulatory Domain Governing Agencies

A regulatory domain is defined as a geographically bounded area that is controlled by a set of laws or policies. Currently, governing bodies exist at the city, county, state, and country-level within the United States forming a hierarchical regulatory domain system. In other countries, governments exist with similar hierarchies or with a single-level of authority at the top level of the country or group of countries. In many cases, these governments have assigned the responsibility of managing communications to a specific organization that is responsible to the government. In the United States, this managing organization is the Federal Communications Commission. In the UK, it is the Office of Communications. In Australia, it is the Australian Communications and Media Authority. The following sections outline four such governing bodies and the roles they play in the wireless networking industry of their respective regulatory domains.

FCC, IC, and CE

The Federal Communications Commission (FCC) was born out of the Communications Act of 1934. Charged with the regulation of interstate and international communications by radio, television, cable, satellite, and wire, the FCC has a large body of responsibility. The regulatory domain covered by the FCC includes all 50 of the United States as well as the District of Columbia and other U.S. possessions, like the Virgin Islands and Guam. In Canada, the Industry Canada (IC) organization certifies RF products for use. The CE mark is given by the European Commission, which allows for RF products to be sold in the European Union.

OfCom and ETSI

The Office of Communications (OfCom) is charged with ensuring optimal use of the electromagnetic spectrum, for radio communications, within the UK. OfCom provides documentation of and forums for discussion of valid frequency usage in radio communications. The regulations put forth by the OfCom are based on standards developed by the European Telecommunications Standards Institute (ETSI). These two organizations work together in much the same way the FCC and IEEE do in the United States.

MIC and ARIB

In Japan, the Ministry of Internal Affairs and Communications (MIC) is the governing body over radio communications. However, the Association of Radio Industries and

Businesses (ARIB) was appointed to manage the efficient utilization of the radio spectrum by the MIC. In the end, ARIB is responsible for regulating which frequencies can be used and such factors as power output levels.

ACMA

The Australian Communications and Media Authority (ACMA) replaced the Australian Communications Authority in July of 2005 as the governing body over the regulatory domain of Australia for radio communications management. Like the FCC in the United States, the ACMA is charged with managing the electromagnetic spectrum to minimize interference. This is done by limiting output power in license-free frequencies, and by requiring licenses in some frequencies.

ITU-R

The International Telecommunications Union — Radiocommunication (ITU-R) is a Sector of the International Telecommunications Union (ITU). The ITU, after an extended history, was designated as a United Nations specialized agency on October 15, 1947. The constitution of the ITU has stated its purposes as:

- To maintain and extend international cooperation between all its Member States for the improvement and rational use of telecommunications of all kinds
- To promote and enhance participation of entities and organizations in the activities of the Union and foster fruitful cooperation and partnership between them and Member States for the fulfillment of the overall objectives embodied in the purposes of the Union
- To promote and to offer technical assistance to developing countries in the field of telecommunications, and also to promote the mobilization of the material, human and financial resources needed for its implementation, as well as access to information
- To promote the development of technical facilities and their most efficient operation with a view to improving the efficiency of telecommunication services, increasing their usefulness and making them, so far as possible, generally available to the public
- To promote the extension of the benefits of new telecommunication technologies to all the world's inhabitants
- To promote the use of telecommunication services with the objective of facilitating peaceful relations

- To harmonize the actions of Member States and promote fruitful and constructive cooperation and partnership between Member States and Sector Members in the attainment of those ends
- In order to promote at the international level, the adoption of a broader approach to the issues of telecommunications in the global information economy and society cooperate with other world and regional intergovernmental organizations, and those non-governmental organizations concerned with telecommunications.

The ITU-R, specifically, maintains a database of the frequency assignments worldwide and helps coordinate electromagnetic spectrum management through five administrative regions. These five regions are:

- Region A: The Americas
- Region B: Western Europe
- Region C: Eastern Europe
- Region D: Africa
- Region E: Asia and Australia

Each region has one or more local regulatory groups such as the FCC in Region A for the United States or the ACMA in Region E for Australia. Ultimately, the ITU-R provides the service of maintaining the Master International Frequency Register of 1,265,000 terrestrial frequency assignments.

In the end, regulatory agencies typically control wireless use in unlicensed spaces in the following important areas:

- Allowed frequencies
- Channel bandwidth
- Area usage, such as indoor and outdoor
- Maximum transmission power at the radiator
- Maximum transmission power at the antenna

IEEE

The Institute of Electrical and Electronics Engineers (IEEE) states its mission as being the world's leading professional association for the advancement of technology. They provide standards and technical guidance for more than just the wireless industry. In this section, I focus on the specific standards developed by the IEEE that impact and benefit wireless networking. These standards include wireless-specific standards, as well as standards that have been implemented in the wired networking domain, which are now being utilized in the wireless networking domain. First, I provide you with a more detailed overview of the IEEE organization.

The IEEE is a global professional society with more than 423,000 members in 160 countries. The constitution of the IEEE defines the purpose of the organization as scientific and educational, directed toward the advancement of the theory and practice of electrical, electronics, communications and computer engineering, as well as computer science, the allied branches of engineering, and the related arts and sciences. Their mission is stated as promoting the engineering process of creating, developing, integrating, sharing, and applying knowledge about electro and information technologies and sciences for the benefit of humanity and the profession. Ultimately, the IEEE creates many standards for many niche disciplines within electronics and communications. In this book, the focus is on computer data networks and specifically wireless computer data networks. In this area, the IEEE has given us the 802 project and, specific to wireless, the IEEE 802.11 standard.

Compatibility and Certification groups

The primary compatibility and certification groups are 3GPP, Bluetooth SIG, CBRS Alliance, CTIA, GSMA, Wi-Fi Alliance, WBA, WiMAX Forums, and Zigbee. This is not intended to be a complete list; however, these groups deal strictly in wireless. Below are several of these organizations and who they are in their own words.

3GPP

According to the 3GPP (https://www.3gpp.org), they are:

- The 3rd Generation Partnership Project (3GPP) unites [Seven] telecommunications standard development organizations (ARIB, ATIS, CCSA, ETSI, TSDSI, TTA, TTC), known as "Organizational Partners" and provides their members with a stable environment to produce the Reports and Specifications that define 3GPP technologies.
- The project covers cellular telecommunications technologies, including radio access, core network, and service capabilities, which provide a complete system description for mobile telecommunications.
- The 3GPP specifications also provide hooks for non-radio access to the core network and interworking with non-3GPP networks.

- 3GPP specifications and studies are contribution-driven, by member companies, in Working Groups, and at the Technical Specification Group level.

The three Technical Specification Groups (TSG) in 3GPP are:

- Radio Access Networks (RAN),
- Services & Systems Aspects (SA),
- Core Network & Terminals (CT)

Bluetooth

According to the Bluetooth Sig (https://www.bluetooth.com), they are:

- Our Vision - A connected world, free from wires. We believe in a world where everything and everyone that wants to connect, should be able to do so in a simple, secure, and wireless way.
- Our Mission - In support of our vision and member companies, the Bluetooth SIG expands Bluetooth technology by fostering member collaboration to create new and improved specifications, drives global Bluetooth interoperability through a world class product qualification program, and grows the Bluetooth brand by increasing the awareness, understanding, and adoption of Bluetooth technology.

CBRS Alliance

According to the CBRS Alliance (https://www.cbrsalliance.org), they are:

- Support the common interests of members, implementers and operators for the development, commercialization, and adoption of LTE solutions for the US 3.5 GHz Citizens Broadband Radio Service. Mission:
- Evangelize LTE-based OnGo technology, use cases, and business opportunities
- Drive technology developments necessary to fulfill the mission, including multi-operator LTE capabilities
- Identify required advocacy steps (e.g., marketing, promotion, certification, branding, regulatory, etc.) and catalyze action in these areas
- Establish an effective product certification program for LTE equipment in the US 3.5 GHz band ensuring multi-vendor interoperability

Cellular Telecommunications and Internet Association (CTIA)

According to the CTIA (https://www.ctia.org), they are:

- The CTIA represents cellular, personal communication services, mobile radio, and mobile satellite services over wireless WANs for service providers and manufacturers.
- CTIA represents the U.S. wireless communications industry. From carriers and equipment manufacturers to mobile app developers and content creators, we bring together a dynamic group of companies that enable consumers to lead a 21st Century connected life.

As the voice of America's wireless industry, CTIA:

- Advocates for legislative and regulatory policies at federal, state, and local levels that foster the continued innovation, investment and increasing economic impact of America's wireless industry. CTIA is active on a wide range of issues including spectrum policy, wireless infrastructure, and the Internet of Things, among others.
- Convenes the industry to tackle our most difficult challenges and coordinates voluntary best practices and initiatives. CTIA works with members to develop test plans and certification processes for mobile devices, coordinates with members and other industry leaders to ensure the security of mobile networks and devices, and leads industry initiatives to enhance accessibility, improve 9-1-1 location accuracy, deter phone theft and encourage safe driving.
- Promotes our members through numerous campaigns aimed at building awareness among policymakers and the general public, as well as through industry-leading events on topics ranging from cybersecurity to 5G.

GSMA

According to the GSMA (https://www.gsma.com), they are:

- The GSMA represents the interests of mobile operators worldwide, uniting more than 750 operators with almost 400 companies in the broader mobile ecosystem, including handset and device makers, software companies, equipment providers and internet companies, as well as organizations in adjacent industry sectors. The GSMA also produces the industry-leading MWC events held annually in Barcelona, Los Angeles and Shanghai, as well as the Mobile 360 Series of regional conferences.

Wi-Fi Alliance

According to the Wi-Fi Alliance (https://www.wi-fi.org/), they are:

- Wi-Fi Alliance® is the worldwide network of companies that brings you Wi-Fi®, one of the world's most valued communications technologies. Our vision is to connect everyone and everything, everywhere.
- Wi-Fi Alliance drives global Wi-Fi adoption and evolution through thought leadership, spectrum advocacy, and industry-wide collaboration. Our work helps ensure that Wi-Fi devices and networks provide users the interoperability, security, and reliability they have come to expect.

With the global economic value of Wi-Fi expected to reach nearly $3.5 trillion USD by 2023 and billions of devices shipped each year, Wi-Fi is one of the greatest success stories of the technology era. The Wi-Fi Alliance commitment to expanding Wi-Fi uses and availability brings with it a relentless focus on delivering the great user experience that underpins the success of Wi-Fi today.

Wi-Fi Alliance Vision: Connecting everyone and everything, everywhere

Wi-Fi Alliance Mission:

- Fostering highly-effective global collaboration among member companies
- Delivering excellent connectivity experiences through interoperability
- Embracing technology and driving innovation
- Promoting the adoption of our technologies worldwide
- Advocating for fair worldwide spectrum rules
- Leading, developing, and embracing industry-agreed standards

Wireless Broadband Alliance

According to the WBA (https://wballiance.com), they are:

- WBA's vision is to drive the seamless and interoperable services experience via Wi-Fi within the global wireless ecosystem. The WBA's mission is to enable collaboration among service providers, technology companies and organizations in the industry who share the same vision. We undertake programs and activities that aim to address business & technical issues, as well as opportunities for member companies.

- WBA work areas include advocacy, industry guidelines, trials and certification. Our key programs include NextGen Wi-Fi, 5G, IoT, Roaming, Testing & Interoperability with Work Groups led by the membership to resolve standards and technical issues that support end-to-end services and accelerate business opportunities. Today, membership includes major operators and leading technology companies.

WiMAX

According to the WiMAX Forum (http://wimaxforum.org/), they are:

- The WiMAX Forum® is a worldwide consortium chartered to deliver certification that achieves global interoperability, develop technical specifications based on open standards, pursue a favorable regulatory environment and promote the vision.

WiMAX Forum Strategic Objectives

- Establish cost-effective and timely certification processes and certification infrastructure for WiMAX that achieve device and network interoperability.
- Ensure that the WiMAX Forum certification process is valued and trusted by network providers, service providers and consumers worldwide.
- Ensure on-time availability of test specifications and certification requirements.
- Ensure test infrastructure is in place to meet the global need for cost-effective WiMAX certification.
- Publish technical specifications to achieve a commercially viable global ecosystem for WiMAX.
- Deliver high-quality technical specifications based on the IEEE 802.16 standard to enable a high-performance, end-to-end Internet network architecture supporting fixed, portable, nomadic and mobile users.
- Establish a WiMAX technology road map to support a wide variety of applications and use case scenarios, and foster a robust ecosystem.
- Enable global roaming for WiMAX-to-WiMAX networks and across networks that meets market demand for ease of use.
- Enable interworking for WiMAX networks with other wireless networks.
- Ensure WiMAX supports coexistence with other wireless technologies to provide access to a broad set of frequency bands.

- Promote the brand and technology to establish WiMAX as the worldwide market leader for broadband wireless.
- Promote attractive services and economic value propositions to foster user demand.
- Promote WiMAX to ensure spectrum availability and a favorable regulatory environment.
- Promote the advantages of WiMAX to facilitate growth of the ecosystem worldwide.

Zigbee

According to the Zigbee Alliance (http://www.zigbee.org/), they are:

- The Zigbee Alliance is the standard-bearer of the open IoT. Established in 2002, our wide-ranging global membership collaborates to create and evolve universal open standards for the smart networks in our homes, businesses, and neighborhoods
- Our Vision - All objects can interact to improve the way the world lives, works, and plays.
- Our Mission - The Zigbee Alliance ignites creativity and collaboration in the Internet of Things, by creating, evolving, and promoting universal open standards that enable all objects to securely connect and interact.
- Our Principles - Striving to create new opportunities and a larger market for all stakeholders in the IoT.
- Bringing together the world's most innovative companies and individuals to create and evolve technologies that enable all IoT devices to connect and interact — regardless of country, brand, market, or network.
- Delivering on our technologies' brand promise of performance and interoperability with robust certification programs and tools.
- Creating and fostering opportunities for collaboration amongst our members, and all IoT stakeholders to drive consensus in the industry.
- Solving industry-wide challenges, by partnering with leading organizations, to enable a more unified Internet of Things.
- Enthusiastically promoting Zigbee Alliance technologies, and members' achievements and solutions built on our technologies.

- Championing the thought leadership of our members to highlight today's opportunities, inspire tomorrow's innovation, and influence the future of the Internet of Things.
- Aiming to be the most trusted IoT consortium by continually working to make it easier for companies to develop and deploy interoperable IoT products quickly, effectively and at a low cost.

As a CWSA, understanding these various organizations, and others not listed here, will help you to grasp the immensity of the wireless industry. It will also assist you in locating the right information at the right time for that next project.

Chapter Summary

This chapter provided a broad overview of wireless networks and an introduction to the industry organizations that are shaping the future of wireless technologies. You were introduced to many topics that will be expanded upon throughout the remainder of this book. The next chapter will go deeper into the various wireless network types, such as WBAN, WLAN, WMAN, and more. It will also address various vertical markets and their use of wireless solutions.

Review Questions

1. What organization is an independent consortium of wireless network products manufacturers whose goal is the interoperability of WLAN technologies, architectures and wireless protected access?
    a. The Wireless LAN association (WLA)
    b. The WICA
    c. The Wi-Fi Alliance
    d. The WLAN SIG

2. What constrains the power levels that can be transmitted and the channels that can be used in wireless communications?
    a. The SSID
    b. The time zone
    c. The regulatory domain
    d. The country domain

3. What organization is behind the standardization of BLE and Bluetooth?
    a. The BLE Alliance
    b. The BLE SIG
    c. The Bluetooth SIG
    d. The ZigBee SIG

4. What does the frequency of an electromagnetic wave express?
    a. The distance between two waves
    b. The power of a wave
    c. The modulation of a wave
    d. The number of cycles per second for a wave

5. What organization is responsible for creating various wireless standards?
    a. The WLA
    b. The FCC
    c. The IEEE
    d. The CE

6. Which standard does not fall under the IEEE 802.15 standard in any way?
    a. BLE
    b. ZigBee
    c. Bluetooth
    d. Wi-Max

7. Which organization has defined a 7-layer model aiming at better understanding the process of sending data over a network?
    a. NATO
    b. OSI
    c. ISO
    d. SISO

8. What does a standard followed by a year represent, for the IEEE? Ex: 802.11-2012
    a. The 802.11ac Wave 1 standard
    b. The 802.11 standard including all its amendments as of that year
    c. The 802.11 standard which was submitted to a new vote in 2011
    d. The standards always carry the style ###.##+capital letters format

9. What unit expresses the number of radio wave cycles per second?
    a. Hertz
    b. Ohm
    c. Volt
    d. Bel

10. Who is the woman called the Mother of Wi-Fi by many today?
    a. Heda Gabler
    b. Hedy Lamarr
    c. Nicolette Tesla
    d. Helen Hertz

Review Answers

1. The correct answer is **C**. The Wi-Fi Alliance is a consortium of manufacturers who aim at making Wi-Fi interoperable and secure.
2. The correct answer is **C**. The regulatory domain identifies the area in which a wireless device is used as well as following the area's laws regarding the use of frequencies, channels, and power.
3. The correct answer is **C**. Although Bluetooth has ties to the IEEE and other RF related standards associations, it is the Bluetooth SIG which is in charge of the standard and its uses.
4. The correct answer is **D**. The frequency of an electromagnetic wave refers to the number of cycles it goes through per second.
5. The correct answer is **C**. The IEEE defined wireless standards among many others.
6. The correct answer is **D**. Wi-Max, a standard that is covered under the 802.16 standard.
7. The correct answer is **C**. The organization's name is ISO, although the model is called the OSI model.
8. The correct answer is **B**. A standard followed by a year for the IEEE represents the standard including all revisions and amendments as of the time of ratification.
9. The correct answer is **A**. The Hertz represents the number of cycles of a radio wave per second.
10. The correct answer is **B**. Hedy Lamarr, a Hollywood starlet who was also involved in the war efforts during the Second World War Who is called the Mother of Wi-Fi. Oh, and it was Nikola Tesla, not Nicolette who was the famous inventor and scientist, though not the mother of Wi-Fi.

Chapter 2: Wireless Network Use Cases

Objectives Covered:

1.3   Define wireless network types

3.6   Administer the wireless solution while considering the implications of various vertical markets

The focus of this chapter is on the general description of various network types and how they are used in different vertical markets. First, the network types will be described and then various vertical markets with their common challenges will be explored.

Wireless BANs

A Wireless Body Area Network (WBAN) is sometimes referred to as a Body Area Network (BAN) or Body Sensor Network (BSN) and includes Medical Body Area Network (MBAN) within its definition. These are the smallest of the networks we will focus on for this certification, except a brief note about Near Field Communication (NFC), the 10-centimeter wireless solution. These wireless networks include devices within the body (implantable, injectable, and ingestible) and wearable electronics. WBAN devices usually communicate to a companion device like a phone, tablet, or laptop to provide a user interface or report into server-based systems on the LAN or Internet.

In May of 2012, the FCC adopted a proposal to allocate 2360 - 2400 MHz (40 MHz of spectrum) for MBAN devices. The purpose of the approval was to provide flexibility for MBAN devices to measure, record, and transmit physiological or other patient information to the applications that process this data. Devices that operate in the 2360 - 2390 MHz range must be within a healthcare facility (indoors) and must send specific data to a frequency coordinator (MedRadio programmer/control transmitter), if this facility qualifies under Section 95.1203 and the facility intends to operate the MBAN devices. When the devices lose communication with the frequency coordinator, they must shut down their radios.

Operations in the 2390 - 2400 MHz band is not subject to registration or coordination and may be used in all areas including residential. These devices may be utilized outside of the medical facilities and wherever the individuals go, which is helpful for other use cases, like athletes.

Wearable and implant technologies usually operate on other frequencies. RFID technologies are commonly used for access and identification purposes. RFID chips can be implanted into pets or humans. These RFID chips usually operated within the ISM bands. Wearable technologies (e.g., FitBit, AppleWatch) are also being used by owners and medical teams to track the movement, health and wellbeing of the users.

These items often utilize Bluetooth or other protocols to communicate with the companion device. WBANs cover spaces of about 1 meter.

Wireless PANs

A WPAN (Wireless Personal Area Network) provides hands-free connectivity and communications within a confined range and limited throughput capacity. They also provide for small-scale mesh-type wireless networks, like those implemented with ZigBee technology. Also, RFID systems are frequently categorized as WPAN technologies because they have a short communications range. Bluetooth is also a good example of a WPAN technology that is both beneficial and in widespread use. Everything from Bluetooth mice to headsets and speakers is being used daily throughout the world.

Operating in the 2.4 GHz band, Bluetooth technologies can cause interference with WLAN technologies like DSSS (802.11-Prime), HR/DSSS (802.11b), ERP (802.11g), and HT (802.11n) when operating in the 2.4 GHz band. However, the newer adaptive frequency hopping technology helps to reduce this interference if not completely removing it. Adaptive frequency hopping is a feature found in Bluetooth 1.2 devices and higher, which makes up most of the Bluetooth device population today. WPANs cover spaces of about 10 meters.

Wireless LANs

WLANs are the primary focus of the CWNA certification and the entire certification track leading up to CWNE. However, it is still a critical component for the CWSA as many IoT devices connect using Wi-Fi.

WLANs are designed to cover homes, office buildings, or campus environments. They provide mobility (moving around with active use during the move), nomadic ability (moving around without active use during the move) and unwired fixed connectivity (no movement). Mobility is provided because the user can move around within the coverage area of the access point, or even multiple access points, while still maintaining connectivity. Nomadic ability — the ability to move from place to place and use the network, although active communications do not take place while moving — is provided because you can power on a wireless client device from any location within a coverage area and use it for a temporary period of time as a fixed location device. It is a given that unwired fixed connectivity must exist if the nomadic ability is provided.

Three primary roles exist, and WLANs play these roles in today's enterprise organizations:

- Access role
- Distribution role
- Core role

In the access role, the wireless network is used to provide wireless clients with access to wired resources. The access point remains fixed while the clients may move. The access point is usually connected to an Ethernet network where other resources, such as file servers, printers, and remote network connections, reside. In this role, the access point provides access to the wireless medium first and then, when necessary, provides bridging to the wired medium or other wireless networks (such as in a mesh network implementation). Figure 2.1 illustrates the access role of a WLAN.

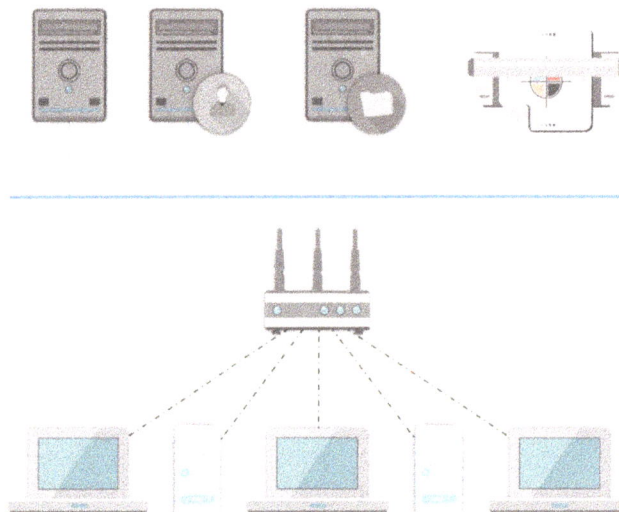

Figure 2.1: WLAN Access Role

In the distribution role, illustrated in Figure 2.2, wireless bridges provide a backhaul connection between disconnected wired networks. In this case, each network is connected to the Ethernet port of a wireless bridge, and the wireless bridges communicate with each other using the 802.11 standard. Once these connections are made, network traffic can be passed across the bridge link so that the two previously disconnected networks may act like one.

The final role is the core role. In the core role, the WLAN is the network. This may be suitable for small networks built on-the-fly, such as those built at construction sites or in disaster areas; however, the limited data throughput will prohibit the WLAN from being the core of the network in a large enterprise installation. Future technologies may change this, but for now, WLAN technologies play the access and distribution roles most often. WLANs cover a building or campus.

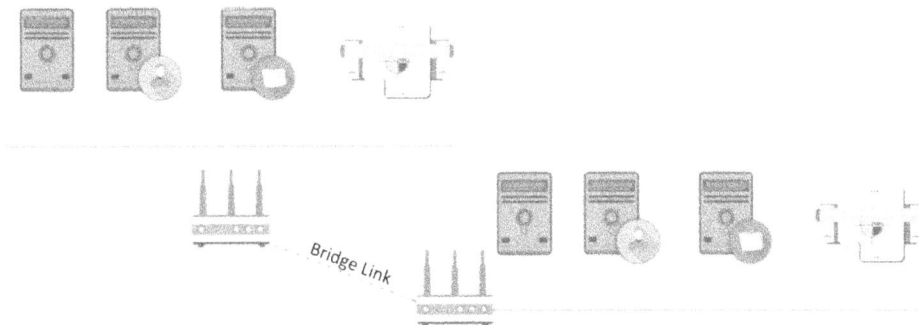

Figure 2.2: Wireless APs in a Distribution Role Creating a Bridge Link

Wireless MANs

WMANs (Wireless Metropolitan Area Networks) differ from WLANs in that they are not usually implemented by the organization that wishes to use the network. Instead, they are generally implemented by a service provider, and then access to the network is leased by each subscribing organization. However, unlike wireless Wide Area Networks, this does not have to be the case. For example, 802.16-compliant hardware could be purchased, and frequency licenses could be acquired to implement a private WMAN, but the expense is usually prohibitive.

WiMAX is a commonly referenced WMAN technology. WiMAX is based on the IEEE 802.16 standard and provides expected throughput of approximately 130 Mbps in the latest specifications. In addition to the throughput speeds, WiMAX incorporates QoS mechanisms that help to provide greater throughput for all users and important applications using the network.

Private LTE networks are starting to appear using MulteFire and CBRS technologies. MulteFire operates mostly within the 5 GHz band in the FCC region; however, for IoT, it leverages 800-900 MHz and 2.4 GHz for long-range connectivity. Some deployments are on 1.9 GHz, also known as sXGP in Japan. At the time of this writing, CBRS is still in the approval process with the FCC for use; however, this operates on 3.5 GHz. WMANs cover a city.

Wireless WANs

Wide Area Networks (WANs) are usually used to connect Local Area Networks (LANs) together. If the LANs are separated by a large distance, WAN technologies may be employed to connect them. These technologies include Frame Relay, analog dial-up lines, cable, DSL, ISDN, and others. What they have traditionally had in common is a physical wire connected to some device that is connected to some other device (usually across a leased line) that is eventually connected to the remote LAN. The wireless WAN (WWAN) is completely different because there is no wire needed from your local LAN to the backbone network or from the backbone network to your remote LAN. Wireless connections are made from each of your LANs to the backbone network. WWANs cover a region.

Examples of WWAN technologies include Free Space Optics, licensed and unlicensed radio, and hybrids of the two. For WAN links that span hundreds of miles, you may need a service provider such as AT&T microwave. For shorter links of a few miles, you may be able to license frequency bands or use unlicensed technology to create the links. The key differentiator of WWAN technologies from WLAN, WPAN, and WMAN is that the WWAN link is aggregating multiple communication channels together (multiplexing) and passing them across the single WAN link.

A WBAN usually covers about 1 meter. A WPAN usually covers about 10 meters. A WLAN covers a building or campus. A WMAN covers a city or an area around a city. A WWAN covers a region or spans the globe.

Wireless Sensor Networks

A Wireless Sensor Network (WSN) is a term used for a group of dedicated sensors that monitor and record their environment. These devices typically process the data with a central server that is part of the collection network. WSN's are utilized for monitoring sounds, temperature, moisture, wind, soil conditions, etc.

Wireless sensor networks range in size from a few sensors to hundreds or even thousands of nodes. Each node has its own wireless radio and electronics that are used to sense the environment it is monitoring. In many cases, these sensors are low power and operate on a battery; however, they may be connected to a dedicated power.

WSN's use a variety of communication methods to communicate amongst each other and to the central server. Many of these are wireless mesh technologies, and they operate in different bands. These mesh networks may be multi-hop or single hop, just like in WLANs. The most common standards in use for WSN's are Thread, ZigBee, Z-Wave, and LORA's LPWAN. Thread and ZigBee operate on within the 2.4 GHz ISM band at data rates up to 250 kbps. Z-wave operates at 915 MHz in the US and 868 MHz in the EU at lower data rates around 50 kbps.

New Network Driver - Internet of Things

The Internet of Things (IoT) is a term widely used for connected devices that do not utilize a traditional user interface, like a laptop, smartphone, or tablet. These devices can be found in industry, wearables, healthcare, retail, education, transportation, smart buildings, agriculture, and smart cities. These devices are changing the use cases and requirements for wireless network deployments; however, the impact they are having is massive, and they need to be accounted for.

IoT devices range in size from tiny circuit boards to large vehicles. These devices use a range of standards for communication amongst each other and to the network. For Internet communications, IoT devices utilize IPv4 and/or IPv6 and connect to some form of a directory server or cloud server for the user of the service to manage them.

Some of the standard IoT devices utilized for wireless communications are as follows.

Short-range wireless:

- Bluetooth mesh networking – Specification providing a mesh networking variant to Bluetooth low energy (BLE) with an increased number of nodes and standardized application layer (Models).
- Light-Fidelity (Li-Fi) – Wireless communication technology similar to the Wi-Fi standard, but using visible light communication for increased bandwidth.
- Near-field communication (NFC) – Communication protocols enabling two electronic devices to communicate within a 4 cm range.
- Radio-frequency identification (RFID) – Technology using electromagnetic fields to read data stored in tags embedded in other items.
- Wi-Fi – technology for local area networking based on the IEEE 802.11 standard, where devices may communicate through a shared access point or directly between individual devices.
- ZigBee – Communication protocols for personal area networking based on the IEEE 802.15.4 standard, providing low power consumption, low data rate, low cost, and high throughput.
- Z-Wave – Wireless communications protocol used primarily for home automation and security applications

Medium-range wireless:

- LTE-Advanced – High-speed communication specification for mobile networks. Provides enhancements to the LTE standard with extended coverage, higher throughput, and lower latency.

Long-range wireless:

- Low-power wide-area networking (LPWAN) – Wireless networks designed to allow long-range communication at a low data rate, reducing power and cost for transmission. Available LPWAN technologies and protocols: LoRaWan, Sigfox, NB-IoT, Weightless, RPMA.
- Very small aperture terminal (VSAT) – Satellite communication technology using small dish antennas for narrowband and broadband data.

IoT for Industry (IIoT)

The growth of IoT devices in industry continues to drive new network and wireless demands. In many cases, worker safety is increased by allowing the workers to control the devices remotely. This is creating a demand for wireless connectivity in locations that previously have not needed connectivity. Manufacturing plants are purpose-built for creating products and the materials used can create challenges for wireless frequencies. Many short-range wireless communications are used in manufacturing such as NFC, RFID, and Wi-Fi as well as Bluetooth.

IoT for Connected Vehicles

Connected Vehicles is a term utilized for vehicles that are connected to the Internet. A connected vehicle is used for cars, trucks, and SUVs that are on the road. Communications to the network may require a data plan from a carrier to utilize the GSM. Safety applications in the US use a dedicated short-range communications (DSRC) radio that operates in the 5.9 GHz for low latency communications. Some connected vehicles are also offering WLAN services within the vehicle itself.

The communications with the vehicle to others are often referred to with a V2x nomenclature (Source: Wikipedia).

- V2I "Vehicle to Infrastructure": The technology captures data generated by the vehicle and provides information about the infrastructure to the driver. The V2I technology communicates information about safety, mobility or environment-related conditions.
- V2V "Vehicle to Vehicle": The technology communicates information about speed and position of surrounding vehicles through a wireless exchange of information. The goal is to avoid accidents, ease traffic congestions, and have a positive impact on the environment.
- V2C "Vehicle to Cloud": The technology exchanges information about and for applications of the vehicle with a cloud system. This allows the vehicle to use information from other networks, through the cloud connected industries like energy, transportation, and smart homes and make use of IoT.
- V2P "Vehicle to Pedestrian": The technology senses information about its environment and communicates it to other vehicles, infrastructure, and personal mobile devices. This enables the vehicle to communicate with pedestrians and is intended to improve safety and mobility on the road.
- V2X "Vehicle to Everything": The technology interconnects all types of vehicles and infrastructure systems with another. This connectivity includes cars, highways, ships, trains, and airplanes.

Connected vehicle technologies focus on reducing congestion on the road, improving safety, and lowering greenhouse emissions. NHTSA estimates that a connected vehicle safety application that helps drivers safely negotiate intersections could help prevent 41 to 55 percent of intersection crashes. Another connected vehicle safety application that helps drivers make left turns at intersections could help prevent 36 to 62 percent of left-turn crashes, according to NHTSA.

To properly administer the wireless network, you need to understand the use cases for wireless technologies and how they continue to evolve. When 802.11 was created, the use cases were limited as the technology was still very young and expensive to deploy. Over the years wireless technology has become more cost-effective and easier to deploy. Most smartphones are more powerful than desktop computers were when 802.11 was released.

Residential Environments

Residential environments were one of the first adopters for wireless technologies. Over the past decade wireless in the home environment has evolved from a nice to have to become a necessity. Connecting to the internet has never been easier than it is with wireless technologies. So much so, that everything in the home now has uses for that connectivity. Consumers started with laptop computers connecting, then moved into smartphones, and tablets for direct interaction with Internet and cloud services.

Technology has continued to evolve from basic internet access into audio, video, and virtual reality. Wireless devices throughout most homes are utilized as speakers for music from both local and internet sources, many of which respond to audible commands for content. These devices can be programmed wirelessly to play at scheduled times as well.

One-way and two-way video solutions continue to become more prevalent. One-way video systems include doorbells, security cameras, and nanny cams. Some of these send data to a central server in the home, others to the cloud wirelessly. Two-way video systems allow for wireless communication to people at a gate, outside the front door, or even around the world.

The "Internet of Things" (IoT) has continued to bring wireless to more and more devices throughout the home for a truly automated home. From light bulbs that can change the brightness or color by wireless commands to power outlets that can be turned on/off or even scheduled. Even thermostats are utilizing wireless to remote sensors to adjust the temperature in the home and even analyze the usage of electric and save money by adjusting when people are not home. Autonomous vacuum cleaners are now using wireless technology to map floor plans and even be told when to start or how long to run for. Irrigation systems can come equipped with Wi-Fi for remote management as well.

Garage doors and home security systems (alarms) can be managed remotely from your phone. Some refrigerators have options to connect wirelessly to notify you when you are running low on things...the list goes on and on, and these are in everyday consumer homes.

More technically advanced homes have digital displays that change what is shown based on who is in the room, TVs can recommend content based on who is in the room, alerts can be created for the sick or elderly if movement in the home is not detected for some time. These are just a few use cases today, and they continue to evolve, and more uses are created daily.

Many modern cars come with wireless communicates as well bringing the need for wireless in the garage and outside the home to another level. As auto-pilot and driver-assist technologies evolve these cars connect to the internet to download updates in the interest of driver safety.

The sad reality is that many consumer-targeted smart home devices are 802.11 wireless devices that operate only in the 2.4 GHz frequency band. The manufacturers suggest that this decision is based on the extended range of 2.4 GHz signals over 5 GHz signals. Of course, single-band chipsets are cheaper, and this fact certainly impacts their decisions. A wireless solutions administrator must be cautious when selecting any wireless device so that the selected devices provide the wireless connectivity required.

Some of the challenges of operating a network in a residential environment include:

- Construction materials vary by region
- Coverage areas are unpredictable and creeping of new locations

- Co-channel contention (mostly an urban problem)
- Limited Channel Selection - All devices must work
- Inability to disable legacy data rates (802.11b)

Retail

Wireless technologies have been adopted in retail in mass since the early 2000s. Inventory management, location awareness, connecting with customers, and point of sale are just a few of the ways wireless is making the retail experience better. Designing and administering a wireless network in a retail environment requires an understanding of the use cases and the environment around the location(s).

Tracking and knowing where your inventory is located is one of the most important things in retail. The entire goal is to get product into the hands of the customers. Wireless technologies make this easier in retail environments. Wireless is used to track inventory as it moves into the store, onto the shelves and as it is sold to consumers. This can be done using barcode scanners or RFID tags.

Wireless also enhances the retailer's ability to interact with the customer in more meaningful ways. Consumers bring their own wireless devices into retail environments. The interactions can now happen via applications on these consumer devices. The retailer can utilize this application to push advertisements, sales, coupons and more to the users. These devices can also be utilized to understand the location in the store, so if a customer is in the area around a product, the application may push a discount code for that item. The retailers can also utilize the location information, and the path customers take through the stores to optimize placement of products.

Retail locations are increasingly enabling staff to bring the checkout capabilities to the customer directly and increasing the quality of the interactions by leveraging wireless technologies for point of sale, receipt generation, etc. Consumers are even using wireless technologies to pay from their smartphones, instead of having to bring a wallet with them everywhere.

Restaurants have similar use cases for wireless technologies as other retail environments. Bringing interactions in the drive-through outside on busier times. They have been utilizing public wireless technology to increase the dwell time and likelihood the consumers will purchase more food and beverages for a while now.

Many sit-down restaurants are even putting kiosk at the table for games, entertainment, ordering, and even payment to turn tables over at the customer's pace.

Operating a network in a retail environment in unlicensed spectrum requires an understanding of everything within your location and what may be interfering from locations surrounding your location. Some of the challenges in retail include:

- Co-channel contention
- Changing wireless environments (stock and people are always moving)
- The proximity of neighboring networks
- Compliance / Regulatory - VISA/CISP
- Unknown/constantly changing throughput requirements
- Backward compatibility with legacy devices
- Poor drivers and implementations by partners with no documentation (digital signage, etc)

Education (K-12)

Wireless technologies have been adopted in education over the past decade. Utilizing technology has enabled schools to provide new ways to connect with students while preparing them for the digital world.

Education has extremely different use cases from retail or other environments. Students are already using wireless devices at home and transitioning this familiarity into the learning tools can really help teachers connect with students. Teachers are able to share their screen directly with the entire class, or one to one when reviewing concepts. Seamlessly transitioning from lectures to individual assignments allows a more content-rich environment than what a textbook alone could provide. Each student having their own device means they have the option for personalized content and lesson plans from the teachers as well. These are just some of the classroom uses.

Wireless technology also allows students and teachers to be connected directly to the Internet, allowing the use of cloud services and research from anywhere on campus. This allows the students to engage in educational activities from anywhere life brings them.

Some of the challenges of operating a network in an educational environment include:

- The density of client devices per room

- Co-channel contention
- Applications requirements (multicast, Intra-BSS communications, etc.)
- BYOD

Higher Education

In college and university environments, wireless technologies include everything from K-12 and add in elements of enterprise networks and home networks. In some universities the sports arenas, football stadiums, and hospitals are also part of the campus and wireless network.

Every year when college students come into student housing, they bring with them every device type imaginable, as they tend to be early adopters of technology. The latest gaming consoles, lighting, speakers, and more. These devices worked at home, so naturally the expectation is that they will work on the college campus as well. This can be a challenge for the administrators of these networks in some environments, as they are bringing the home devices onto an enterprise-grade network build for a high density of devices and they tend to have a lot more interference than in residential neighborhoods.

When hospitals are part of the campus, they bring all of the HIPAA and hospital requirements as part of the campus. Stadiums, Arenas, and public areas also bring their own requirements as well. Both of these are covered in their respective sections; however, these are also large parts of the campuses they belong to.

Some of the challenges to operating a network in Higher Educational environment include:

- BYOD
- Security
- The density of client devices per room
- Co-channel contention
- Rogue Access Points
- Applications requirements (multicast, Intra-BSS communications, etc.)
- Diverse environments and varying requirements by area of campus

Agriculture

Agriculture historically has not been thought of as having a use for wireless, since when do plants or cows need wireless? Even in agriculture, wireless technologies are becoming more and more valuable. Enabling network communications around the property using point to point or cellular technologies allows for devices to talk to each other. For crops rain or soil sensors can be placed to know when the irrigation system needs to run, where it needs to run and when it should stop. This environmental data can be used to save on water usage or optimize growing conditions. The soil sensors will also let the farmers know when the soil needs to be enriched with fertilizer.

Wireless technologies are utilized in tractors for automation. John Deere released a tractor that can navigate fields without a driver steering it. This eliminates the possibility of error and allows the farmers to focus on other important tasks while the tractor plows the feel or harvests the crops.

Wireless cameras are often used to monitor livestock activities and locations on farms; however, this is not the only wireless system in place. RFID tags can be used on horses, cows, or other livestock to track their locations throughout the day, which field they are in, and their movements.

Drones are also commonly used as a means to put eyes over large areas quickly. This helps the farmer know if where to spend time on a given day.

Some of the challenges of operating a network in agricultural environments include:

- Device Support
- Coverage
- Weather
- Mounting locations

Smart Cities / Public Access

Wireless technologies have enabled communications in cities like never before. Cities are leveraging these technologies for utilities, public safety, automation, and public access. Cities are taking advantage of public/private partnerships for deployments with carriers, service providers, and in some cases building out their own networks. Smart city applications include demand-based road tolling, pollution monitoring, and even city-wide municipal Wi-Fi in some cases.

When traveling around, you may notice people are using wireless technologies to stay in communication with friends and family. Wireless technologies in urban areas are used for nearly everything, as running cables can be expensive and in some cases not possible. In addition to connecting consumer devices, public safety is one of the most common use cases. Public safety has evolved from connecting the police and fire department radios into providing laptops to the police with direct access to the tools to validate identities and even look up individuals for outstanding violations or safety concerns. Cameras can be connected for real-time monitoring as well. (The world of RoboCop is becoming a reality.)

Utility companies are able to install meters that communicate back wirelessly, either by utilizing Wi-Fi in the home, on the pole, or cellular, to eliminate the need for a manual reading every month. In some markets, the power companies will offer incentives to businesses at peak times to consume less electricity. The company sees demand on the grid and will wirelessly trigger an alert to participating entities to reduce consumption. Many of these locations then use wireless technology (ZigBee) to tell the thermostats to go into power save mode if they can. All seamless to the individuals on the property and without human involvement.

Cameras are utilized to monitor traffic conditions and traffic flows. Traffic flows can also be monitored wirelessly counting the number of devices passing by a specific location. This enables cities to more effectively time traffic lights, plan public transportation routes, schedules, and methods of transport.

Some of the challenges of operating a network in urban environments include:

- Backhaul - How & to where
- Environmentals (Weather)
- Co-channel contention
- High Noise Floors
- Equipment mounting locations

Health Care

Hospitals have wireless workstations and tablets similar to office environments and many that are unique to them. Nurses require the power of desktop computers in a mobile platform, many times with other systems on them. Engineers have created a workstation on a cart that has wheels custom-built for this purpose. Communication within the hospitals is absolutely critical, and almost all of these are wireless. Voice communications using VoIP phones or badges and secure text messages between the care teams.

Medical records are, in many cases digital and accessed from wireless devices. This enables the care team to quickly pull up a patient's medical history to ensure a proper care plan is in place.

Health Care has continued to use wireless technology to collect information and get this information to those who need it. Patient monitoring systems are able to directly notify the hospital staff if a patient's vital signs change, in the room or if they are mobile. This enables the care team to know how to respond in a critical event and increase the likelihood of saving a life.

In some cases, doctors can work remotely. Tests can be run, and the results can be transmitted to the patient's doctor in another location or even around the world. Doctors can utilize two-way video technologies with patients to communicate the results of these tests, diagnosis and recovery plans. Most of this was not even possible a few years ago; however, today wireless and smartphones have enabled this revolution.

Wireless is also utilized by the patient's family and guests while waiting within the hospital. Many times, bringing their technology from their home or office into the hospital to facilitate work or entertainment while the patient is treated, asleep, or cared for.

Some of the challenges of operating a network in medical environments include:

- Device Support
- Proper Design / Repurposed building spaces
- Co-channel contention
- Access to areas for site survey/installation
- Compliance / Regulatory - HIPPA
- Coexistence of point solutions

Office Buildings

Office buildings are getting smarter, increasing automation, and measuring their performance. Wireless connectivity, sensors, and devices are deployed throughout many buildings, and ones that have pervasive coverage are known as "Smart Buildings."

The sensors in these smart buildings collect data from various points, communicating it back to a database, and an analytics system runs on this information to automate action. These include lighting, HVAC, electric, cameras and much more. This enables the building to only provide services while people that need them are present and this can save property owner's money.

Security systems in smart buildings are wireless as well. NFC, RFID, and Wi-Fi are commonly used to grant access into the building or specific areas within the building. They can also be used to track the location of people within the building, which is helpful in the event of an emergency to let first responders know how many people are in the building and where they are.

Some of the challenges of operating a network in office environments include:

- Neighboring APs / Co-channel contention
- Device Support
- High bandwidth demands

Hospitality

Wireless technologies have been adopted in hospitality in mass since the early 2000s. It started with guest access and then transitioned into more use cases. Designing and administering a wireless network in hospitality requires an understanding of the use cases and the environment around the location(s).

Hotels have been on the bridge between home and office buildings for a while, bringing technology from both environments into these locations for safety and convenience. The wireless experience within a hotel starts when you walk into the building with Wi-Fi being available throughout the building. This Internet access allows guests to connect back to their home, office, or to any other resources they may desire.

Hotels utilize wireless for the door locks within the room. In many cases, you can utilize an app on your phone to by-pass the front desk entirely to get to your room, saving you time. Once you are in the room though, wireless can be used to put content from your device onto the TV, to print documents, or even to charge your phone with wireless power (Qi).

The hotels themselves use wireless location services to track luggage carts. Wireless communication is also used to know when the maids are complete with a room or when supplies are needed. The same wireless communication system can be utilized by guest services to get the closest person to assist a guest in need. Check-in kiosks are also able to be moved around for both conferences and hotel check-in purposes.

Some of the challenges of operating a network in office environments include:

- Construction materials
- Neighboring APs / Co-channel contention
- Device Support

Industry

Industry is extremely diverse, and wireless is used in most of them. In manufacturing, machines continue to automate manual tasks. Wireless is used to control many of these machines and wireless sensors monitor environmentals of the machines and let engineers know when maintenance may be required. This may be a change in how the machines sound, heat output, or that they have stopped working.

Automated carts and forklifts are used in many cases to move materials around the plant. These can be equipped with wireless cameras so that someone central can monitor the activities and stop the devices if they are malfunctioning. In some environments wireless controls of vehicles can be utilized with the cameras to keep workers safe in dangerous environments; this is extremely helpful in mining activities.

Wireless tracking tags have multiple uses. They can be used to track inventory through a warehouse or manufacturing plant. This enables the business to know how much of any given item it has on hand and when supplies need to be refilled or are overstocked. Location can be for worker safety, for example, if a mine collapses tags for each worker will quickly let the safety team know how many people and who was working in a given area.

Some of the challenges of operating a network in office environments include:

- Environmentals (Extreme hot / cold)
- Getting backhaul to the wireless bridge

- Building materials
- Competing wireless systems (Co-Channel)
- Device Support

Stadiums, Arenas, and Large Public Venues

Stadiums have historically been a problem area for wireless technologies. When 20,000 - 100,000 new people show up to an area they are not there every day it stresses the network capabilities. Stadiums added DAS (Distributed Antenna Systems) to help offset this problem through the early 2000s. With the use of smartphones, data demands rapidly increased, and these venues brought in Wi-Fi, BLE, and other wireless systems to complement the DAS.

These locations tend to offer free Wi-Fi to the end-users to better support in venue applications and league-sponsored content. In-venue applications range from marketing for sponsors, wayfinding (maps to help you find where you are going), instant replay, and highlights from the event. Food and drink can be purchased, then brought directly to your seat, without the need to stand in line, this can be done from your mobile device or by an employee of the venue.

In building communications are done wirelessly as well, allowing security to stay in contact with each other. This allows information to flow to all areas within the building within milliseconds instead of requiring people running from location to location.

Some of the challenges of operating a network in stadium environments include:

- Limited usage times
- Environmentals (Weather)
- Co-channel contention
- Equipment mounting locations & cabling
- High duty cycles to contend with

Chapter Summary

In this chapter, you learned about the different wireless network types or categories and the planning and administration tasks common in various vertical markets.

Review Questions

1. What kind of network includes personal medical monitoring devices directly participating in the network?
    a. WBAN
    b. WWAN
    c. WMAN
    d. Hospitality

2. Which one of the following is an example of a WPAN technology?
    a. NB-IoT
    b. LoRaWAN
    c. Bluetooth
    d. 802.11

3. When all clients connect to a WLAN through an access point, what role is used?
    a. Access
    b. Distribution
    c. Ad-hoc
    d. None of these

4. What best describe the area covered by a WWAN?
    a. City
    b. Building
    c. 10 meters
    d. Region

5. Approximately what space is covered by a WBAN?
    a. 1 meter
    b. 10 meters
    c. A building
    d. A campus

6. Which one of the following is an example of a long range IoT solutions?
   a. LoRaWAN
   b. Bluetooth
   c. NFC
   d. Z-Wave

7. What is a somewhat unique challenge in implementing retail wireless networks?
   a. Guideline compliance
   b. Very high density
   c. Lead-lined walls
   d. None of these

8. What is a challenge faced in both higher education and K-12 education?
   a. Density of client devices per room
   b. Rogue access points
   c. Compliance with PCI-DSS
   d. None of these

9. What is a challenge faced in agricultural wireless not seen in indoor wireless deployments?
   a. Coverage problems
   b. Varied client devices
   c. Weather-related issues
   d. Legacy clients

10. What vertical market would have to consider HIPPA regulations during wireless application implementation?
    a. Retail
    b. Industry
    c. Healthcare
    d. Home

Review Answers

1. The correct answer is **A**. Personal medical devices are often categorized as WBAN devices.
2. The correct answer is **C**. Bluetooth is a WPAN technology.
3. The correct answer is **A**. The WLAN access role is described.
4. The correct answer is **D**. A WWAN covers a region.
5. The correct answer is **A**. WBANs generally have a range of about 1 meter to cover the human body. In some cases, they communicate with other devices and may have a longer range.
6. The correct answer is **A**. LoRaWAN is a long-range IoT solution. The others work in building or in even shorter ranges.
7. The correct answer is **A**. The best answer is guideline compliance because payment card processing standards (like PCI-DSS) are guidelines imposed on retail.
8. The correct answer is **A**. Both K-12 and higher education struggle with the number of client devices per room between the devices provided to the students and the devices brought by the students.
9. The correct answer is **C**. Many agricultural networks are deployed outdoors and weather-related issues can impact communications.
10. The correct answer is **C**. HIPPA is a regulation related to health and patient records.

Chapter 3: Planning Wireless Solutions

Objectives covered:

3.1   Identify and document wireless system requirements

3.2   Identify and document system constraints

3.3   Select appropriate wireless solutions based on requirements and constraints

3.4   Implement effective project management practices for wireless solution planning and implementation

As engineers, we love to just jump right in and start designing or building something when presented with a problem to solve. In school, the professors give the students well-defined problems that usually are limited in scope and have one right answer, in order to teach particular concepts. Alas, real-world projects are never that neat and tidy.

Most of the time, the problems are more complex, not very well defined up front, suffer from scope creep during (or even after) the project, and specific conditions such as budget and cabling access may make the theoretically *"right"* answer an impractical one.

The reality of system design for complex systems is that there is never one "right" solution. However, there will inevitably be "better" and "worse" solutions. But how does one evaluate the quality of design options and solutions? Fortunately, as much as the design remains something of an *art*, there is also a *science* for selecting better vs. worse solutions. The science of complex system design to arrive at the "best" solution, or at least a "better" solution, is to approach complex system using a methodical process to guide one's thinking both qualitatively as well as quantitatively.

Once a system design engineer can frame the design problem properly, he or she can actually properly process and interpret the information provided by the customer, stakeholders, equipment vendors, and their own organization. The solution administrator can identify the right questions to ask when information is missing, understand the ramifications when the scope is changed, and do all of the other tasks required to create a functional system that meets the needs of the stakeholders.

A system design engineer likely will not have detailed expert-level knowledge of all of the subsystems and components that comprise an overall system. However, a system design engineer will need to understand these systems sufficiently to understand how they interact, relying upon subject matter experts (internal engineers or external equipment vendors) for specific details. A system design engineer is therefore analogous to an orchestra conductor; the conductor need not necessarily know how to play any of the instruments, though does need to know the types of sounds that each instrument creates and how those instruments work with each other to create a cohesive and tonal musical work.

Identifying Use Cases and Applications

Over the last several years, there has been a lot of hype about the *Internet of Things (IoT)*, which is projected to be an array of billions of individual networked appliances. Despite a slight amount of irrational exuberance, this is a field that is clearly growing and for which there are a lot of companies investing a lot of money to propose a diverse array of use cases and applications. Some use cases that have been proposed for IoT applications include home monitoring and automation, commercial asset monitoring, public safety, industrial factory, and warehouse monitoring and automation, city-wide network access and surveillance, etc. Integrating these types of devices with new or existing wireless networks is one of the core challenges for a Certified Wireless Solutions Associate (CWSA).

IoT essentially describes mechanisms and devices that consist of a combination of *sensors* and *actuators* that communicate over and/or are controlled by a wired or wireless network, as opposed to simply operating autonomously. In this context, a *sensor* is a device that measures something in the environment, such as a video camera, microphone, thermometer, barometer, motion detector, etc. An *actuator* is a device that does something to the environment, such as a speaker, a flashing alarm light, a gate opening mechanism, etc. One or more sensor and/or actuator functions are often bundled into the same physical device to perform a given task. A thermostat is a good example of this, as a thermostat performs both the function of measuring the temperature of the environment and the function of activating or deactivating the HVAC system to adjust and maintain the temperature to the desired setpoint. A "smart thermostat" is a thermostat that also communicates with a server on the Internet, which can be accessed from a web browser or smartphone app so that the desired temperature setpoint can be manipulated from a remote location, and not just physically at the thermostat itself.

Connecting a series of sensors and actuators to either a private network or the Internet clearly creates numerous potential use cases that were previously impossible or impractical. Sensors and actuators no longer need to be physically co-located or directly "wired" to each other in order to work collectively. Furthermore, the algorithmic logic (i.e. the software) of how to use sensor data to determine how to adjust the actuators need not be physically located with the sensors and actuators themselves, giving developers the freedom to dynamically refine their algorithms from a central "cloud" location without ever having to touch the physical location.

Alas, many current IoT offerings are *solutions in search of a problem*, and many more make wild and grandiose marketing claims that cannot (at least yet) be supported by the technology. There are also a whole new set of potential problems that IoT solutions introduce, especially in the realms of security and privacy, as early adopters of these technologies have unfortunately learned the hard way.

The IoT offerings that will ultimately prove successful are those that understand the *requirements* and *constraints* of an actual use case and have selected the most appropriate *design parameters* that address those requirements and constraints.

> Don't start with the shiny new cool gadget. Instead, identify the actual problems to be solved, and then pick the best solutions to solve those problems.

This is, naturally, a chicken and egg problem, as new technologies open up new use cases that were never conceived of before. Prior to the launch of the iPhone in 2007, only fiction writers envisioned having pocket-sized devices for accessing the combined knowledge of 6000 years of human civilization. Could we have envisioned watching TV on such devices, or using such a device to write a chapter in a book about them? Marketing hype should therefore not be ignored, as it opens up a new world of potential applications, getting people to start imagining and envisioning new possibilities. Nonetheless, the marketing claims should be screened with a very critical eye, so that when a new use case is identified and a project is launched, realistic expectations can be set and achieved.

A CWSA, therefore, needs to start with identifying *requirements* and *constraints*. In this context, the *requirements* dictate <u>what</u> the system has to do in order to work properly, while the *constraints* dictate what the system has to <u>*work around*</u> in order to meet the requirements. Thus, requirements define the needed functionality, and constraints limit the viable choices of potential design solutions. The difference between requirements and constraints is often subtle, and the stakeholders inevitably fail to distinguish these and therefore present both requirements and constraints simultaneously. Nonetheless, it is important to get these properly distinguished up front. Getting requirements and constraints properly identified and categorized at the outset will enable the ability for both identifying and evaluating the quality of alternative design options, as well as having a mechanism to properly manage scope creep.

Once the requirements and constraints are appropriately characterized, then begins the process of identifying the *design parameters*, which are <u>how</u> the system meets the requirements and constraints. There are typically multiple options, so the options need to be evaluated in a systematic way to determine which solutions will provide a more robust (i.e. "better") design. A more robust design can more easily accommodate changes (i.e. scope creep), which is an inevitable part of the process.

Common Wireless Requirements and Constraints

For wireless systems, there are a set of requirements and constraints that are encountered on virtually every project. A CWSA must be attuned to seeking out these specific requirements and constraints to ensure they are properly captured up front.

Security

Any wireless system will have some type of *users* that are being serviced. These users may be human (e.g. Wi-Fi client devices) or machine (i.e. surveillance cameras, IoT sensors, gate openers, panic buttons, etc.), depending on the application. More complex applications may have multiple types of human and/or machine users. It is therefore essential to define the *access control* methodology of how each type of user shall be authenticated as being "valid", as well as the *client isolation* policies of how associated users of each type are allowed to interact, or not interact, with other users on the same network. The access control and client isolation policies are commonly distinct for different classes of users (e.g. guests, staff, in-house IoT devices, external consumer appliances, etc.).

Security is always about defense-in-depth. At every point in the design, security needs to be considered. This potentially means having multiple and redundant layers of security in different sections of your system so as to ensure that a breach of one line of defense doesn't compromise the whole system. Nonetheless, there is an inevitable tradeoff between security and ease-of-use. The more secure a wired and/or wireless network is made, the harder it becomes for a user to connect to the network and to perform its intended tasks. Depending on the application and the sensitivity of the data, more or less security may be required. For certain types of data, especially financial (e.g. FINRA, PCI-DSS (a data security standard that may impost security constraints on a retail payment card processing network – it is not a regulation, it is a standard)) and health care (e.g. HIPAA), there are government and industry standards of security that must be conformed to and which are periodically audited for compliance.

Access control on a wireless network is usually implemented in one of the following manners:

- **<u>Open access:</u>** This method requires no credentials, so any client device can connect to the network. This method is generally not advisable, as there is no control over what client devices can connect. Additionally, the wireless traffic is unencrypted, so messages can be intercepted in the air, or the connection may be subject to man-in-the-middle (MitM) attacks. Generally, open access is only used for guest access in hospitality Wi-Fi or a hotspot Wi-Fi environment, where the intention is to encourage client devices to access the network. In these cases, client isolation is usually implemented to prevent users from intercommunicating with each other, and access control to the Internet is often implemented centrally by a separate appliance.

- **<u>Personal (Password/Passphrase):</u>** This method requires the connecting client device to possess a particular password or passphrase. This may be pre-configured into the device. The password/passphrase is used as a seed to set up symmetric encryption between the client device and the wireless access point. In Wi-Fi, this is generally implemented with WPA2-Personal, which uses a passphrase along with the MAC addresses of the access point and client device to establish a unique symmetric 128-bit AES encryption key between the Wi-Fi access point and the client device. This methodology is often useful for IoT appliances but has some intrinsic weaknesses. If the association traffic between the access point and the Wi-Fi client is intercepted, it is possible to derive the encryption key for that client device as well as potentially deriving the passphrase in a dictionary attack. Additionally, passphrases often are accessible by human users and thus may be shared intentionally or unintentionally.

- **<u>Enterprise (Authentication Server):</u>** This method requires the connecting client device to authenticate to a separate authentication server before being allowed on the network. In Wi-Fi, this is generally implemented with WPA2-Enterprise, which uses digital certificates or other variations of pre-loaded credentials on the client device (supplicant) to allow an authentication server to authorize the device to access the network through the access point (authenticator) and establish a unique symmetric 128-bit AES encryption key between the Wi-Fi access point and the client device. While this method provides excellent security, it can be laborious to set up and may be impractical in networks with large numbers of guest devices. Additionally, many IoT appliances still don't currently support this method of authentication.

When the network needs to support different classes of users simultaneously, it is common to segment the one physical network into multiple virtual local area networks (VLANs). VLANs are typically (though not always) blocked from interacting with each other, so as to keep the different classes of users isolated from each other. Within each VLAN, the access control and client isolation policies will be different. For wireless connections, a separate SSID for each VLAN is often used, though some systems support *dynamic VLANs* where one SSID is used but the client device is placed on a particular VLAN based on feedback from the authentication server. As an example, a hotel Wi-Fi network may consist of the following types of VLANs:

- Guests: This VLAN is for guests staying at the hotel, and will generally have open access, and client isolation enabled.

- Staff: This VLAN is for staff members of the hotel, and generally will utilize WPA2 Personal or WPA2 Enterprise to only allow pre-configured staff devices on to the network. This network will generally allow connected devices to intercommunicate.

- VoIP / VoWi-Fi: This VLAN is for wired or wireless phones, and generally will utilize WPA2 Personal or WPA2 Enterprise to only allow pre-configured phone devices to the network. This network must allow connected devices to intercommunicate.

- Appliance: This VLAN is for IoT appliances used to detect room occupancy, room temperatures, and other potential applications. WPA2 Personal is typically used, as the appliances likely will not support WPA2 Enterprise.

Client intercommunication may be allowed or not allowed, depending on the nature of the application and whether the appliances are interacting with each other or just reporting to an external cloud server.

Determining Appropriate Bandwidth Based on Expected Usage Requirements

One of the critical sub-requirements of network usage is to understand how much Internet bandwidth is required, and what maximum level of service, commonly known as the service level agreement (SLA), is to be provided to each client device. The amount of bandwidth will generally be critical to actual performance of the system, as well as the perceived performance of the wireless since the wireless portion of the system inevitably gets blamed whenever there are performance issues, even when the root cause of the problem lies elsewhere. Internet bandwidth is also usually the highest ongoing operational expense (OPEX); thus, it is critical to be in the Goldilocks zone, such that there is enough bandwidth available for the application with reasonable margin, without having too much bandwidth that unnecessarily drives up OPEX.

In any telecommunications system, there is always one part of the system that will be the slowest, dictating the overall throughput capacity. This is known as the *bottleneck* of the system. Since budget is virtually always a constraint, the bottleneck should always be the most expensive part of the system. For wireless systems, the most expensive component is always Internet bandwidth to the system, due to very high ongoing operational costs. Ironically, Internet bandwidth is typically one of the easiest items to upgrade after the system is deployed, as continually increasing demand means that service providers are generally investing in their own infrastructure over time to both increase overall capacity and decrease subscription costs.

Internet bandwidth into the system should always be designed to be the bottleneck. The overall internal wireless capacity should always exceed Internet bandwidth capacity.

If the application is to transmit small amounts of data from IoT sensors once every several hours, only minimal bandwidth is required. Conversely, if the application is Wi-Fi in a student housing environment, where individual residents are likely to be streaming different 4k videos to multiple devices simultaneously, investment in adequate bandwidth is critical.

The first step, therefore, is to determine the appropriate SLA per user. As will be seen later, this topic generally will be a critical topic to discuss with various stakeholders and becomes a sub-requirement of both usage and capacity. For machine clients (e.g. IoT sensors, video surveillance cameras, etc.), the amount of bandwidth per client and the expected quantity of simultaneous clients on the network should be straightforward to quantify. For human users (e.g. Wi-Fi for guest access), this becomes more subtle and complex, as the requirement has components that are both quantitative (e.g. bandwidth required for Netflix streaming an HD movie to a client device) and qualitative (e.g. the network is perceived to be fast enough by guest users). Many hospitality environments have tried to monetize this with tiered service plans, where customers are offered a free service at a low SLA but can purchase a higher SLA optionally. In practice, very few customers ever purchase the higher-level SLA, yet will complain bitterly if the "free" SLA is not good enough. Another complication is that the required SLA is likely to increase over time during the life of the wireless system, as applications like 4k and 8k video streaming to client devices become more prevalent. For typical guest access applications, 2-3 Mbps per user (both upstream and downstream) is minimal, with 5-10 Mbps per user is typical. For student housing, specifications of 20-30 Mbps per user are not unusual. These thresholds are inevitably going to increase over time.

Some requirements, like bandwidth, are likely to change during the life of the wireless system. These need to be understood and identified up front such that the system as-deployed is capable of meeting those evolving needs with no or relatively minimal changes.

Once the SLA is determined, the amount of bandwidth to be supplied to the system needs to be quantified. In early telephony, engineers realized that not everybody uses their telephone to be on a phone call simultaneously, and thus they could share the capacity across their subscriber base. Hence, an *oversubscription ratio* is defined to quantify how much real capacity is needed to provide service. For modern networking, the same concept applies; statistically, not every client will consume its maximum SLA simultaneously. Accordingly, the required total bandwidth needed to adequately meet the needs of the use case can be shared, and thus is significantly less than simply multiplying the number of users by the SLA per user. Granted, this is a single "fudge factor" based on what is truly a fairly complex statistical analysis, but it turns out to be a reasonably accurate guideline. For Wi-Fi in residential apartments and hotels, typically a 25:1 or 20:1 oversubscription ratio is used.

As an example, let's take a hotel where up to 500 simultaneous client devices are expected during peak usage times, and the property wants to provide a maximum SLA of 10 Mbps / device. With a 20:1 oversubscription ratio, the required bandwidth to the property is:

500 users x 10 Mbps per user / 20 = 250 Mbps

For Wi-Fi in student housing, which is significantly more bandwidth-demanding, a 10:1 ratio is more appropriate. Conversely, Wi-Fi in an assisted living facility usually does not get a lot of usage, so a 30:1 or even 40:1 oversubscription ratio is appropriate. A student housing property and an assisted living property are often architecturally similar, so consider comparing two such properties with the same number of residents, same SLA, and even the same number and layout of APs. Nonetheless, the student housing property requires approximately 3x – 4x the amount of bandwidth as an analogous assisted living property, simply due to the differences in how the network shall be used.

Constraints in terms of both the potential bandwidth available at the system location as well as the OPEX budget for the bandwidth will influence the results of this calculation. In the example above, it was determined that 250 Mbps was required to provide a 10 Mbps SLA to 500 simultaneous users. However, if the service provider can only get a 100/10 cable circuit for this property, the SLA of 10 Mbps is not realistically achievable. One can sometimes shave a little from the oversubscription ratio assumption (e.g. use 25:1 vs. 20:1), yet the SLA would need still to be lowered to at most 5 Mbps downstream and 1 Mbps upstream. For a hotel, this might be an acceptable compromise. For a student housing property, this would likely lead to unhappy residents. In the latter case, a more expensive bandwidth alternative would be needed, such as using 2-3 cable circuits with a WAN load balancer or a fiber connection from a different service provider, both of which may add significant CAPEX and OPEX costs. The requirement for what speed to provide per user will need to be balanced against the budget constraints of your customer.

Remember that the throughput the user experiences will be based on the airtime available to that user (or device) on the wireless medium. Therefore total airtime utilization required by the various devices and applications becomes the most important factor related to capacity in the channel.

Performing a Wireless System Design

The general process for correctly generating a successful design is as follows:

1. Identify the use case
2. Capture the needs of all of the stakeholders
3. Translate the stakeholder needs into actionable engineering requirements and constraints
4. Evaluate potential design parameters for meeting the requirements and constraints
5. Iterate the design based on changes to the requirements and constraints (i.e. scope creep)

Identify the Use Case

The use case itself is usually a high-level conceptual vision, driven by an organizational *need* that is not being fulfilled. A particular solution may or may not be achievable, especially once constraints enter the discussion, but that doesn't change the *need* for it.

Often, the barrier to such applications is either too much cost and/or the lack of appropriate technology. In wireless projects, the barriers are often related to the costs of infrastructure backhaul wiring. For instance, if the need is to create a city-wide surveillance network, the cost in terms of money, time, and manpower of ripping up streets and sidewalks to run cable or fiber to each location are likely prohibitive. Wireless technologies eliminate the need for much of that data cabling, but the cost of wireless backhaul may still be significantly higher than the costs of the cameras themselves. As technology progresses in terms of both higher capabilities and reduced costs, such wireless applications become more cost-effective.

Once the overall need of a use case is identified, the typical first steps are to perform feasibility analysis, using a very rough set of initial requirements and constraints and evaluating these against a shortlist of potential technology solutions and their costs. This enables the generation of a rough estimate of costs, time, and manpower. The goal of this analysis is to determine whether or not the project is even feasible. This is typically done as a budgeting exercise, though may also involve performing some initial technology evaluation and working with equipment vendors to understand current and potential capabilities and limitations.

Capture the Needs of the Stakeholders

If the high-level use case seems feasible, the next step is to gather the complete set of needs of all of the potential stakeholders, with the eventual goal of converting such needs into requirements and constraints. This process is much harder than it seems, yet it is also the most important. The total set of stakeholders is not always obvious at the outset, but even when all of the stakeholders are identified, do not expect them to be able to fully identify, understand, or articulate their own needs.

Nonetheless, up-front investment in capturing the needs of the stakeholders is critical to the eventual success or failure of the project. One of the biggest sources of scope creep results from a failure to identify all the stakeholders up front, or a failure to adequately capture all their needs. Stakeholders that are late to the table are the least likely to have their needs addressed, and thus ae most likely to be dissatisfied with the result. Furthermore, attempting to satisfy such needs late in the design cycle, which impose new or altered requirements and/or constraints, can often compromise how well other requirements are satisfied and thus the success or failure of the endeavor overall.

Identify all of your stakeholders up front. Don't short-change this step. An extra few days or weeks up front can save months or years of pain later. This will dictate the ultimate success or failure of the project.

To further complicate matters, stakeholders will have different and potentially conflicting needs. It is the job of the system designer to sort this out, potentially necessitating a negotiation process to get different parties to compromise on their conflicting needs.

The stakeholders will obviously be unique for every project, though stakeholders generally span the following functional roles. Depending on the scale of the project and the organization, there may be multiple individuals responsible for a particular role, or conversely a single individual may be responsible for multiple roles.

- **System Owner:** This is the entity (person or organization) that owns the wireless system, or at least the property in which it will be installed and operated, such as a building owner. For the example of a city-wide surveillance network, this is the municipal government. For a hotel Wi-Fi network, this is the hotel management company. The system owner is responsible for funding the initial and ongoing investment in the system. Accordingly, the needs of this stakeholder are largely going to be driven by budget limitations, aesthetics concerns, and branding / customer-facing portions of the system.

- **System Operator:** This role encompasses the persons or organizations responsible for installing, operating, and maintaining, the wireless system. This may be supervised by an IT department of the owner's organization or outsourced to one or multiple external service providers. For large projects, there may be several organizations or departments involved in different phases, including external contractors. In the example of a city-wide surveillance network, the physical installation and physical maintenance may be performed by the Department of Public Works, whereas the day-to-day monitoring and management of the system may be performed by the Police Department. The needs of these stakeholders will generally be operational, focusing around the ease of installation and system turn-up, ease of monitoring and managing the wireless system activity from either a central location or from multiple decentralized locations, ease of maintenance, and system uptime and availability. There may also be constraints imposed on the choice of equipment vendors, to simplify integration with existing systems.

- **Integrated Systems:** These are the machines and/or systems in which the wireless system needs to work with (or around). For co-located wireless systems, this requires frequency coordination to ensure that the wireless systems don't interfere with each other. For surveillance applications, the integrated systems are the cameras, and there will be needs related to the image quality (number of pixels, frame rate, compression, etc.) and the number of cameras at each location, which will ultimately drive the necessary throughput requirements at those locations. For a network of IoT devices, this will drive needs based on the types, quantity, and location of the devices using the wireless network as backhaul infrastructure. Aside from video surveillance cameras, most IoT devices generally do not need much bandwidth, but a high density of devices may influence the number of access points needed. Furthermore, a Wi-Fi network may need to be established only for the integrated systems; one may not need a Wi-Fi access point in a boiler room for human user access, but the sensors and actuators to regulate the building machinery will certainly require wireless access. The needs here are primarily going to be centered around functionality, such as throughput, capability to pull data from sensors and/or push data to actuators, availability, security, and privacy.

- **Clients / Subscribers:** These are the people who are using the system, i.e. the people for whom the use case applies. In a building Wi-Fi network, these are the users who are connecting their devices to the Wi-Fi. For city-wide surveillance, this is the public being observed. The needs here will be centered around functionality, including throughput, usability, availability, security, and privacy.

> *Integrated systems* could potentially be considered as clients/subscribers, but there is a critical distinction between human users and machine users. Integrated systems do not have perceptions and emotional needs, whereas managing the perceptions and emotions of human users is critical for success.

- **Indirect Organizational Entities:** These are the people or organizations that don't have a direct interest in the system yet still have some level of involvement, especially during the procurement and installation phase. Examples of this could include the accounting and marketing departments of the owner, the building manager, etc. These stakeholders are likely to impose project constraints, but may also present additional functionality needs (e.g. coverage areas not otherwise identified) or even *opportunities* (e.g. an unused/underused ring of dark fiber in the building / municipal area left over from a separate project).
- **Government or regulatory authorities:** Wireless systems are constrained to meet regulatory constraints on the frequencies used and transmit power levels.

There may also be local, state, or Federal laws regarding specific applications, especially related to privacy in video and audio surveillance use cases.

It is usually necessary to interview the stakeholders to solicit their inputs, either in person, by phone, or through email or a web survey. It may be impractical or cost-prohibitive to speak directly to all stakeholders, but in-person or video/voice calls are always better than email or an online survey. Such conversations are often revealing, as it is easier to get a stakeholder to verbalize otherwise unstated, controversial, or "delicate" needs in a conversation than in a written email, even when the person knows the interview is being recorded. One common example of a "delicate" need is whether or not Wi-Fi coverage is necessary in the bathrooms. (Bathrooms are typically high-use areas for smartphones, but rarely does anyone acknowledge this.)

It is important when soliciting needs from the various stakeholders to assess their relative *importance* and *elasticity*. *Importance* is a measure of which functions are absolutely essential for success, versus functions that are "nice to have" if not a lot of extra effort or cost is involved. Different stakeholders are likely to have different perspectives on this. *Elasticity* is a measure of how much flexibility there is in satisfying particular needs. For example, in some circumstances the maximum cost budget is fixed, whereas other times the budget is a target value, and there may be some willingness to increase it, especially in return for adding some "nice to have" functionality.

Even if the stakeholders don't get everything they are asking for, it is important for the stakeholders to feel heard and understood. There will be much wider buy-in to the final design solution if everyone has had the opportunity to provide their input and perspective.

While we design and build machines and systems, our stakeholders are human beings. We need to satisfy our stakeholders' emotional needs and perceptions as well as their technical needs. Never underestimate the human element.

Identifying Requirements and Constraints

Once the needs are captured from the various stakeholders, they need to be sorted into *requirements* and *constraints*. The needs identified by the stakeholders will ultimately be used to judge whether the system as implemented is successful. However, needs are usually qualitative items that rarely can be directly designed to.

Recall that *requirements* dictate <u>what</u> the system has to do in order to work properly, while the *constraints* dictate what the system has to <u>work around</u> in order to meet the requirements. In order to properly characterize a stakeholder need as either a requirement or a constraint, define the requirements to be intrinsically independent of each other, so as to capture all of the core functionality needed for performance of what the system has to <u>do</u>, without (yet) considering how those requirements will be satisfied and what may constrain your ability to satisfy them. A use case will generally dictate a list of multiple independent requirements.

> *Requirements* are a summation of what the system has to do, ignoring any limitations. Thus, when defining requirements, assume no limits: you can spend whatever money you like, use whatever vendor equipment you like, have access to whatever expertise or manpower you need, can run cables anywhere you want, etc.

For more complex systems, it commonly makes sense to structure requirements hierarchically. This does not violate independence, though does allow for the requirements to be grouped logically. For example, in a hotel Wi-Fi network, there is typically a high-level requirement to provide client access to all users, which may consist of guests, staff, and appliances. This requirement can be broken down into requirements for isolating the different groups of users (e.g. separate SSIDs on separate VLANs), access control (e.g. open network with captive portal for guests, WPA2 Personal encryption for staff), client isolation (e.g. yes for guests, no for staff), upstream and downstream bandwidth limits (dependent upon number of expected users per VLAN / SSID and amount of bandwidth available at the property), etc.

Unlike requirements, constraints can be highly interdependent with each other, and inevitably impact one or more requirements. Constraints ultimately limit the design options available to be selected and implemented, and thus ultimately drive how well the design can satisfy the requirements.

A system is *overconstrained* if one or more constraints either directly conflict with each other or conflict with one or more requirements. For example, a system is overconstrained if there is a requirement to provide video surveillance of a particular region, but local government regulation makes such surveillance illegal. A system is also *overconstrained* if no viable solutions can be devised. For example, the budget on a project is so small and inelastic that no equipment vendor can meet the cost targets while satisfying other performance requirements. It is best to identify such overconstraints as early as possible in the process. Resolving such overconstraints means circling back with the relevant stakeholders to compromise to remove, or at least relax, the affected requirements or constraints, such that a design solution is possible.

Constraints are generally unique to each project but generally fall into the categories of either physical or organizational.

Physical Constraints

Physical constraints are those dictated by the environment. For wireless systems, such constraints generally include the following:

- **Interior and exterior building materials:** When radio frequencies (RF) interact with objects in the environment (e.g. walls, buildings, furniture, people, etc.), they are subject to several different physical effects, generically known as attenuation. Most importantly, RF propagates differently through different building materials, and generally lower frequency signals tend to propagate better through building materials than higher frequency signals, enabling lower frequency signals to transmit farther. Additionally, building materials like metal are far more reflective than wood or drywall, which will influence where the RF signal propagates. The building materials, in combination with the layout of the property, will serve to dictate how many wireless transmitters and receivers are necessary.

- **Lack of Cable Path Availability Within Buildings:** It is much easier and less expensive to pull power and/or Ethernet cabling in new construction before the walls are installed vs. an existing facility, and easier to pull cables in environments with drop ceilings vs. hard ceilings. In existing buildings with hard ceilings, one is often limited to pulling cable down hallways in cable chases (i.e. crown molding at the top of the wall near the ceiling), and sometimes even that isn't possible. If telephone closets aren't stacked on adjacent floors, one may have to core drill through stairwells or other rooms to get cabling where it needs to go. Cable paths, or lack thereof, can significantly limit where wireless transmitters can be installed, and thus the resulting signal coverage. Being limited to hallways can also significantly limit certain applications, such as Wi-Fi, due to decreased in-unit coverage and higher co-channel interference. In extreme cases, the application may not be doable, as the lack of cable path availability means that wireless transmitters cannot be placed where they are needed for an application. For an example, indoor positioning via Wi-Fi cannot be performed accurately if APs can only be installed in the hallways since trilateralization algorithms would not be able to determine which side of the hallway a signal is coming from. In such an over-constrained case, it may be necessary to relax the positioning accuracy requirement to the general area of a floor, and not a specific room, because of the inability to put wireless sensors in the right locations. It may also be necessary to change the design approach from a Wi-Fi-based system to a BLE-based system using battery-powered sensors that use the Wi-Fi as backhaul, as these would provide positioning accuracy but increase maintenance costs and operations since batteries would need to be periodically changed.

Whenever a design is over-constrained, *SOMETHING* has to give. It is best to control what functionality must get compromised and get buy-in for it during the design process. Otherwise, it will inevitably show up later as an operational failure.

- **Lack of Cable Path Availability Between Buildings:** It is common to need to interconnect multiple buildings for networking or surveillance applications, but it can be cost-prohibitive and/or logistically impractical to rip up roadways and sidewalks to run Ethernet or fiber cable. This is especially true if the buildings are not all on the same private property, such as being across a public street from each other. This type of constraint can usually be worked around by adding in point-to-(multi)point links, though this itself adds expense and channels need to be more carefully managed if co-locating PTP/PTMP links and Wi-Fi APs, as strictly speaking these are co-located wireless systems serving different functions, even though they may be part of the same overall local area network (LAN).

- **Overlapping Wireless Systems:** This is a frequent constraint in Wi-Fi networks, but is applicable to multiple wireless systems on the same frequency band that overlap with each other. It is important to distinguish between *co-located* systems and *neighboring* systems, as well as who owns and controls the overlapping wireless systems. Co-located wireless systems are two systems that intentionally cover the same area. A common example would be a Wi-Fi network for client access and a Zigbee network for IoT or Bluetooth beacons for indoor positioning. For co-located wireless systems, it is best to try to coordinate the channels used (e.g. limiting Zigbee to specific channels at the edge of the Wi-Fi range). Such frequency coordination can be done if the system owner actually owns all of the co-located systems, as this gives the system designer complete control over all of the operating channels. However, if a third party (non-stakeholder) owns and controls that co-located system, that can be devastating to satisfying your key system requirements. Neighboring systems are intended to cover adjacent physical areas, but due to the physics of RF propagation, there is usually some unintentional signal overlap. A very common example of this is two neighboring Wi-Fi systems in adjacent apartments/office suites/buildings, and usually these are owned and controlled by separate entities. In this case, there isn't much that can be done – one can try to do frequency coordination, but with unlicensed spectrum there is no legal recourse if both parties are following the regulatory rules and neither party is intentionally jamming the other. Auto-channel is usually recommended for such scenarios, but auto-channel has its own limitations and can compromise the quality of the rest of your own wireless network. Sometimes, you need to design your network to your own needs and hope your neighbors adapt to you, as opposed to you adapting to your neighbors.

Organizational Constraints

These are constraints dictated by the customer or other stakeholders, based on their organizational needs. Common types of organizational constraints are as follows:

- **Budget:** All projects have some constraint on the amount of money that can be spent, though the level of elasticity in the budget can vary widely by vertical market, by customer, and even for specific projects. Generally, the lower the overall budget available and the more inflexible the elasticity of that budget, the more restricted the potential design options and the more likely that some requirements will not be adequately satisfied. Unrealistic expectations on budgets will overconstrain a design, and particular requirements or other constraints may need to be relaxed or eliminated if the budget cannot be compromised upon. Note that many customers will have specific constraints on budget related to capital expenditures (CAPEX) vs. operational expenditures (OPEX). Some organizations will have funds allocated for a particular project for the initial install (high CAPEX), but will not want be able to make substantial changes to their ongoing operational budget (low OPEX). Such projects generally need equipment that doesn't require ongoing licensing or support contracts during operation or at least will need to pay for such fees in advance. The analogy here is purchasing a new car with all cash up front (i.e. no financing). Other organizations may be structured where they cannot provide all of the funds for equipment up front (low CAPEX) but would prefer to spread out their costs over time (high OPEX). In these situations, the equipment is often leased to the customer, either permanently (e.g. an automobile lease) or for a certain time frame (e.g. an automobile purchase with a car loan from the dealer paid off over a series of years).

- **Schedule**: Like money, there is usually never enough time, though schedules can also vary widely in their elasticity. When deploying a wireless system for a particular event, such as temporary Wi-Fi for a concert, then the schedule absolutely cannot be compromised on by definition. The schedule can emerge as a devastating constraint late in a project. For example, in new construction, a wireless system generally cannot be installed until the building is nearly complete, both for equipment security as well as practical limitations; if you are installing Bluetooth beacons on walls, the walls have to be physically in place before the beacons are installed. Weeks or even months of construction delays are all-too-common, which can significantly compress the installation time allotted to a wireless system in a rush to get the building open to tenants on time. This can lead to installation shortcuts (e.g. skipping Ethernet cable testing, not properly verifying wireless system configuration settings, etc.) which can easily lead to degraded system performance.

- **Aesthetics:** Except for hard-core wireless engineers, nobody likes to see antennas. Some high-end hospitality or apartment complexes, as well as historical museums and landmarks, don't even want to see access points or IoT appliances at all and will insist that they are placed in the walls or ceilings. In addition to making installation and serviceability more difficult, this can impede functionality as additional structure now exists between the APs and the client devices that was not accounted for in the design. There are some creative ways around aesthetics issues, such as painting or even skinning access points to make them virtually invisible. Projects with significant aesthetics constraints are usually willing to spend additional funds to meet such constraints. Conversely, lower-end residential and commercial environments, as well as most industrial environments, are likely not to impose any aesthetic constraints.

- **Equipment Security:** This is a limitation on where wireless system components need to be physically located to protect the hardware assets themselves. On some properties, potential theft of equipment can be a problem, so the equipment needs to be either locked down, hidden, or located out of reach. Damage protection may also be a significant issue, either accidental (e.g. a stray basketball in a gymnasium) or intentional (e.g. concerns about people shooting at equipment). Equipment placed in harsh environments (e.g. extreme heat or cold, saltwater beaches, dirty industrial facilities, etc.) may also require protection from the environment, as well as temperature regulation from fans, heaters, or coolers. Some vendors have equipment in hardened enclosures, or equipment can be placed in locked and/or hardened cases, though usually this incurs additional expense and installation effort. Even for properties where physical equipment security is not a significant concern, it is always a good idea for the supporting equipment in racks in the main distribution frame and intermediate distribution frames (e.g. switches, routers, servers, etc.) to be placed in locked rooms or rack cabinets with limited access, so as to minimize the risk of a system outage.

- **Use of Particular Vendors:** Most equipment vendors want to enforce brand loyalty, and do this by making their product offerings *sticky*, to make it more challenging to switch to a competitor. In wireless applications, this is often done with a combination of sales tools such as cloud-based AP controllers, training, volume purchasing incentives, licensing, and, most importantly, interpersonal relationships. If a service provider is heavily invested in a particular vendor's technology and has already deployed it elsewhere for several other projects, they are generally not going to switch to a new vendor. Sometimes, the customer will be the one to dictate the allowed systems to be used; it is extremely common in hospitality Wi-Fi for hotel brands to dictate the particular AP vendors and models that their franchisees can deploy for "brand consistency." Alas, not all properties and projects are created equal, and different equipment vendors tend to have different niche specialties, so a vendor that works well in one type of vertical or environment may not be the optimal choice for others with respect to performance and/or price.

- **Accessibility:** Many properties may have restricted or secure areas that are only accessible at certain times, with advanced notice, and/or with an escort. Such constraints generally complicate installation and servicing, which can put pressure on scheduling constraints and/or availability and uptime requirements. That said, be careful to avoid the temptation to have an inconvenience justify a compromise to core functionality and performance. In hospitality and other multi-dwelling unit Wi-Fi deployments, accessibility constraints are usually cited as justification for putting access points in hallways instead of in the units, even though hallway deployments generally compromise the coverage and co-channel interference of the Wi-Fi, and thus its ultimate performance. Most such properties would consider overall tenant satisfaction with the Wi-Fi network far more important than a temporary localized outage issue because they need to give a tenant advanced notice to enter their unit to conduct a repair. Even the tenant would most likely welcome the technician in right away to fix the Wi-Fi. If this is properly explained up front, the system owner and the system users are quite likely to relax the accessibility constraints and/or the uptime requirements.

> Understand the relative importance of your requirements and constraints, and design accordingly. Do not compromise on must-have functionality to attempt to avoid some potential inconveniences.

- **Unique Property Constraints:** Some specific verticals or geographic locations may impose unique constraints. Unique property constraints are generally manageable but can require some out-of-the-box creativeness or additional design effort to create a customized solution vs. using off-the-shelf components. As an illustrative example, deploying a wireless system in a hazardous waste processing facility can be especially challenging, since all electronics need to be in explosion-proof enclosures to protect the facility itself. This constraint goes well beyond normal outdoor ingress protection ratings and typically necessitates that all active electronic components are sealed in thick steel boxes, which makes it hard to propagate wireless signals. Such a constraint may necessitate a custom design where external antennas are mounted to the outside of the box with appropriately sealed antenna connectors and perhaps specialized shielded and sealed cabling.

> Whenever installing in an unfamiliar vertical or environment, ask probing questions to ensure all physical and organizational constraints are captured. It is always better to understand these constraints up front vs. being surprised late in the project.

Scope Creep

Inevitably, most projects suffer from some level of *scope creep*, where the requirements and/or constraints are changed at some point during, or even after the completion of, the project. Scope creep usually occurs because the full set of stakeholder needs were not properly identified and captured up front. However, sometimes needs change: budgets are cut, certain hardware isn't available or perform as expected requiring some additional workaround elsewhere, or project scopes are enhanced, such as adding more cameras to a wireless surveillance network, necessitating more bandwidth.

In surveillance projects, additional cameras are almost always introduced late in the design cycle, sometimes even after the wireless backhaul network is fully deployed and operational.

Scope creep will occur, and therefore it needs to be managed in the design process. There are three basic techniques to *design for scope creep,* namely (1) include excess margin, (2) minimize complexity, and (3) maintain functional independence. Note that all of these techniques may not be possible or only of limited applicability due to specific requirements and constraints of a project.

- *Design Margin:* Never design to the limits of your hardware. In a wireless link, it is usually best practice to design for only 60% - 70% of the rated capacity. Links may not be perfectly aligned, may be subject to external interference, and may be forced to carry additional data load, for example because of more client devices than specified. There may also be *surprises*, such as building materials in the environment not being what was expected. Designing with margin will inevitably lead to additional access points, but this is well worth doing up front if there is elasticity in the budget. Similarly, on the wired side of the network, never consume every port in a switch, but always leave about 20-25% of your ports unused – this allows for adding devices later, accommodating bad switch ports, etc. Note that there is a risk of *over-designing* the system in pursuit of design margin, which should be avoided. From a capacity standpoint, this may mean designing for excess capacity that the system will never use, which naturally drives up both cost and system complexity. There is always a fine line to balance between designing with sufficient margin vs over-designing the system.

- *Minimize complexity:* While harder problems will naturally necessitate more complex solutions, it is generally advisable to minimize complexity wherever possible. The greater the complexity, the harder it is to install, operate, and maintain, and the less robust it is to changes when (not if) the scope creeps. Many vendors are beginning to offer solutions leveraging machine learning and artificial intelligence to attempt to manage the complexity of modern wireless deployments. While these algorithms are necessary in certain specific use cases and can provide enhancements for both operations and diagnostics, they add their own additional layer of complexity to the system and may lead to unexpected behavior.

Minimizing complexity usually can be achieved with the following techniques:

- *Minimize the number of components:* In general, the fewer pieces of hardware you have, the easier a system is to install, operate, and maintain. Note, this may need to be balanced against the need to have additional components for margin as well as redundant equipment if high availability / short mean-time-to-repair (MTTR) is a critical requirement.

- *Standardize the design:* While a wireless system is always customized to its environment, complexity can be significantly reduced if standard components (e.g. APs, switches, routers, sensors, etc.) are used and standard configurations are implemented, such as a standardized IP address and VLAN scheme used across multiple projects, even if not every VLAN / IP range is used on specific projects. The more that multiple similar projects are configured to look like each other, the simpler a deployment will be to implement and the easier it will be to troubleshoot. Note that some projects may impose constraints on using particular equipment vendors and/or conforming to a particular IP addressing and VLAN scheme, in which case the design must conform to those constraints.

- *Avoid excess features:* In general, higher-end enterprise hardware for wired and wireless network components (e.g. APs, switches, routers, etc.) have additional features and subsystems that may not be used on your projects. Examples of these from Wi-Fi networking include wireless intrusion detection and/or prevention systems (WIDS/WIPS), layer 7 firewalls for stateful packet inspection, integrated antenna beam steering, artificial intelligence (AI) / machine learning (ML), etc. These features add additional complexity in both initial setup and troubleshooting, especially if they are enabled when they are not intended to be. Additionally, the cost of these features is embedded in the price of the hardware, whether you use them or not. Such features are available for particular use cases and applications; if such hardware and/or software features are applied to satisfy particular requirements or constraints, by all means deploy them. If not, however, consider going with simpler systems that don't provide those features to save on both cost and complexity.

Note that some unused features may be standard in all equipment, either by industry-standard or by de facto practice, or the choice of equipment vendor may be a constraint in the design. Accordingly, it is recommended that you understand all of the features in the components you select, as well as explicitly disabling any excess features not being used. The equipment vendor should have documentation or support that can help answer and clarify these things; else it is worthwhile having access to one or more experts (i.e. in-house or exterior consultants) that understand these features and can indicate where they should and should not be used. Some features may also be available in multiple system components, though may be cleaner to implement in some components vs. others, or in both if those features are critical for performance. For example, client isolation is a typical requirement in guest Wi-Fi networks, as guests should be able to get out to the Internet but not intercommunicate with each other, or with other network users such as staff or security. Virtually all APs allow clients connected to a given SSID to be isolated from each other on the same APs. Blocking clients from intercommunicating on the same SSID but different APs is more challenging, and vendor solutions often require much complex configuration and/or may not do a perfect job of isolating such users, especially if there may be a combination of wired and wireless guest users. However, managed network switches have a feature called access control lists (ACLs), which act as a layer 3 firewall on your network switch. A standardized IP scheme can be implemented to ensure a standard set of ACL rules that allow the intercommunication you want and block the intercommunication you don't.

- *Require installers and maintainers to produce professional installations:* There are several best practices and guidelines on wiring techniques to keep the main distribution frame (MDF) and intermediate distribution frames (IDFs) neat and tidy. This includes using appropriate lengths of patch cables (i.e. so long cables do not clutter the installation), using patch panels and patch cords instead of directly running the wired feeds into a network switch, color coding patch cables by application (voice, video, data, etc.), bundling cables in cable trays, using zip ties or velcro ties to bundle patch cables, etc. This adds some extra time, effort, and thus cost to the original installation, but will save significant effort and money later on during troubleshooting, as well as avoiding mistakes during system maintenance or when (not if) scope creep requires adding additional components to the network.

- *Maximize functional independence:* The requirements of the system are defined independently from each other. The design parameters selected to satisfy the requirements and constraints will dictate whether or not that independence is maintained. The more that independence is maintained in the design, the easier it becomes to accommodate scope creep, as the change in a particular functional requirement will only impact one aspect of the design, and not ripple into the design of how other requirements are satisfied.

Selecting and Evaluating Design Parameters

Once the requirements and constraints are fully captured, then *design parameters can be generated* and evaluated. The specific design parameters need to be matched to the specific requirements and constraints. *Design parameters* dictate <u>how</u> the requirements are going to be satisfied. There are always choices to be made in satisfying the functional requirements, even in the presence of constraints. As stated above, there is no one "right" answer, but there will be several better and worse alternatives, so a systematic method is necessary in order to evaluate different design alternatives and select the best options of all of the available choices. Several methodologies have been proposed that vary in both structure and approach, but all such methods generally serve to maximize independence and minimize complexity. One such method, known as *axiomatic design*, is presented below as it is simple to understand and apply effectively in the design of wireless systems.

Evaluating the Quality of a Design

Functional requirements (FRs) are independent of each other, by definition. Ideally, each functional requirement should have one, and only one, corresponding design parameter (DP), and that DP should only influence its corresponding FR. This ideal case is known as an *uncoupled design*. As design requirements get more intricate and more constrained, this is usually difficult, if not impossible, to achieve in practice. A *coupled design* is the case where all of the DPs impact all of the FRs. This is the situation that needs to be avoided, as the change to any single FR or constraint (e.g. from scope creep), and thus to its corresponding DP, impacts all of the other FRs. The change to the DP requires additional changes to other DPs to compensate, which then ripple back to the original FR. Such a design requires iteration and thus becomes very difficult to optimize. More importantly, the design is very *fragile*, and will, therefore, have difficulty accommodating even minor changes in the requirements. Fortunately, appropriate choices of DP can limit the amount of coupling, and it is frequently possible to select and limit DPs in scope to provide a *decoupled design* so that the DPs can be changed in a particular sequence so as to affect the FRs without requiring further iteration.

In an *ideal design*, the number of FRs equals the number of DPs. An *insufficient design* exists when there are more FRs than DPs, as it is impossible to independently satisfy all of the FRs because there is an insufficient number of DPs to do so. Conversely, a *redundant design* exists when there are more DPs than FRs is known as *redundant*. In this case, one or more of the additional DPs can generally be held fixed or tweaked within minor ranges, to allow independent or sequential manipulation of the other DPs. (The Principles of Design, N. P. Suh, 1990)

A simple illustration of these principles is demonstrated by the design of a water faucet. The two fundamental FRs for a water faucet are as follows:

- FR1: Control the water flow rate
- FR2: Control the water temperature

Water is generally supplied to a faucet via two pipes, one supplying hot water and the other supplying cold water. Accordingly, the seemingly simplest solution, as shown in Figure 3.1, is to have two faucet valves, one to control the volume of hot water flow and one to control the volume of cold-water flow. The hot water and cold-water valves are therefore the two DPs. However, anyone who has ever used such a faucet knows intrinsically that this is a *coupled design* – you cannot adjust either valve by itself without affecting both water flow rate and water temperature. Getting the desired temperature and flow rate, therefore, require iterating the position of both faucets.

Contrast this with a faucet with a mixer tap valve, as shown in Figure 3.2, such that moving the lever vertically controls the flow rate by drawing in water from both pipes equally, while moving the lever horizontally controls the temperature by deliberately changing the valve size (i.e. inlet areas) of each pipe, thus altering the ratio of hot water to cold water. Both parameters can now be controlled independently, making this an example of an *uncoupled design*.

$$\begin{bmatrix} \text{FR1: Flowrate} \\ \text{FR2: Temperature} \end{bmatrix} = \begin{bmatrix} X & X \\ X & X \end{bmatrix} \begin{bmatrix} \text{DP1: Hot Faucet} \\ \text{DP2: Cold Faucet} \end{bmatrix}$$

Figure 3.1: Coupled water faucet design.

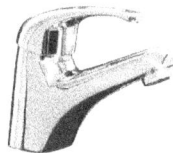

$$\begin{bmatrix} \text{FR1: Flowrate} \\ \text{FR2: Temperature} \end{bmatrix} = \begin{bmatrix} X & 0 \\ 0 & X \end{bmatrix} \begin{bmatrix} \text{DP1: Mixer Valve Vert.} \\ \text{DP2: Mixer Valve Horiz.} \end{bmatrix}$$

Figure 3.2: Uncoupled water faucet design.

A mixer tap valve generally assumes that the pressure and size of the hot and cold-water feeds, and thus their volumetric flow rates, are equivalent. What would happen, however, if one of the feeds had a disproportionately higher flow rate, such as a system constraint where the cold-water pipe provides cold water at a higher pressure, and thus a higher flow rate than the hot water pipe. In this scenario, adjusting the vertical lever of the mixer tap valve impacts both the flow rate and the temperature, as the volumetric flow rates from each supply pipe are not equal. However, adjusting the horizontal lever of the mixer tap valve still only impacts the temperature (i.e. the ratio of hot to cold water) and not the flow rate. By adjusting these two parameters in sequence, i.e. first the vertical lever for flow rate and then the horizontal lever for temperature, both parameters can be adjusted without further iteration. This is an example of a *decoupled design:*

$$\begin{bmatrix} \text{FR1: Flowrate} \\ \text{FR2: Temperature} \end{bmatrix} = \begin{bmatrix} X & O \\ X & X \end{bmatrix} \begin{bmatrix} \text{DP1: Mixer Valve Vert.} \\ \text{DP2: Mixer Valve Horiz.} \end{bmatrix}$$

Wireless System Requirements and Constraints

This technique can readily be extended to larger systems, such as the high-level functional requirements that apply to a typical wireless network, whether it is Wi-Fi, Bluetooth, Zigbee, Z-Wave, or even cellular. The typical functional requirements for a wireless network are as follows:

- **FR1: Connect human and/or machine client devices and applications to a wireless network.** This requirement dictates understanding the intended use case(s), so encompasses what types of client devices that connect to the network and what types of applications those devices will be running. For a Wi-Fi network, this may consist of smartphones, tablets, and laptops in an environment that are doing email, web browsing, video streaming, etc. For a wireless in-building positioning network, this could consist of Bluetooth beacons getting signals from smartphone apps. This requirement also encompasses how clients will be authorized and if multiple types of client devices need to access the same network, such as guests connecting to an open Wi-Fi network via a captive portal, IoT devices and/or camera connecting to a secured network using WPA2 Personal, and staff devices connecting to the network with WPA2 Enterprise. Thus, this requirement will serve to dictate the number of SSIDs that need to be broadcast and the security, access control, and bandwidth requirements of each SSID. The bandwidth required per device type is also encompassed here.

- **FR2: Provide adequate wireless signal coverage to all areas of the facility.** This requirement dictates understanding where client devices are going to be accessing the wireless network in the facility and the appropriate signal strengths to ensure at least minimal acceptable levels of performance.

- **FR3: Provide adequate capacity.** As more and more devices are connecting to wireless networks, this requirement reflects the need to provide enough capacity to handle all of the simultaneous devices that will be connected to the network. This will be satisfied both in terms of the technology of the APs (i.e. for Wi-Fi this would consist of the choice of 802.11n vs. 802.11ac vs. 802.11ax) as well as the quantity of the APs, as more APs are required to handle high client density environments beyond simple signal coverage. Note, this is intentionally separated from the client devices and applications; if three SSIDs are required on a Wi-Fi network because of the different client device types and applications to be used on the network, those three SSIDs are still required independently of whether there are 30 client devices, 300 client devices, 3,000 client devices, or 30,000 client devices connecting to the network simultaneously. Naturally, different SSIDs may require accommodating different quantities of client devices.

- **FR4: Manage the network.** This requirement dictates the need to monitor and maintain the network, and potentially to be able to make changes to the network over time. This may dictate the need for a network management system, a vendor's cloud controller, a custom framework integrating APIs from multiple sources to get disparate systems to intercommunicate properly, or simply configuring everything in standalone mode and only look at it when something breaks. This mechanism will be different whether it is an internal IT team vs. an external vendor that is maintaining the network, whether multiple vendor systems need to be integrated, etc.

- **FR5: Integrate with the backhaul infrastructure.** There are no wireless networks without wires, and the quality of the wireless network is only as good as the quality of the wired network infrastructure that it relies upon. Thus, this requirement encompasses the need for all of the cabling infrastructure, wireless PTP/PTMP backhaul links, network switches, and routers necessary to establish communication both through and outside the network. This requirement may also require integration with a data analytics engine to capture data and to process it to accommodate the use case, such as for a wireless in-building location network.

Note that these requirements have been defined independently of each other. The design parameters selected may (and usually will) break that independence. Design *coupling* occurs when a particular design parameter influences multiple requirements. The degree to which the design parameters allow you to satisfy these requirements independently will ultimately dictate how well one can accommodate changes to particular requirements (i.e. scope creep) without sabotaging the overall functionality of the system.

In terms of constraints, budget is extremely common, though other constraints such as aesthetics, cable paths, co-located RF networks, etc. need to be taken into account.

The corresponding design parameters for a generic Wi-Fi network are as follows:

- **DP1: Access Point Model(s).** This dictates the choice of a particular AP vendor and the model(s) of access point, which dictate its technological capabilities needed to satisfy the client device types and applications (FR1) as well as its environmental and mounting needs. Depending on the coverage area requirements (FR2), multiple compatible models may be selected, such as indoor access points vs. outdoor access points, models with external antenna ports to accommodate directional antennas, etc. If there is a constraint to using a particular vendor, either due to budgetary constraints and/or constraints to use a particular AP vendor to accommodate existing knowledge and remote management infrastructure, there may be limits as to the choices of models are available or limits on the availability of particular features that may be desirable to fine-tune the network.

- **DP2: Access Point Locations.** This dictates where the APs are placed throughout the facility to satisfy both coverage (FR2) and capacity (FR3) requirements. If there are constraints on aesthetics or where cables can physically be run to, this may constrain where the APs are located and ultimately the quality of coverage in particular areas.

- **DP3: Access Point Channels.** This dictates the channel settings on each radio frequency band that the AP operates on. Note that APs on the same or overlapping channels on neighboring APs (FR2) will result in interference, which can significantly impact throughput performance (FR1) and capacity (FR3). While most access point vendors offer some auto-channel capability to select channels automatically, no vendor does this perfectly (despite grandiose marketing claims to the contrary), as the selection of channels is often dictated by the location of the APs relative to each other and the physical structure (e.g. walls, warehouse aisles, etc.) in between them. Frequency coordination can be a significant issue, based on constraints to accommodate other wireless systems in the environment on the same radio frequencies. For example, if an independent ZigBee system for an IoT application is co-located in the environment, this may further limit what channels are available to ensure that the Wi-Fi network and the ZigBee network don't interfere with each other.

- **DP4: Access Point Transmit Power.** This dictates the transmit power settings on each radio frequency band that the AP operates on. The transmit power dictates the effective coverage area (FR2). In many applications, the maximum transmit power of the client devices is significantly lower than the maximum transmit power of the access point, meaning that the access point transmit power needs to be lowered so as to not create "false" coverage areas where the client device can hear the AP (since the AP is shouting), but the AP cannot hear the client device (since the AP is whispering). In high-density environments (FR3), transmit power may also be intentionally lowered, to create smaller coverage cells and accommodate more APs within a specific area with minimal self-interference to handle larger densities of client devices. Note also that if the transmit power is too high, an AP is more likely to interfere with its neighboring APs due to channel re-use, which will impact overall performance (FR1).

- **DP5: Network Management System.** This dictates the system that will be used to monitor and manage the network. Many AP vendors have controller appliances and/or cloud-based controllers for remote monitoring and configuration of their networks, with some systems being a lot more sophisticated and expensive than others. If the responsible party for managing the network is already invested in a particular AP vendor's management system (or a third party or custom system that has been tuned to accommodate particular vendor equipment), they will generally constrain the choice of AP vendor and AP model to in current and future applications so that it more easily integrates into their monitoring and management system.

- **DP6: Wired Network Infrastructure.** This dictates the backbone infrastructure that is used to support the wireless network and to provide backhaul from the access points to wherever the data needs to go, whether that is the Internet, an on-site server for analytics, etc. Thus, this is the system of cabling, network switches, servers, routers, wireless PTP/PTMP bridge links, etc. that make up the core of the network. Budgetary and physical cabling constraints may limit the effectiveness of the wired network infrastructure. In many complex system designs, the support infrastructure often gets less attention, though the performance of the overall system absolutely depends on the quality and robustness of the underlying infrastructure.

In this case, we have six DPs and five FRs, resulting in a redundant design, as shown below. Generically, this is also a coupled design.

$$
\begin{bmatrix} \text{FR1: Usage} \\ \text{FR2: Coverage} \\ \text{FR3: Capacity} \\ \text{FR4: Control} \\ \text{FR5: Integration} \end{bmatrix} = \begin{bmatrix} X & X & X & X & X & X \\ X & X & X & X & X & X \\ X & X & X & X & X & X \\ X & X & X & X & X & X \\ X & X & X & X & X & X \end{bmatrix} \begin{bmatrix} \text{DP1: AP Model} \\ \text{DP2: Location} \\ \text{DP3: Channel} \\ \text{DP4: Tx Power} \\ \text{DP5: Network Mgmt} \\ \text{DP6: Wired Network} \end{bmatrix}
$$

Evaluating the Design Parameters Against the Requirements

There are certain choices we can make in the design parameters above to ensure that we result in a decoupled design. Often, these choices are characterized by vendors and other engineers as *best practices* in deployments. Best practices are usually techniques that are empirically learned over time and recommended because they seem to work in many applications. In actuality, best practices work because they actually serve to maximize the independence (i.e. reduce the overall coupling) in a design, making deployments simpler and easier to manage.

In Wi-Fi, it's considered best practice to put the APs in rooms instead of hallways, and to stagger their position on neighboring floors. Why? We know that the client devices (FR1) tend to have significantly weaker transmitters than the access points (DP1). Accordingly, the AP is generally shouting whereas the client devices is whispering. Optimal performance is therefore achievable by placing the APs (DP2) as close as possible to the clients with the minimum number of physical obstructions (e.g. walls). With respect to other APs, we want to discourage overlapping communications due to channel conflicts (DP3), so the APs should be placed as far apart as possible with as many intermediate obstructions as possible. In practical terms, this means staggering the position of APs both horizontally and vertically. APs should be placed in rooms on alternating sides of the hallway and staggered from floor to floor. By positioning APs so that they have minimal AP-to-AP interference, the impact of the AP position (DP2) on ultimate performance is minimized if not eliminated.

For channel settings (DP3), a static staggered channel pattern on both the 2.4 GHz band and 5 GHz band is recommended, based on the AP positions (DP2). Additionally, unless throughput requirements (FR1) dictate otherwise, it is generally best to use the smallest channel sizes available so as to maximize the number of independent channels. This allows the largest amount of space and the largest amount of internal building structure between APs that repeat the same channel. For high-density deployments (FR3) with dual-band client devices (FR1), it may be necessary to disable 2.4 GHz radios to minimize co-channel interference, such that high-density capacity is handled on the 5 GHz band and the 2.4 GHz band is primarily used for lower-capacity coverage.

The transmit power settings (DP4) impact the effective range at which an AP can be heard by a client device (though not the range at which a client device can be heard by the AP). Furthermore, the laws of physics (effectively a constraint) dictate that 5 GHz travels less far than 2.4 GHz, especially through walls and other structure. To simplify AP location (DP2) and channelization patterns (DP3), it is common best practice to set a fixed transmit power on all APs, which makes the coverage area roughly the same and allows the APs to be located (DP2) in an evenly spaced manner, which also allows the channel pattern (DP3) to be simpler to formulate. Furthermore, a fixed offset of 6-9 dB between the 2.4 GHz and 5 GHz bands to ensure a roughly equivalent coverage area (DP2) on both bands. If particular areas require more or less coverage, due to building layout and structure, transmit power can be tweaked on individual APs.

This is illustrated in the following example of a Wi-Fi design for a multi-level hotel. In Figure , the APs are located in the hallways, and while staggered in position from floor to floor, they still result in coverage problems in particular guest rooms (FR2), as well as high co-channel interference between APs, despite having proper channel (DP3) and transmission power (DP4) settings.

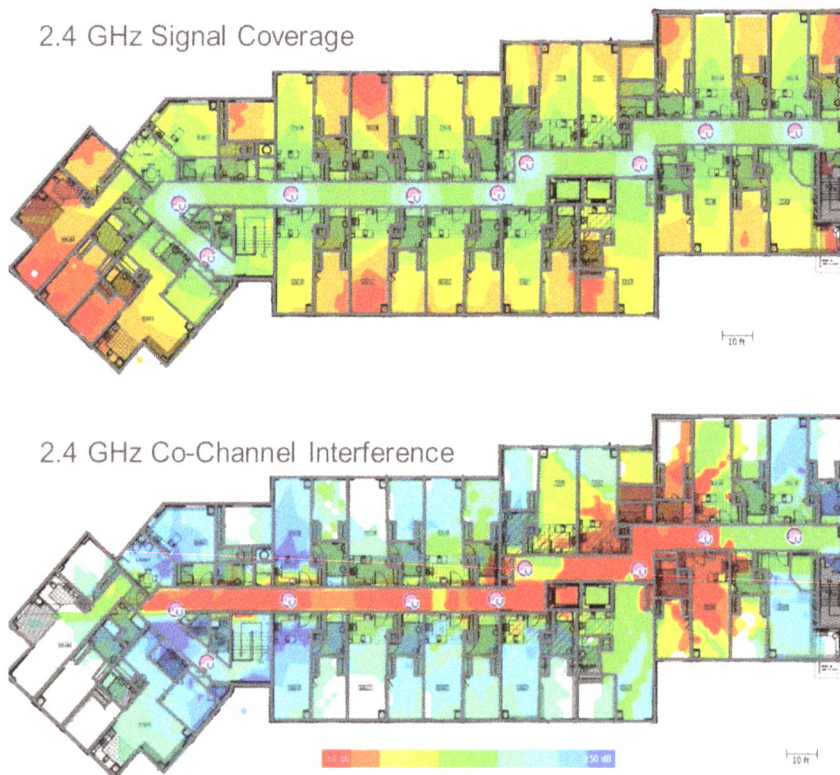

Figure 3.3: Worse design locations: Predictive coverage model (top) and co-channel interference (bottom) for a hotel with APs in hallways (faded APs are on neighboring floors)

By contrast,

Figure 3.4 shows the result of placing the same number of APs in alternating rooms on each side of the hallway. Coverage in the guest rooms (FR2) is dramatically improved, and co-channel interference between APs, while not eliminated because of the limited channel choices on the 2.4 GHz band, is drastically reduced (DP3).

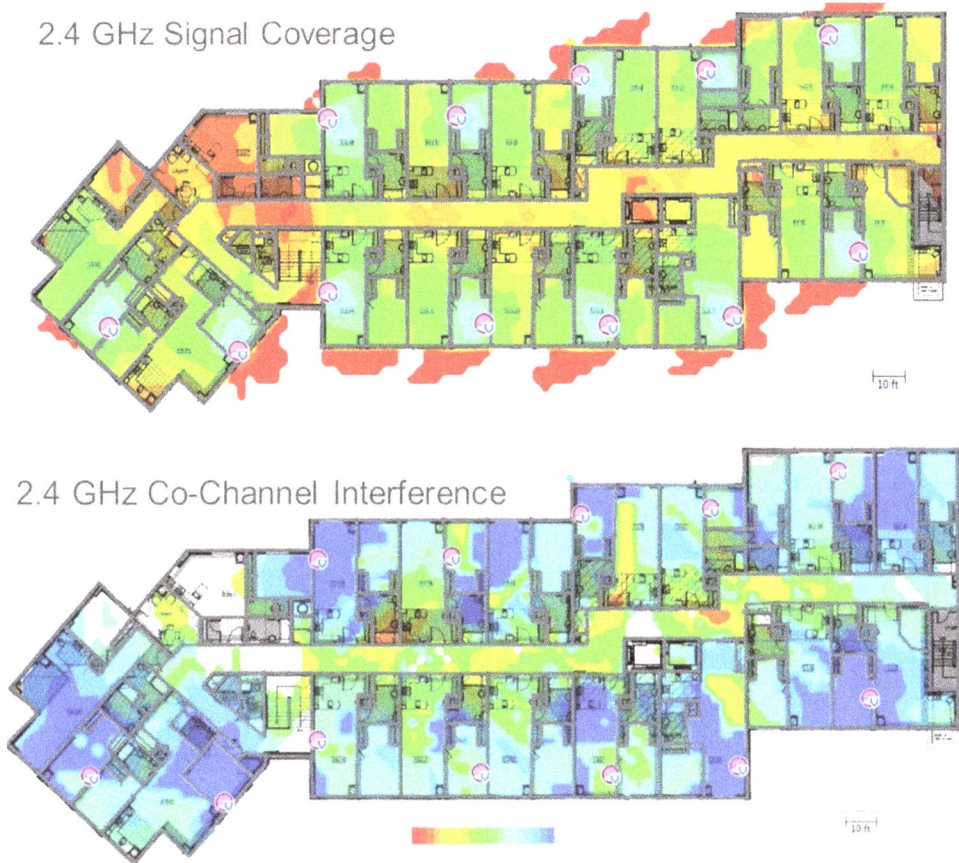

Figure 3.4: Better design locations: Predictive coverage model (left) and co-channel interference (right) for hotel with APs in rooms (faded APs are on neighboring floor)

Thus, we can decouple our design matrix by using the following best practices in sequence:

- Set a fixed transmit power (DP4) on all of the APs on both the 2.4 GHz and 5 GHz band to ensure an equal coverage area (DP2) from all APs on all bands. This resolves the redundant design problem by selecting one DP to have a fixed value.

- Select the APs (DP1) for the application. If using APs with internal antennas, which is common to meet basic aesthetics constraints, the antenna pattern and gain is fixed based on the AP vendor and model. If using APs with external antenna ports, the antennas must be selected at this step.

- Select the AP locations (DP2). This can be done starting with predictive modeling as shown in the figures above, and/or reinforced with passive site surveys to understand how the AP model selected will propagate through the walls in the environment. Placing the APs within rooms instead of hallways also simplifies channelization (DP3) and minimizes the potential for co-channel interference.

- Select the AP channels (DP3). The antenna (DP1), locations (DP2), and transmit power (DP4) are now established, so a channelization pattern can now be created to minimize the potential for co-channel interference.

- Select the AP controller (DP5). The AP vendor and model selected will often dictate what control options are available. Some vendors may offer multiple methods of control (e.g. standalone, local controller, cloud controller), which can be selected based on the use case (FR1), the capacity (FR3), and the monitoring and management needs (FR4).

- Establish the wireless infrastructure (DP6). The locations of necessary MDF and IDF closets, horizontal cable runs to APs (DP2), any wireless point-to-point / point-to-multipoint wireless connections, and other infrastructure needs get established here. Some AP vendors may also allow for control of switches and routers, which may dictate specific hardware choices in order to be compatible with the monitoring and management requirements (FR4).

By following these best practices, in sequence, the design can now be decoupled, as shown below:

$$
\begin{bmatrix} \text{FR1: Usage} \\ \text{FR2: Coverage} \\ \text{FR3:Capacity} \\ \text{FR4: Control} \\ \text{FR5:Integration} \end{bmatrix} = \begin{bmatrix} X & X & 0 & 0 & 0 & 0 \\ X & X & X & 0 & 0 & 0 \\ X & X & X & X & 0 & 0 \\ X & X & X & X & X & 0 \\ X & X & X & X & X & X \end{bmatrix} \begin{bmatrix} \text{DP4: Tx Power} \\ \text{DP1: AP Model} \\ \text{DP2: Location} \\ \text{DP3: Channel} \\ \text{DP5: Network Mgmt} \\ \text{DP6: Wired Network} \end{bmatrix}
$$

This same process can be used to evaluate alternatives in a systematic way. For example, in the case of selecting AP locations (DP2), placing the APs in the hallways will create much self-interference between neighboring APs, which will degrade the ability to satisfy the use case (FR1). This clearly creates a coupling term between the AP locations (DP2) and the use case (FR1), which we want to avoid. Putting the APs in the rooms minimizes or eliminates the co-channel interference, and thus removes (or at least minimizes) that coupling term. Similarly, selecting channels poorly, or allowing an AP with an insufficient algorithm to do it for you (DP3), will potentially create self-interference and thus impose a coupling term between channel setting (DP3) and the use case (FR1).

Thus, when evaluating design alternatives, the CWSA needs to evaluate all of the requirements in sequence and understand what the potential impact of that design choice is on the other requirements. The larger the impact on the other requirements, the more coupled and thus more fragile the design.

Chapter Summary

When designing a wireless system, it is essential to fully understand your requirements and constraints before diving into the generation of a system design. This means understanding the use case and engaging with all stakeholders to collect their needs. Once the needs are collected, they need to be sorted into requirements (i.e. what the system needs to do) and constraints (i.e. what the system needs to work around). After the requirements and constraints are quantified, each design parameter can be selected and evaluated for its ability to satisfy its own intended requirement and the constraints while minimizing its impact on the other requirements. Going through this process systematically will provide a design that is more robust to scope creep, as it will more readily accommodate changes to requirements and constraints as the project progresses.

Review Questions

1. What is defined as *how* the system meets the requirements and constraints?
    a. Design parameter
    b. Functional requirement
    c. Business constraint
    d. Regulatory constraint

2. What is the slowest part of a network system known as?
    a. Firewall
    b. Bottleneck
    c. Router
    d. Port aggregator

3. Why is oversubscription used on key network links such as WANs and Internet connections?
    a. Because CPUs cannot keep up with modern network speeds
    b. Because Internet links are always faster than required
    c. Because WAN links are always faster than required
    d. Because devices do not require full throughput at all times

4. What role is responsible for maintaining, administering and securing a wireless solution?
    a. System operator
    b. System owner
    c. System user
    d. None of these

5. Building materials impose a physical constraint on wireless solutions?
    a. True
    b. False

6. Which one of the following is a common organizational constraint?
    a. Overlapping wireless systems
    b. Budget
    c. Lack of cable paths between buildings
    d. Lack of power at mounting locations

7. What term or phrase is used to reference an increase in requirements after project execution has started?
    a. Annoying growth
    b. Scope creep
    c. Irregular intrusion
    d. None of these

8. What term or phrase defines a design wherein each functional requirement has one and only one corresponding design parameter?
    a. Coupled design
    b. Grouped design
    c. Uncoupled design
    d. Inferior design

9. What is often built into a project plan to address scope creep?
    a. ROM estimates
    b. Margin
    c. Detractions
    d. Retractions

10. Requirements dictate what a system must do in order to achieve the desired results.
    a. True
    b. False

Review Answers

1. The correct answer is **A**. A *design parameter* specifies *how* the system meets the requirements and constraints.

2. The correct answer is **B**. The bottleneck is the slowest component in an interconnected system.

3. The correct answer is **D**. Oversubscription is used because not all devices require all the available throughput all the time. Some ratio is used to provide that which is actually required on WAN links and Internet connections.

4. The correct answer is **A**. The system operator is responsible for installing, maintaining, operating and securing a wireless solution.

5. The correct answer is **A**. It is true that building materials impose a physical constraint on wireless solutions. Outer walls made of metal may reflect signals inward. Internal walls will cause attenuation. Ceilings and floors also impact propagation.

6. The correct answer is **B**. Of those listed, budget is an organizational constraint. Additional organizational constraints include schedule, aesthetics, vendor lock-in, and accessibility.

7. The correct answer is **B**. Scope creep involves changes in requirements and constraints after project execution has begun.

8. The correct answer is **C**. Uncoupled design is achieved when each functional requirement is matched with a single design parameter.

9. The correct answer is **B**. Margin is often built into project schedules, budgets or requirements to allow for some scope creep throughout the project.

10. The correct answer is **A**. Requirements dictate what the system has to do in order to work properly and achieve the desired results.

Chapter 4: RF Communications

Objectives Covered:

2.1   Explain the basic RF wave characteristics, behaviors and measurements used for wireless communications

2.2   Describe the fundamentals of modulation techniques used in wireless communications

2.4   Describe the basic use and capabilities of the RF bands and other wireless carriers (light and sound) used for communications

In today's connected world, there is a multitude of wireless technologies that connect us and our systems. Think about all of the "wireless" systems that you use daily: Wi-Fi, Bluetooth, cellular networks, GPS, perhaps AM radio if you want to listen to some sports updates or political commentary. Also, let's not forget about television, whether it's being delivered via satellite or terrestrial (ground-based) antennas. You might even have a set of handheld radios that you use for work or recreation.

All of these systems serve wildly different purposes, including pinpointing your location on the globe, browsing online auctions, or just simple voice communication with a friend or coworker who is a couple of kilometers away. So, we've only named a handful of systems with which everyday people interact. What about all of the radio systems that are used by government, military, emergency responders, corporations, the aerospace industry, and the other countless verticals that use wireless technologies? The amount of radio systems in use today is staggering - the list provided above puts but the tiniest, most inconspicuous scratch on what humans do with radio technology daily.

Hopefully, we've made our point: there's a lot of wireless stuff out there, and although they're all completely different types of systems, they all have something in common: they all utilize radio waves to transmit information. At their most basic form, all of the systems that we mentioned above including Wi-Fi, Bluetooth, GPS, AM and FM radio, broadcast and satellite television, and even handheld radios all use the same basic concepts to wirelessly move information. As a result, all of the most fundamental concepts of how they work are the same. Don't misunderstand us: each technology has a tremendous amount of complexity in the layers above physical aspects of radio waves, but they are the same in their most basic form.

All radio systems are subjected to the same physical limitations and share behaviors when interacting with objects in the physical realm. That said, there are many variables such as frequency, which wildly changes the behavior of radio waves on a sliding scale. If this all sounds complicated, it is, but don't worry. The purpose of this chapter is to lay the groundwork of how radio waves work, define important terminology around radio communications, and cover some of the different methods used to move information via radio.

RF Wave Characteristics

Later in this chapter, we'll cover some of the modulation schemes that wireless technologies use to convey information, such as FM, Amplitude Shift Keying, and frequency hopping. To fully understand how data is modulated, it's important to what a radio frequency (RF) wave is, and what it's characteristics and variables are. You don't need to know everything about the physics behind electromagnetic waves, but this guide will serve as a starting point.

Waves

The first thing we must define is a *wave*. A *wave*, in the realm of physics, can be defined as a motion traveling through matter or space. Note that the wave is not necessarily a movement of matter, but it is a motion — such as oscillation — traveling through matter or space (non-matter). To visualize this, think of the waves in the ocean bobbing up and down. Now imagine a beach-ball placed on top of the waves: as the waves pass by, the ball moves up and down (vertically), but the ball won't move *with* the waves (horizontally). If you investigate even more closely, you'll notice that the water doesn't travel with the waves, either. Instead, the waves are *passing through* the water.

Similarly, an electromagnetic wave is an oscillation traveling through space. A specific form of electromagnetic waves (wavelength and frequency, which we'll get to shortly) is used for radio communications and is known as a *radio frequency wave*, with *radio frequency* often being shortened to *RF*. RF systems rely on the phenomenon of electromagnetic waves to wirelessly transmit information.

Waves of all types are often represented using a *sine wave*, which represents the complete cycle of a radio wave: starting at zero, moving to a positive, going back to zero, and down to a negative, and then back to zero again where the whole cycle will repeat as seen in Figure 4.1.

Frequency

Frequency refers to the number of wave cycles that occur in a given window of time. Measured in one-second intervals, a frequency of 1 kilohertz (kHz) would represent 1,000 cycles of the wave in one second. To remember this, keep in mind that a wave cycles frequently, and how frequently it cycles determines its frequency. Table 4.1 shows the relationship between hertz, kilohertz, megahertz, and gigahertz.

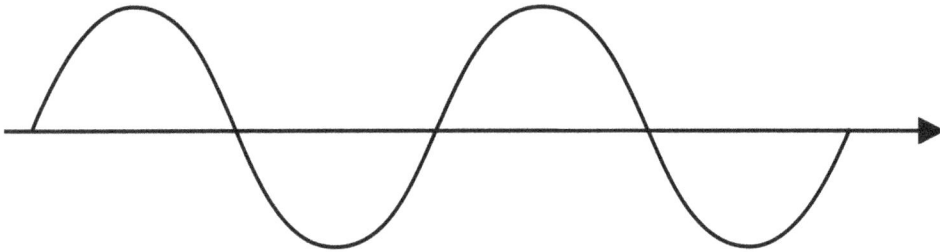

Figure 4.1: A sine wave

1 hertz	1 cycle per second
1 kilohertz	1000 cycle per second
1 megahertz	1,000,000 cycles per second (one million)
1 gigahertz	1,000,000,000 cycles per second (one billion)

Table 4.1: Hertz

Because all electromagnetic waves, including radio waves, move at the speed of light, the frequency is related to the wavelength. We observe that wavelength and frequency are interdependent. Higher frequencies have shorter wavelengths, and lower frequencies have longer wavelengths. An AM radio station at 670 MHz has a lower frequency than an AM station at 1400 MHz; the station at 670 MHz is a lower frequency, and thus longer wavelengths. Figure 4.2 shows two sine waves at differing frequencies.

The concept of frequency exists not only in RF engineering, but sound engineering as well. Figure 4.3 shows a piano keyboard and the sound frequencies to which the keys are tuned. While radio waves and sound waves are not the same phenomena, they do share characteristics such as amplitude, frequency, and wavelength (more on these terms shortly). Because of the similarities between sound waves and radio waves, sound waves make a great starting point for understanding radio waves.

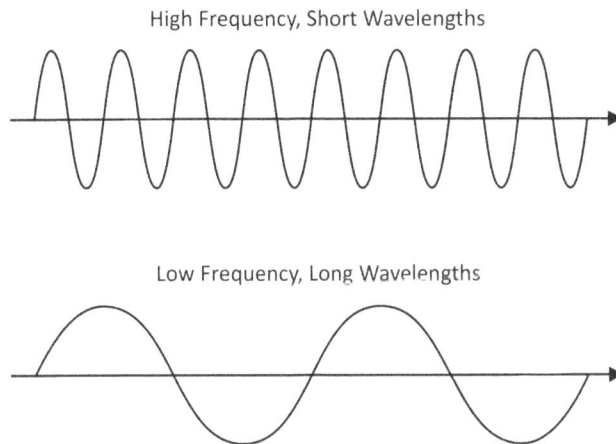

High Frequency, Short Wavelengths

Low Frequency, Long Wavelengths

Figure 4.2: Two sine waves, each with different frequencies and wavelengths.

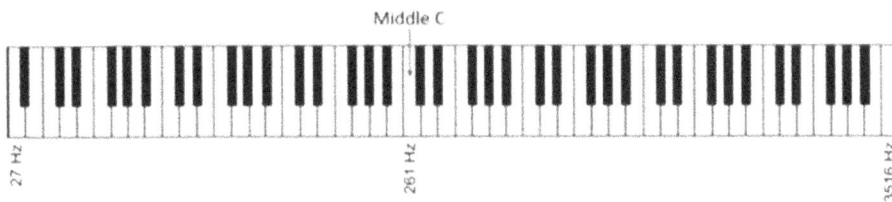

Middle C

27 Hz   261 Hz   3516 Hz

Figure 4.3: A piano keyboard, with the lowest, middle, and highest keys labeled by sound frequency

Looking at the piano keyboard again in Figure 4.3, you can see that the keys to the left produce sound waves at a frequency as low as 27 Hz. A "middle C" key plays a frequency of 261 Hz, and the key to the right produces a frequency of 3516 Hz - a much higher frequency than the 27 Hz we started at! Remember that these aren't on the electromagnetic spectrum, instead being vibrating air in the audio spectrum, but audio spectrum still creates sine waves in varying frequencies, so the piano works well for illustrating how radio waves can be transmitted and received at differing frequencies.

Wavelength

The wavelength of a radio frequency (RF) wave is calculated as the distance between two adjacent identical points on the wave. Figure 4.4 shows a standard sine wave. Note that Point A and Point B mark two identical points on the wave; the distance between them is defined as the wavelength. Notice that you can mark any two identical, recurring points on the wave, but the wavelength is frequently measured as the distance from one crest of the wave to the next.

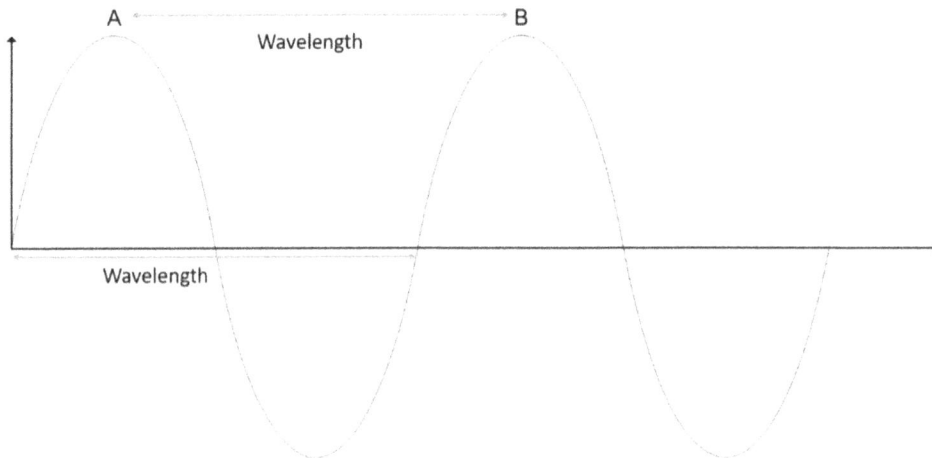

Figure 4.4: Examples of wavelength measurement.

Wavelength is a very important factor in wireless communications, as it dictates optimum antenna lengths for specific frequencies, and determines how the RF wave will interact with the environment that it is in. For example, an RF wave is more likely to reflect when it strikes an object that is larger than the wavelength, and it will be more likely to scatter if the object is smaller than the wavelength. We will discuss reflections and scattering more later in this chapter.

The wavelength at any given frequency is related to the speed of light. If you know the frequency, you can calculate the wavelength. Inversely, if you know the wavelength, you can calculate the frequency, since the speed of the wave is constant, being roughly the speed of light.

When the frequency is known, you can calculate the wavelength in meters, where λ (lambda) is the wavelength, and f is the frequency in hertz:

λ = 299,792,458 / f

Therefore, 2.45 GHz (converted into Hz) would have a wavelength that is calculated with the following formula:

λ = 299,792,458 / 2,450,000,000 = 0.123

The result is 0.123 meters, which is approximately 12.3 centimeters, or 4.8 inches. So we know that a 2.45 GHz radio wave has a wavelength of 4.8 inches.

Alternatively, if you know the wavelength, you can calculate the frequency. Here is the formal equation for reference (where c is the constant for the speed of light):

λ = c / f

Amplitude

To describe amplitude, let's continue to lean on examples in the realm of sound. You've probably interacted with a sound amplifier of some kind, whether an amplifier in a car audio system, an amplifier for music equipment (like a guitar amp), or even an amplifier for a vinyl record player. Put simply, an audio amplifier takes an audio signal and makes it louder. In other words, the audio signal is *amplified*. You could also say that the loudness or *amplitude* of the audio signal is increased.

With both audio and radio frequency, amplitude defines how loud a sound is, or how strong a radio frequency signal is. A signal with low amplitude is weak, and a signal with high amplitude is strong. On a sine wave, amplitude is represented with the height of the wave (and this is true for in both sound and radio frequency). Figure 4.5 shows two sine waves, one with low amplitude, and one with high amplitude.

Coming back to the sound analogy: higher amplitude (louder) sounds can be heard from much further away than lower amplitude (quieter) sounds. When trying to eavesdrop on a conversation that is far away, you'll notice that it can be very difficult to differentiate between the conversation (signal) and the background noise. The same is true for radio frequency communications: a radio receiver will have an easier time understanding high-amplitude signals than a low-amplitude or quieter signal. At some point, the signal will become lost in the noise. We'll discuss this concept later in this chapter when we discuss RF Noise and Noise Floors.

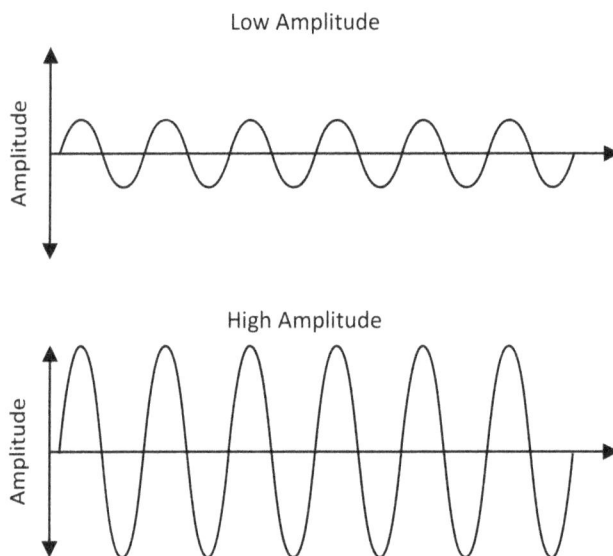

Figure 4.5: Two sine waves of differing amplitudes

Phase

Unlike frequency, wavelength, and amplitude, the *phase* is not a characteristic of a single RF wave but is instead a comparison between two RF waves. Think back to our first discussions about sine waves, where we noted that single wavelength could start at zero, transition to full positive, go back to zero, transition to full negative, and go back to zero again. This full cycle of a wave can be mapped to degrees on a circle, such as $0°$, $180°$, $270°$, and $360°$ as the cycle completes. Think of the circle as representing a sine wave that is standing still, and not propagating through space. You can see both the circular and wave representations of this in Figure 4.6.

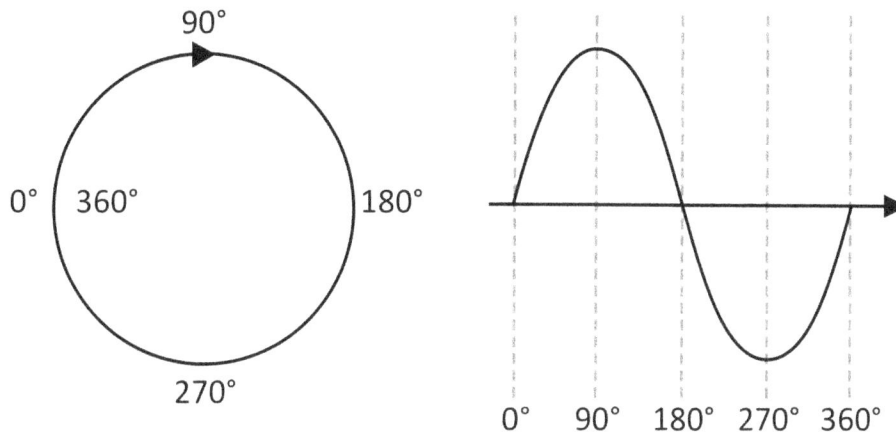

Figure 4.6: Degrees of a circle and degrees of a single wavelength in a sine wave

Now that you understand the different phases of a wavelength let's discuss how two sine waves can be compared. If two radio waves arrive at a receiver perfectly aligned with each other, then they're said to be *in-phase*, since their phases match. In Figure 4.7, you can see two sine waves that are in-phase with each other. When two radio waves arrive in-phase at the receiver, they will have the effect of combining and increasing the received amplitude of the signal.

Next, you can see another radio wave that is shifted ¼ out-of-phase with the initial phase. This wave is 90 degrees out-of-phase with the initial phase. Finally, you can see the sine wave that is 180 degrees out-of-phase with the original. This will have an especially destructive effect on the incoming signal, as when the two sine waves are received 180 degrees out-of-phase at the receiver, they will cancel each other out, and the receiver won't be able to discern a sine pattern so that no signal will be derived from the carrier wave.

Traditionally, out-of-phase signals were an especially destructive activity for wireless communications. Two signals arriving at 180 degrees out-of-phase were especially harmful since the signal would effectively cancel itself out. However, relatively recent developments in wireless technology now harness phases to increase the amount of data transmission that can be performed. We'll discuss this later in this chapter, specifically in the section about Phase Shift Keying.

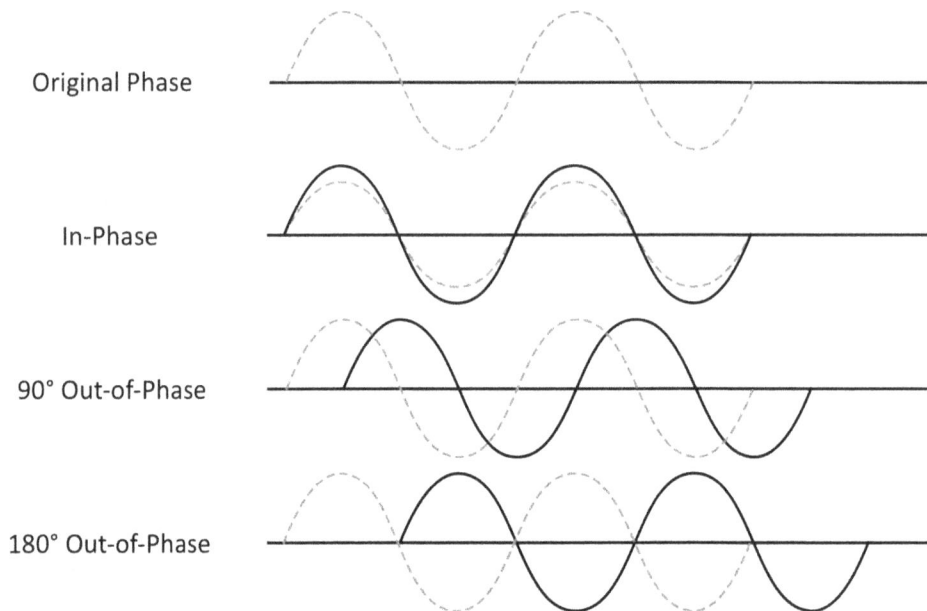

Figure 4.7: RF waves in various phases

RF Propagation Behaviors

As RF waves propagate through space, they will encounter a variety of different solids, gasses, liquids, and also space with none of the above. All of these physical objects (or in the case of space, non-objects) will affect RF waves in different ways, in the same way that light would react. In fact, much as we leaned on examples from the world of sound, we can now rely heavily on behaviors that we see in light to understand how RF is affected by the environment that it is propagating in.

Amplification

Before we look at how physical objects can interact with and affect RF, it's important for us to have a quick discussion about amplification. The analogies about visible spectrum (light) will have to wait a moment, as we need to compare the similarities between audio and RF once again as we cover Amplification.

Amplification is an increase in the amplitude of an RF signal. You probably remember when we discussed audio amplifiers for use in car stereo systems or guitars. These amplifiers do precisely what you'd think they do: they take a low-amplitude incoming signal, and with an external power source and the appropriate electronics, export a higher-amplitude outgoing signal. Audio amplifiers and radio amplifiers are entirely identical in this regard, but remember, any noise in the audio stream will also be amplified. The same is true for radio amplifiers.

Amplification, where an external power source is involved, is known as *Active Gain*. There is another type of gain called *Passive Gain*, which increases amplitude with no external power. Passive gain is usually accomplished with an antenna that provides more focus. An omnidirectional antenna is a type of antenna that radiates RF energy evenly in all directions (with some limitations due to the structure of the antenna). In this scenario, the energy is very "spread out."

A directional antenna can provide passive gain by radiating energy in a more specific direction, and thus *not* radiating RF energy in all directions. Think of the radiation pattern of an antenna as a ball of clay. With an omnidirectional antenna, the ball of clay is perfectly round, representing an antenna that radiates omnidirectionally.

If the ball of clay was to be flattened out as much as possible, it now represents a higher-gain antenna - one that radiates RF energy out horizontally, but less vertically, providing passive gain. The same amount of clay has been used to create the shape, but the shape of the clay is very different.

Attenuation

Attenuation is what occurs when an RF signal's amplitude is reduced. Attenuation usually occurs after the RF signal has been transmitted, and is passing through objects that it encounters as it propagates. Essentially, attenuation is the technical term for blocking or reducing signal strength.

A common *attenuator* for radio signals is the walls in buildings. For high-frequency signals like Wi-Fi in the 5 GHz frequency band, attenuation happens very quickly as the signal passes through walls, refrigerators, or shelves full of books. Figure 4.8 shows an example of an RF wave experiencing attenuation as it passes through an object (such as a wall).

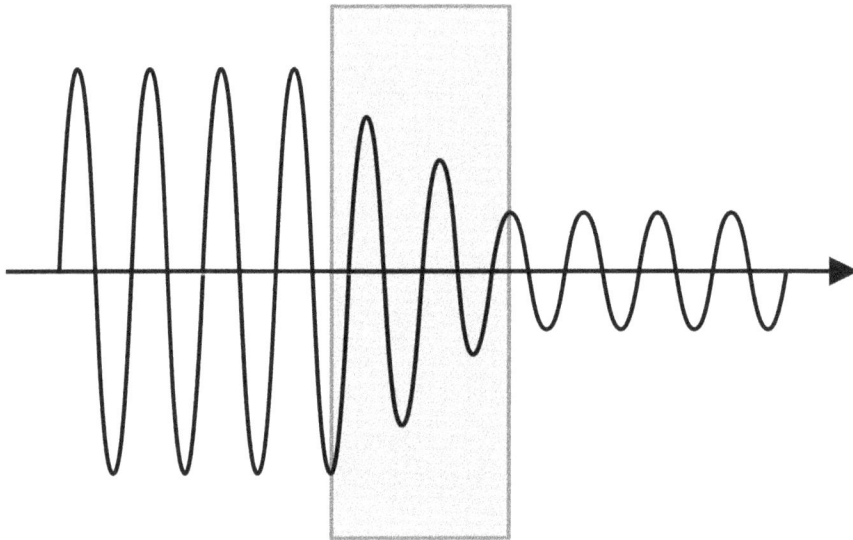

Figure 4.8: An obstacle in the environment causing a radio wave to attenuate.

Lower frequency signals experience less attenuation than high frequency signals. For example, have you ever noticed the impressive range of AM radio? It's not uncommon to listen in on a station that is up to 160 kilometers (that's just over 100 miles) away. Part of this is due to the lower frequency, and thus the longer wavelengths of AM radio. AM radio experiences far less attenuation from buildings, terrain, and atmosphere than Wi-Fi does, which explains why you can listen to AM radio in the desert in the middle of nowhere, but Wi-Fi barely works in your backyard.

Free Space Path Loss

Free space path loss (FSPL), sometimes simply called *free-space loss* (FSL) or just path loss, is a weakening of the RF signal due to a broadening of the wavefront. The broadening of the wavefront is known as signal dispersion. Consider the concentric circles in Figure 4.9 as representing an RF signal propagating out from an omnidirectional antenna (which we discussed briefly earlier in this chapter). Notice how that the wavefront becomes larger as the wave moves out from the antenna. The broadening of the wavefront causes a loss in amplitude of the signal at any specific point in space because the energy is spread over a larger area. Therefore, the signal is weaker at point B than it is at point A.

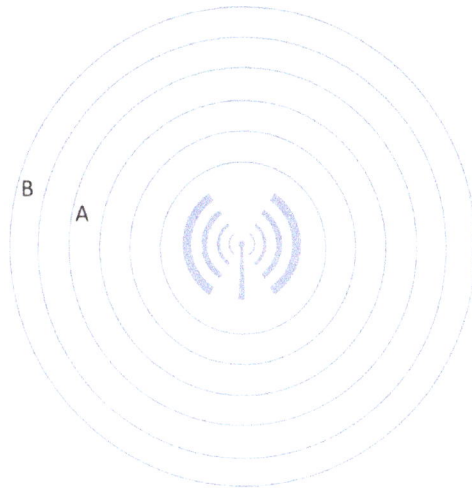

Figure 4.9: FSPL Illustrated

Absorption

Microwave ovens use the 2.4 GHz frequency range to heat food. While Wi-Fi devices (which work in the same frequency band) have output levels of around 30 milliwatts (mW), a microwave oven usually has an output power between 700 and 1400 watts (W). What does this have to do with wireless engineering? The microwave oven works because RF waves are absorbed by materials that have moisture (molecular electric dipoles) in them. The absorption converts the RF wave energy into heat energy and therefore heats your food.

As a result, if you've ever used a microwave oven, you've experienced absorption first-hand! Another place where you can experience absorption is in your closet, especially if it is a walk-in closet. Clothes on hangers in closets do a fantastic job of absorbing sound. Next time you go into a closet, note how "dead' the closet sounds. Closets are a great place to record voice-over work, or even record your first hit album!

Back to RF: Liquids are especially absorptive, so expect water tanks, terrain, or even large groups of people to absorb radio frequency signals significantly. Fortunately, most RF systems that we use and interact with don't concentrate power as a microwave oven does, so there's no danger of being heated up a measurable amount by RF absorption. What does measurably warm up a person includes going outside for some sunshine, or putting a couple more pieces of wood on the fireplace. Figure 4.10 shows RF signal absorption.

Reflection

When an RF signal bounces off a smooth, non-absorptive surface, the signal sharply changes direction in a process is known as reflection. Reflection is probably the easiest RF behavior to understand because we see it frequently in our everyday lives. You can shine a light on a mirror at an angle and see that it reflects off that mirror in relation to the angle. When you look in the mirror, you are experiencing the concept of visible spectrum reflection, which is essentially the same as RF reflection.

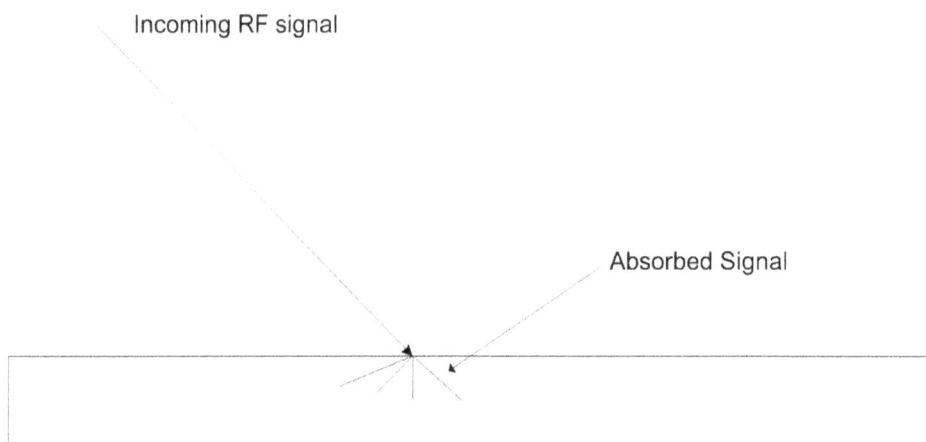

Incoming RF signal

Absorbed Signal

Figure 4.10: RF Absorption

Figure 4.11 illustrates this concept. As you can see, the light waves, which are electromagnetic waves are similar to RF signals first reflecting off the object and travel toward the mirror. Next, the light waves reflect off the mirror and travel toward your eye. Finally, your eye acts as a focusing device and brings the light waves together at the back of the eye, giving you the sense of sight. However, the critical thing to note is that what you are "seeing" is the light reflected off the object onto the mirror, and off the mirror into your eyes. The ability to see objects all around us is driven by the reflective properties of the materials and the light waves striking against them.

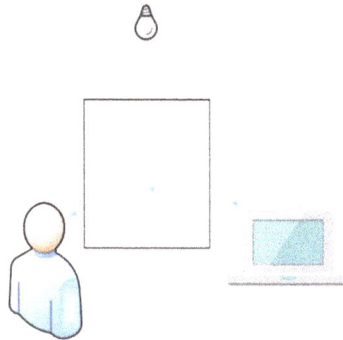

Figure 4.11: Reflection Illustrated with a Mirror

RF signals also reflect off objects that are smooth and larger than the waves that carry the signals. Earlier, it was noted that the wavelength impacts the behavior of the RF wave as it propagates through space. Reflection is an example of the relationship of the wavelength and the space through which the wave travels. If the space were empty, there would be no reflection, but since all space we operate in (earth and its atmosphere, at least, for now) contains some elements of absorption, reflection, refraction, and scattering are to be expected.

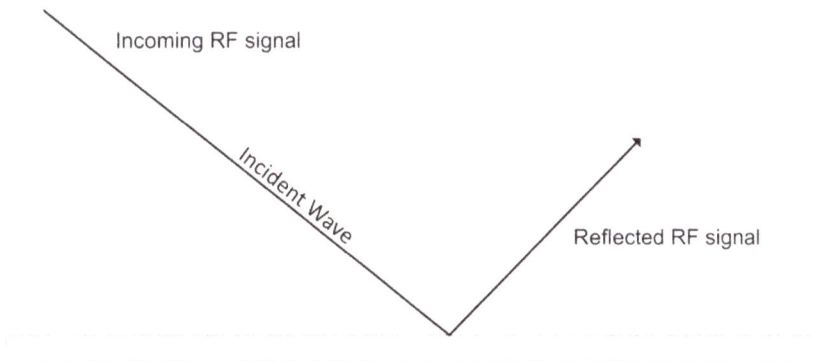

Incoming RF signal

Incident Wave

Reflected RF signal

Figure 4.12: RF Reflection

Typically, the object that causes reflection will be smooth and larger than the wavelength. For example, waves that interact with Wi-Fi radios in the 2.4 GHz band are about 13 centimeters in wavelength. As such, it follows that smooth objects greater than 13 centimeters in size will have a tendency to cause reflections for 2.4 GHz Wi-Fi, any other RF activity occurring in the 2.4 GHz frequency band.

Refraction

Refraction occurs when an RF signal changes speed and is bent while moving through media of different densities. Different mediums, such as drywall, wood or plastic, will have different *refraction indexes*. The refraction index helps in determining how much refraction will occur.

Let's go back to the light analogy for a moment. If you wear glasses, you are wearing a refraction device. The lens refracts or bends the light, to make up for the imperfect lens in your eye. The glasses you to see clearly again because the lack of focus in the eye is corrected by the refraction caused by the lens in the glasses.

Figure 4.13 shows an RF signal being refracted. As you can see, when refraction occurs with RF signals, some of the signal is reflected, and some is refracted as it passes through the medium. Of course, as with all mediums, some of the signal will be absorbed as well.

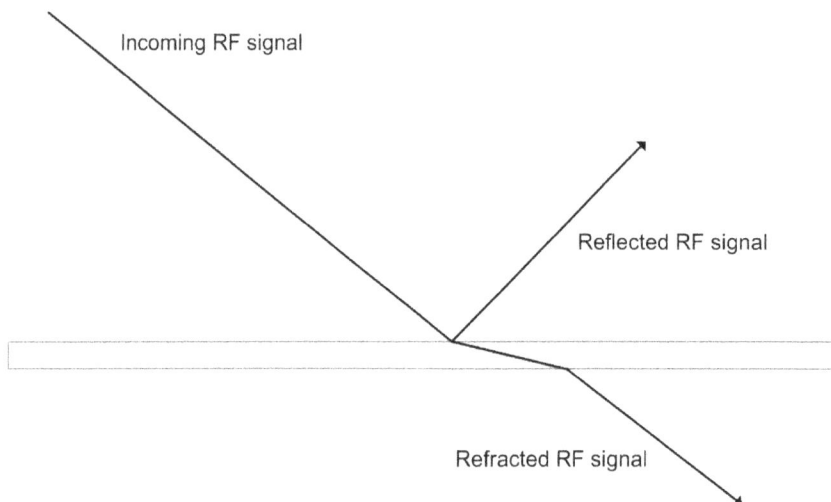

Figure 4.13: Refraction

Usually, significant refractions don't occur in indoor-only wireless systems. Instead, they're more common in outdoor systems, especially site-to-site links using 5 GHz, 6 GHz, and higher-frequency links using bands like 24 GHz. Site-to-site or long-distance wireless links typically use directional antennas with a narrow beam of focus, and as the narrow beam of RF passes through different atmospheric conditions such as changes in air pressure, or varying amounts of water vapor in the air.

The issue here is simple: if the RF signal changes from the intended direction as it's traveling from the transmitter to the receiver, the receiver may not be able to detect and process the signal. The result can be a broken connection or an increase in error rates if the refraction is temporary, or sporadic due to fluctuations in the weather around the area of the link.

An excellent experiment can be easily performed that demonstrates the concept of refraction. Take a large clear bowl and fill it with water. Now, place a spoon (or another piece of flatware) into the water at an angle and look through the transparent side of the bowl at the knife. What did the spoon do? Well, nothing other than entering the water; but what did it *appear* to do? It appears to bend. This illusion is because the light waves are traveling slower in the water medium, and this causes refraction of the light waves. It's not the spoon that's bending — It's the light that's bending because it's the light that you see.

Diffraction

Diffraction is defined as a change in the direction or intensity of a wave as it passes by the edge of an obstacle. As seen in Figure 4.14, this can cause the signal's direction to change, and it can also result in areas of RF shadow. Instead of bending as it passes into or out of an obstacle, like refraction, diffraction describes what happens as light travels around the obstacle.

Diffraction occurs because the RF signal slows down as it encounters the obstacle, and this causes the wavefront to change directions. Consider the analogy of a rock dropped into a pool and the ripples it creates. Think of the ripples as analogous to RF signals. Now, imagine there is a stick being held upright in the water. When the ripples encounter the stick, they will bend around it, since they cannot pass through it. A larger stick has a more significant visible impact on the ripples, and a smaller stick has a lesser impact. Diffraction is often caused by buildings, small hills and other larger objects in the path of the propagating RF signal.

The RF shadow caused by diffraction can result in areas without proper RF coverage. If you are in an RF shadow area, you will not be able to receive communications from the wireless network. An example of this phenomenon indoors is an elevator shaft. Often, when the access point is on one side of the elevator, and the client is on the opposite side, the signal will be insufficient for communications in that location. Many times, RF shadow problems can be resolved with very slight adjustments in the location of the antennas used on the access point or wireless router, or by installing additional access points. For example, if you install access points in the areas on both sides of the elevator shaft, one access point can serve one side, and the other can serve the remaining side.

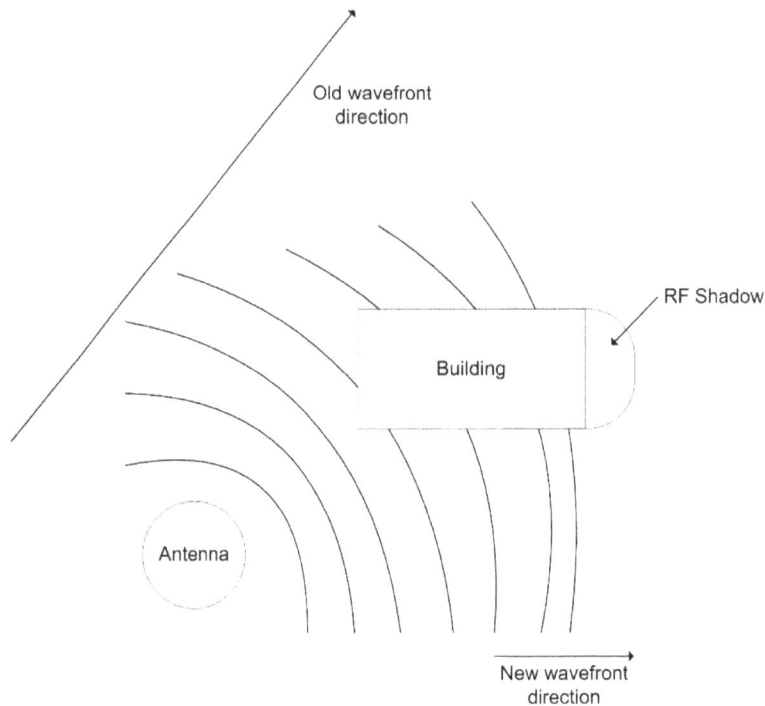

Figure 4.14: RF Diffraction

Scattering

Scattering happens when an RF signal strikes an uneven surface (a surface with inhomogeneities — there's a word you can use around your family to sound smart) causing the signal to be scattered instead of absorbed so that the resulting signals are less significant than the original signal. Another way to define scattering is to say that it is simply multiple reflections. Figure 4.15 illustrates this.

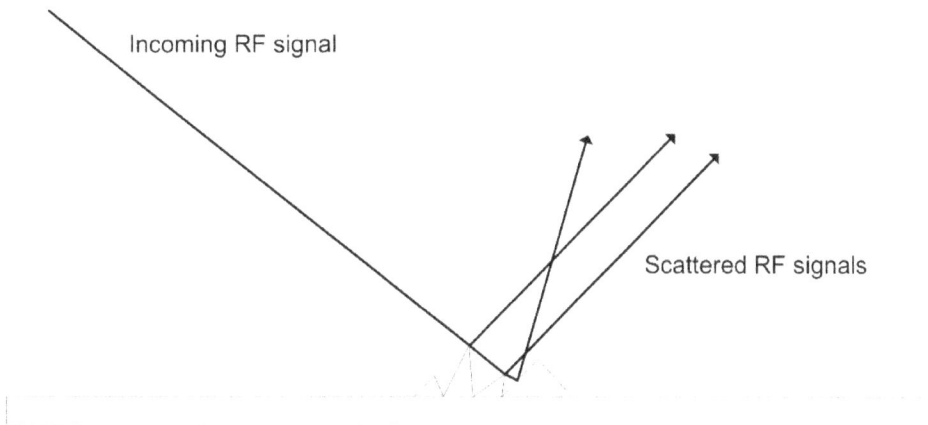

Figure 4.15: RF Scattering

Scattering can happen in a minor, almost undetectable way when an RF signal passes through a medium that contains small particles. These small particles cause scattering. Smog is an example of such a medium. The more frequent and more impacting occurrence is that caused when RF signals encounter things like rocky terrain, leafy trees, or chain link fencing. Rain and dust can cause scattering as well.

RF Signal Metrics

One of the most important aspects of working with wireless systems is to measure RF signal strength, whether measuring a change in power or power at absolute levels. If you begin to investigate signal strength measurements, you might find them to be confusing. In this section, we'll investigate absolute RF signal measurements such as the Watt, Milliwatt, and Microwatt, as well as ways of measuring changes in power such as decibels.

Watt

The watt (W) is a basic unit of power equal to one joule per second. It is named after James Watt, an 18th-century Scottish inventor who also improved the steam engine, among other endeavors. This single watt is equal to one ampere of current flowing at one volt. Think of a water hose with a spray nozzle attached. You can adjust the spray nozzle to allow for different rates of flow. The flow rate is comparable to amperes in an electrical system. The water hose also has a certain level of water pressure — regardless of the amount that is flowing through the nozzle. The pressure is like the voltage in an electrical system. If you apply more pressure or you allow more flow with the same pressure, either way, you will end up with more water flowing out of the nozzle. In the same way increased voltage or increased amperes will result in increased wattage since the watt is the combination of amperes and volts.

In wireless systems, outdoor links often use power levels measured in watts at the transmitter. In indoor wireless systems, the watt is too powerful, so many indoor systems and consumer electronics transmit in milliwatts of power instead of watts of power.

Milliwatts and Microwatts

Most wireless systems do not need a tremendous amount of power to transmit a signal over an acceptable distance. For example, you can see a 7-watt light bulb from more than 83 kilometers (50 miles) away on a clear night with line of sight. Remember, visible light is another portion of the same electromagnetic spectrum, so this should give you an idea of just how far away an electromagnetic signal can be detected. For this reason, many systems "step down" from the Watt, and use a measurement of power that is 1/1000th of a watt, known as a *milliwatt*. 1 watt (W), then, would be 1,000 milliwatts (mW).

A good example of common devices that work in the milliwatt range is Bluetooth devices. Bluetooth devices can implement different classes of transmit power, depending on their intended use, as shown in Table 4.2.

	Max Transmit Power	Expected Range	Used For
Class 1	100 mW	100 meters (328 feet)	Industrial devices
Class 2	2.5 mW	10 meters (33 feet)	Most headphones and headsets
Class 3	1 mW	Fewer than 10 meters	Very low-power devices

Table 4.2: Bluetooth transmit power classes

While a milliwatt represents one-thousandth of a watt, a *microwatt* (μW) is one-millionth of a watt. It represents an incredibly small amount of RF power.

Decibels

Now that we've established watts, milliwatts, and microwatts as units of RF power, why would we want to use any other units to measure signal strength and RF power? Let's take a look at a few common signal strengths that you might observe in the realm of Wi-Fi and indoor IoT networks in the 2.4 and 5 GHz frequency bands:

- **0.01 mW** - A very, very high signal strength

- **0.0000002 mW** - Good signal strength, plenty fast for streaming video

- **0.00000001 mW** - Poor signal strength, it probably won't work well

- **0.000000001 mW** - Signal strength so low that it will typically be useless

All of the values above represent completely normal signal strengths that your Wi-Fi devices work with every day. However, as an engineer, keeping track of so many decimal places is very difficult. Imagine asking someone over the phone, "What is your signal strength?" and hearing, "It's point zero zero zero zero zero zero zero zero milliwatts of signal." That's very difficult for people to work with, isn't it?

This is where *decibels* come in. Now, before we explain what decibels are and how they work, let's look at the exact same signal strength values again:

- **-20 dBm** - A very, very high signal strength

- **-67 dBm** - Good signal strength, plenty fast for streaming video

- **-80 dBm** - Poor signal strength, it probably won't work well

- **-90 dBm** - Signal strength so low that it will typically be useless

Even without knowing how decibels work, those signal strength differences should be much easier to read and relay to other people. Milliwatts are very precise, but decibels are great for representing big changes in signal strength while being relatively easy to read. Let's dive into decibels and learn how they work.

First, it's good for you to know that a decibel is 1/10th of a bel, which was developed by Bell Laboratories to calculate losses in telephone communication power as ratios. For our discussion, we'll be focusing on the decibel.

While milliwatts are an absolute measurement of power, a decibel is a relative measurement that shows changes in signal strength. While milliwatts increase and decrease linearly, decibels increase, and decrease logarithmically. In other words, small numbers can mean very big jumps. For example, a 3 decibel (dB) jump in power means double the signal strength. Going up another 3 dB means the signal strength has doubled again. Moreover, an increase of another 3 dB means we've doubled the signal strength *yet again*. In only 9 dB of increase or gain, we've doubled our signal strength three times! This is the power of a logarithmic scale at work; you can represent big changes with small numbers.

The example above leverages the rule of 3's and 10's as a simple way to understand signal strength changes without having to resort to complex, logarithmic math. As you read the rules, keep in mind that *gain* is an increase in power, and *loss* is a decrease in power. Here are the basic rules:

- +3 dB (a gain of 3 dB) means double the power

- -3 dB (a loss of 3 dB) means half the power

- +10 dB (10 dB of gain) means 10x more power

- -10 dB (10 dB of loss) means 10x less power

- Gains and losses a cumulative (increasing by success additions or subtractions)

Now, let's evaluate what these rules mean, and the impact they have on your RF math calculations. First, 3 dB of gain doubles the output power. This means that 100 mW + 3 dB of gain equals 200 mW of power, or 30 mW + 3 dB of gain equals 60 mW of power. The power level is always doubled for each 3 dB of gain that is added. Rule five stated that these gains and losses are cumulative. This means that 6 dB of gain is the same as 3 dB of gain applied twice. Therefore, 100 mW of power plus 6 dB of gain equals 400 mW of power. The following example illustrates this based on 9 dB of gain (3 dB added three times). Note that both formulas are saying the exact same thing.

40 mW + 3dB + 3dB + 3dB = 320 mW

40 mW * 2 * 2 * 2 = 320 mW

Both formulas are saying the same thing. Now consider the impact of 3 dB of loss; 3 dB of loss halves the output power. Look at the impact on the following formula with 6 dB of gain and 3 dB of loss:

40 mW + 3 dB + 3 dB – 3 dB = 80 mW

40 mW * 2 * 2 / 2 = 80 mW

Let's look at one last example to illustrate the rule of 10's. Remember that 10 dB of gain means 10x more power, so we need to multiply our power by 10. 20 dB of gain means we'd multiply it by 10, and then multiply it by 10 again:

40 mW + 10 dB + 10 dB = 4000 mW (4 W)

40 mW * 10 * 10 = 4000 mW (4 W)

It is also important to know that the 10s and 3s can be used together to calculate the power levels after any integer gain or loss of dB. This is done with creative combinations of 10s and 3s. For example, imagine you want to know what the power level would be of a 12 mW signal with 16 dB of gain. Here is the math:

12 mW + 16 dB = 480 mW

But how was this calculated? The answer is very simple: add 10 dB and then add 3 dB twice. Here it is in longhand:

12 mW + 16 dB = 480 mW

12 mW + 10 dB + 3 dB + 3 dB = 480 mW

12 mW * 10 * 2 * 2 = 480 mW

Sometimes you are dealing with both gains and losses of unusual amounts. While the following numbers are completely fabricated, consider the assumed difficulty they present to calculating a final RF signal power level:

30 mW + 7 dB – 5 dB + 12 dB – 6 dB = power level

At first glance, this sequence of numbers may seem impossible to calculate with the rules of 10s and 3s; however, remember that the dB gains and losses are cumulative, and this includes both the positive gains and the negative losses. Let's take the first two gains and losses: 7 dB of gain and 5 dB of loss. You could write the first part of the previous formula like this:

30 mW + 7 dB + (-5 dB) = 30 mW + 2 dB

Why is this? Because (+7) + (-5) = (+2). Carrying this out for the rest of our formula, we could say the following:

30 mW + 7 dB + (-5 dB) + 12 dB + (-6 dB) = 30 mW + 2 dB + 6 dB

or

30 mW + 8 dB = power level

The only question that is left is this: How do we calculate a gain of 8 dB? Remember, the rules of 10s and 3s. We have to find a combination of positive and negative 10s and 3s that add up to 8 dB. Here's a possibility:

+10 + 10 − 3 − 3 − 3 − 3 = 8

Decibels to Milliwatts (dBm)

So far, we've only discussed dB, and how it is used to show changes in power. However, earlier, you may have noticed that we converted milliwatts to *decibels in relation to a milliwatt* (dBm) to show absolute power. We did this to illustrate that while dB looks complicated, it's very helpful for simplifying measurements of power. Let's take a closer look at *decibel-milliwatts* (dBm).

dBm is an absolute measurement of power where the *m* stands for milliwatts. Effectively, dBm references decibels relative to 1 milliwatt or that 0 dBm equals 1 milliwatt. Once you establish that 0 dBm equals 1 milliwatt, you can reference any power strength in dBm. Depending on the transmit power of the wireless technology that you are working with, you may find yourself using positive numbers such as 2 dBm or 5 dBm, or for low-power, indoor systems, you will see numbers dipping well into the negatives such as -10 dBm, -50 dBm, or even -90 dBm.

Because a wireless receiver can detect and process very weak signals, it is easier to refer to the received signal strength in dBm rather than in mW. For example, a signal that is transmitted at 4 W of output power (4000 mW or 36 dBm) and experiences -63 dB of loss has a signal strength of .002 mW (-27 dBm). Rather than say that the signal strength is 0.002 mW, we say that the signal strength is -27 dBm.

The formula to get dBm from milliwatts is:

dBm = 10 * log10(Power_mW)

For example, if the known milliwatt power is 30 mW, the following formula would be accurate:

10 * log10(30) = 14.77 dBm

The result of this formula would often be rounded to 15 dBm for simplicity; however, you must be very cautious about rounding if you are calculating specific transmit powers that need a high level of accuracy. Table X provides a list of common milliwatt power levels and their dBm values.

One of the benefits of working with dBm values instead of milliwatts is the ability to easily add and subtract simple decibels instead of multiplying and dividing often huge or tiny numbers. For example, consider that 14.77 dBm is 30 mW as you can see in Table 4.3. Now, assume that you have a transmitter that transmits at that 14.77 dBm, and you are passing its signal through an amplifier that adds 6 dB of gain. You can quickly calculate that the 14.77 dBm of original output power becomes 20.77 dBm of power after passing through the amplifier. Now, remember that 14.77 dBm was 30 mW. With the 10s and 3s of RF math, which you learned about earlier, you can calculate that 30 mW plus 6 dB is equal to 120 mW. The interesting thing to note is that 20.77 dBm is equal to 119.4 mW. As you can see, the numbers are very close. While we've been using a lot of more exact figures in this section, you'll find that rounded values are often used in vendor literature and documentation.

RF Noise and Noise Floors

Let's once again think back to our examples of how RF and audio spectrum relate to each other. When you are trying to have a conversation with someone, any other sounds that you hear interfere with your ability to understand what the other person is saying. It could be loud music, a noisy car driving by, or even just other people talking, any sound that you are not able to distinguish individually is noise. If the background noise consistently overpowers the person that you're trying to talk to, eventually you just give up, nod, and smile, because you just can't understand them.

This is exactly what noise is to a radio receiver. Whether the noise is just natural background noise (like wind blowing through the trees, or birds chirping), or another device nearby talking on the same frequency (like someone driving by with their stereo turned up very loud), noise is any signal other than the signal that the receiver is attempting to hear and decode.

Natural background noise in the environment is known as the *noise floor*. It's there in the realm of audio, too. Sit quietly sometime and listen, and you'll notice that even when it's quiet, there's always a distant noise in the background when you dig for it. The same is true for radio receivers.

mW	dBm (rounded)	dBm (rounded to two decimal places)
1	0	0
10	10	10
20	13	13.01
30	15	14.77
40	16	16.02
50	17	16.99
100	20	20
1000	30	30
4000	36	36.02

Table 4.3: mW to dBm conversion table (rounded to two precision levels)

SNR and SINR

Background RF noise, which can be caused by all the various systems and natural phenomenon that generate energy in the electromagnetic spectrum, is known as the noise floor. The power level of the RF signal relative to the power level of the noise floor is known as the signal-to-noise ratio or SNR. It is the difference between the signal strength and the noise floor, so don't let the term "ratio" confuse you. It is not typically referenced as a ratio in wireless communications, but as a dB value. Figure 4.16 illustrates the concept of SNR.

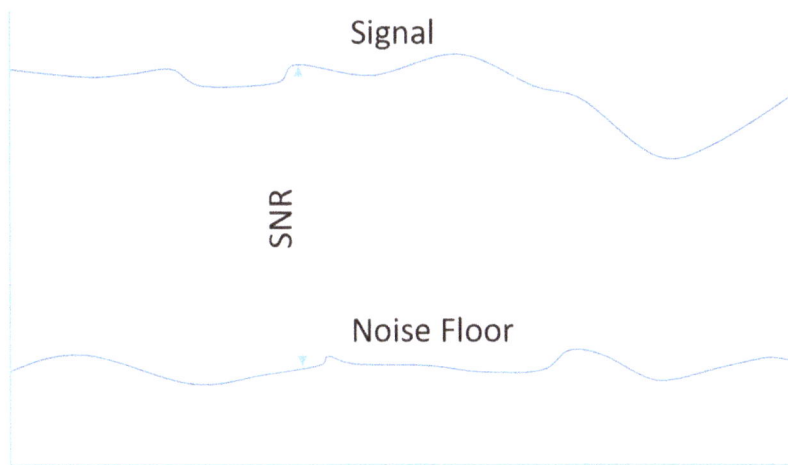

Figure 4.16: SNR Illustrated

When working with radio technologies, SNR is a very important measurement. If the noise floor power levels are too close to the received signal strength, the signal may be corrupted, or it may not even be detected. It's almost as if the received signal strength is weaker than it actually is when there is more electromagnetic noise in the environment. You may have noticed that when you yell in a room full of people yelling, your volume doesn't seem so great; however, if you yell in a room full of people whispering, your volume seems to be magnified. In fact, your volume is not higher, but the noise floor is less than before. RF signals are impacted in a similar way.

Technically, in wireless signal reception, SNR is defined as the difference between the noise floor and signal strength in dB. The formula for calculating SNR is simple:

SNR = signal strength value in dBm – noise floor value in dBm

135

If the noise floor is rated at -95 dBm and the signal is detected at -70 dBm, the SNR is 25.

In addition to the term SNR, the term SINR has become common. SINR is the signal to interference plus noise ratio. Like SNR, it is not expressed as a ratio, but a value in dB. The difference is that SINR is more momentary in nature than SNR. SNR looks at the noise floor at a given point in time and assumes it doesn't change drastically, which is usually a good assumption. However, sporadic interferers may generate RF energy for small bursts of time; during that time window, SINR is a better measurement of reality. Even if the SNR would allow for the reception of an RF signal at a given data rate, the SINR may not because of the temporary interference from other devices, transmitting on the same frequency.

Fundamentals of Wireless Modulation

By themselves, sine waves are pretty boring; they repeat in the same, predictable pattern as long as they pass by our receiver's antenna. They're nice and all… but how do they actually carry data? Earlier in this chapter, you learned the characteristics of an RF wave, such as the wavelength, frequency, amplitude, and phase. In this section, we'll take a close look at how we can manipulate those aspects of a wave to carry data on it.

Amplitude Shift Keying

Earlier in this chapter, you learned that amplitude specifies the height of a radio wave. A radio transmitter can vary the amplitude of a signal by changing the output power, whether that be up to increase the amplitude and make the sine wave "taller", or down to decrease amplitude and make the wave "shorter." In both cases, the frequency and wavelength stay the same - only the amplitude changes.

These variations in amplitude can be used to carry a digital signal with Amplitude Shift Keying (ASK). In other words, a lower amplitude can be used to indicate a 0, and a higher amplitude to indicate a 1. You can see where the word "keying" comes from in Amplitude Shift Keying; the information is "keyed" by changing amplitude, as you can see in Figure 4.17.

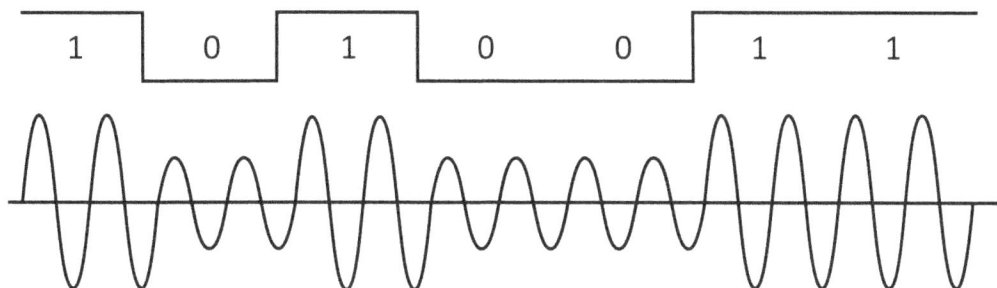

Figure 4.17: Amplitude Shift Keying (ASK)

While Amplitude Shift Keying (ASK) might seem similar to Amplitude Modulation (AM), the latter is typically used for transmitting analog signals, such as AM radio. We'll discuss Amplitude Modulation (AM) later in this chapter.

Frequency Shift Keying

Another concept that was discussed earlier in this chapter was the frequency of an RF wave. So far, every wave we've shown as a consistent frequency, but the reality of an RF wave is that the frequency can be changed on the fly. Because it can be changed, it can be used to modulate a digital signal in a process called Frequency Shift Keying (FSK). For example, a shift to a lower frequency (and thus a longer wavelength) might indicate a 0, and a shift to a higher frequency (shorter wavelength) could indicate a 0. The information is "keyed" to the frequency, as you can see in Figure 4.18.

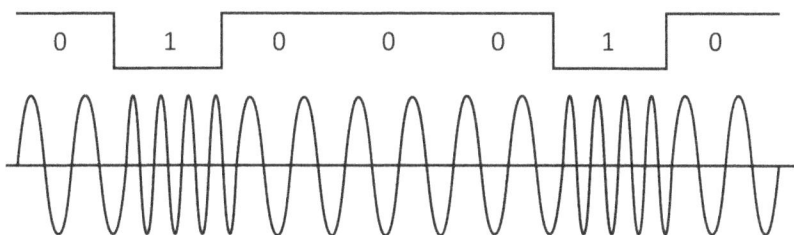

Figure 4.18: Frequency Shift Keying

Frequency Shift Keying (FSK) might seem similar to Frequency Modulation (FM), but FM is typically used for the transmission of analog signals, such as FM radio. We'll discuss Frequency Modulation (FM) later in this chapter.

Phase Shift Keying

Phase Shift Keying (PSK) leverages changes in phase to convey data on an RF signal. For example, a transmitter that is modulating data with Phase Shift Keying (PSK) will change the phase of the sine wave on-the-fly as a means for encoding information. Figure 4.19 shows how changes in the phase can be used to indicate a 0 or a 1.

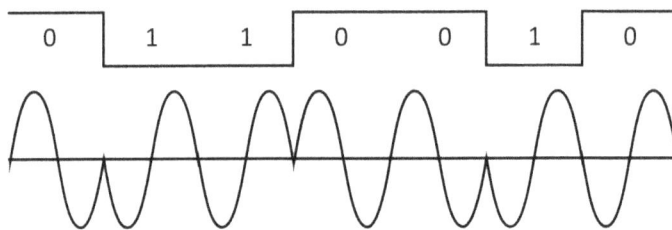

Figure 4.19: Phase Shift Keying, showing how changes in phase can represent binary data

Binary Phase Shift Keying (BPSK) is the simplest form of Phase Shift Keying (PSK). Figure 4.20 shows a *constellation diagram*, which is very basic in appearance in this example but will increase in complexity as we investigate more complex forms of modulation. The circle on the left is the target for a phase of 0° and the one on the right is a target for a phase of 180°. While the circle is the *constellation point*, the reality is that the phase can land anywhere on the left side and register a 0, or it can land anywhere on the right and register a 1. This is known as an *error vector magnitude* (EVM). BPSK, since it is binary, is an elementary form of modulation, but due to the large EVM, it is very forgiving when background noise and interference encroaches on the signal.

But what if we need more speed? To get more speed, we need to transmit and receive more bits of data in the same amount of time. While Binary Phase Shift Keying (BPSK) could only convey a 0 or a 1 (hence the name "binary"), the next "level up" in modulation is Quadrature Phase-Shift keying (QPSK). Figure 4.21 shows the constellation diagram for QPSK, which now has four target vectors instead of two. Each target, instead of representing a single bit, now represents two bits at a time, such as 00, 01, 11, and 10. These groups of bits are known as *symbols*, and they can be

138

transmitted in the exact same amount of time and frequency space, essentially doubling the amount of data we can transmit and receive in a given period.

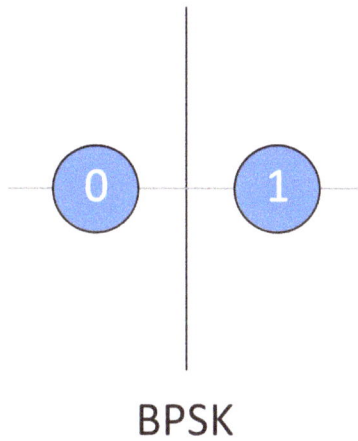

Figure 4.20: A BPSK constellation

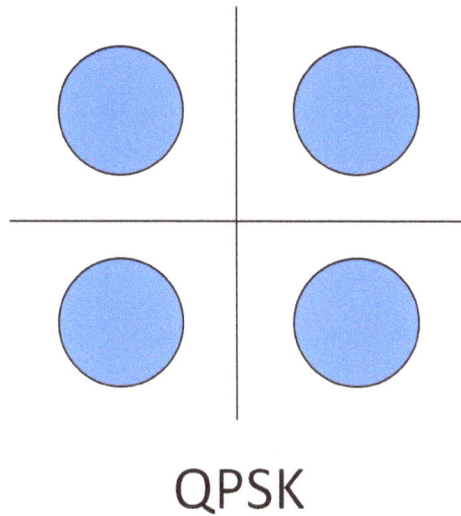

Figure 4.21: QPSK Constellation

This added speed comes at a cost: a smaller *error vector magnitude* (EVM). Now, hitting the targets with phases is twice as hard, because the EVM boxes are smaller. But at this point, this modulation scheme is still very simple, and very resilient to noise and interference.

The specifics about QPSK and BPSK aren't important for the exam. What is important is understanding Phase Shift Keying (PSK). The purpose of explaining QPSK and BPSK is simply to set up for the next section, which is Quadrature Amplitude Modulation.

Quadrature Amplitude Modulation

Now that you understand Phase Shift Keying (PSK) let's take a look at the next level of modulation: Amplitude and Phase Shift Keying (APSK) modulation. This is identical to PSK, except it adds the variable of Amplitude to PSK. With APSK, symbols (that is, groups of bits) are no longer represented solely with changes in phase, but now with changes in phase and changes in amplitude.

Quadrature Amplitude Modulation (QAM) is a form of APSK, and it is a type of modulation that you'll see in many wireless technologies such as Wi-Fi, digital broadcast television, satellite television, DSL in plain old telephone lines, and many others. Let's first look at 16-QAM, which is a relatively simple version of QAM that sees extensive use in technologies that you use every day. Figure 4.22 shows at 16-QAM constellation. Note that it looks just like a Quadrature Phase Shift Keying (QPSK) constellation, except now there are more than four targets. Now, there are 16, and they cannot be hit by changing phase alone. This is where the Amplitude in Amplitude and Phase Shift Keying (APSK) comes into play; amplitude determines the distance from the center of the constellation. As you can see, APSK, and thus 16-QAM uses both phase and modulation to hit the targets but gives the benefit of four bits per symbol. For example, a symbol might contain 0000, 0001, 0011, 0111, etc.

Just like before, this added complexity increases the number of bits we can transmit and receive in the same time span, but it also decreases the error vector magnitude (EVM) boxes, making it more susceptible to noise and interference.

64-QAM, 256-QAM, and 1024-QAM work exactly the same way to add more symbols, smaller EVM boxes, thus increasing speed potential but decreasing reliability. While Figure 4.23 only shows up to 1024-QAM, some technologies are known to use 4096-QAM, which provides a whopping 12 bits per symbol (for example, 011001000111).

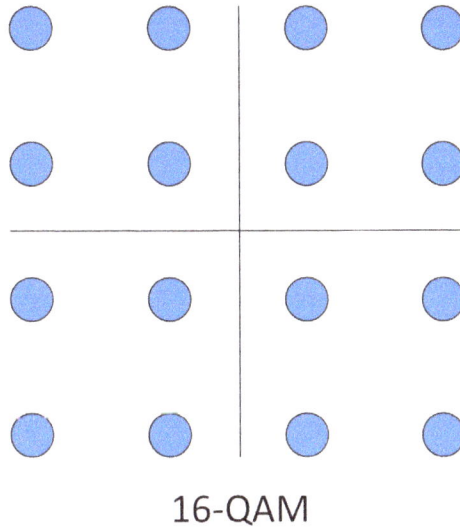

16-QAM

Figure 4.22: A 16-QAM constellation.

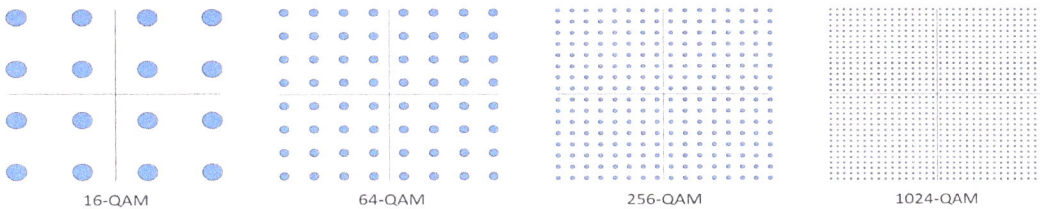

Figure 4.23: 16-QAM, 64-QAM, 256-QAM, and 1024-QAM

Orthogonal Frequency Division Multiplexing

Orthogonal Frequency Division Multiplexing (OFDM) divides a certain amount of frequency space into *subcarriers*, which are small divisions in the operating channel that are far apart enough from each other to avoid self-interference. There are usually three types of subcarriers:

- **Pilot Subcarriers**, which the receiver uses to synchronize with and tune to the incoming transmission

- **Data Subcarriers**, where data transmissions are carried out

- **Null subcarriers**, which don't carry data, and can function as guards at the outer edges of the transmission channel

All of the data subcarriers work together to simultaneously move data, but each one can move a different piece of data to increase transmission speeds, or copies of the same data to ensure reliability, depending on the coding scheme in use. Each subcarrier is commonly occupied by either Phase Shift Keying (PSK) or Amplitude Phase Shift Keying (APSK) modulation, with all subcarriers active at the same time with whatever modulation scheme is in use.

Think of subcarriers like strings on a guitar. When all of the strings are strummed on a guitar, it plays five notes simultaneously, but they are usually all different notes. The same is true for subcarriers in OFDM: they're all transmitted simultaneously, but are arranged at slightly different frequencies so each one can be heard and understood, carrying its own small piece of data. The device in the receiving end of the transmission will demodulate all of the subcarriers and reassemble the data.

Orthogonal Frequency Division Multiple Access

Orthogonal Frequency Division Multiplexing (OFDMA) is a variant of OFDM that allows data transmission to multiple, separate receivers at the same time. With OFDM (Orthogonal Frequency Division Multiplexing), a transmitter that needed to send unique pieces of data to multiple receivers would need to transmit data to each receiver, one at a time. This could create a performance bottleneck in the time domain, as a large amount of time could be consumed as the transmitters perform each unique transmission. Figure 4.24 shows how an OFDM transmitter would need to send separate pieces of data, one at a time.

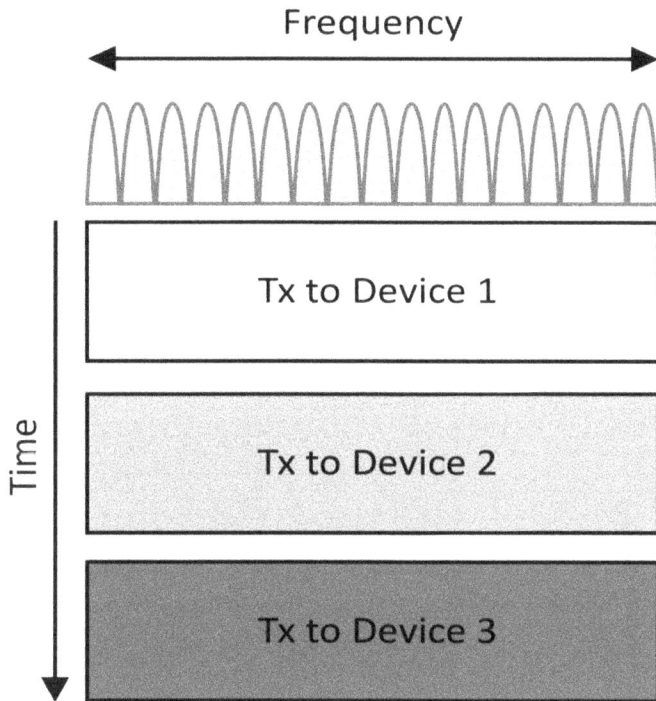

Figure 4.24: An OFDM transmitter sending unique data to three different receivers, one at a time

OFDMA alleviates this problem by allowing a transmitter to split it's channel up into multiple pieces in the frequency domain, and transmit data to multiple receivers at the same time. This is accomplished by dedicating a block of subcarriers to one receiver, another block of subcarriers to another receiver, and so on. These blocks of subcarriers are called *subchannels*, *resource blocks*, or *resource units* (all terms are equally acceptable). Figure 4.25 shows a transmitter simultaneously transmitting data to two or three devices concurrently, depending on what transmissions need to occur.

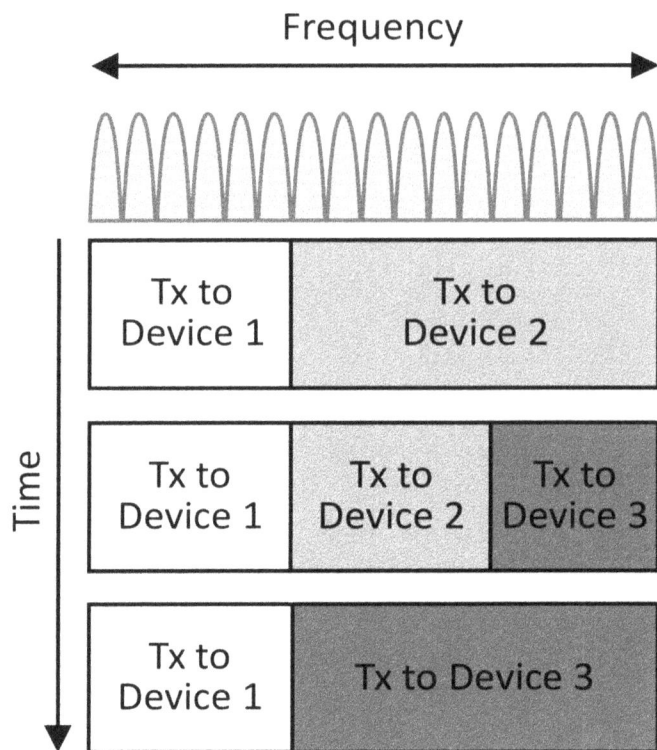

Figure 4.25: A single transmitter utilizing OFDMA to transmit data to multiple receivers at the same time

Frequency Hopping

Frequency-Hopping Spread Spectrum (FHSS) devices avoid interference from other devices by rapidly moving from channel to channel as they transmit information. During data transmission, they usually consume very little channel space, only using 1 the 3 MHz of frequency space. This makes them sound very similar to a narrowband transmitter like an old 2.4 GHz cordless phone, or a 5 GHz analog video camera, but unlike a narrowband transmitter, they do not stay still.

Frequency-hopping (FHSS) devices use varying methods to determine a pattern of channels to hop to and from. The transmitter will tune to a channel, dwell on that channel for a certain amount of time, tune to a new channel, dwell on the new channel for a certain amount of time, and repeat the process over and over, constantly hopping all over the spectrum that they're designed to work in. The amount of time that an FHSS transmitter spends on a single channel is called *dwell time*. Figure 4.26 shows how an FHSS device dwells on a channel for a short time before moving on.

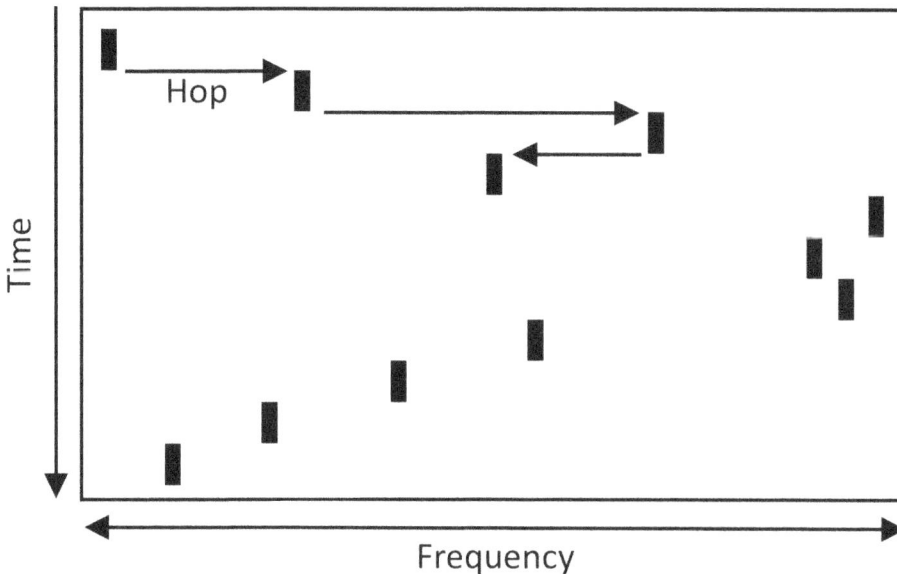

Figure 4.26: A FHSS device, shown over time

Some FHSS devices use a predetermined pattern of channels that both the transmitter and receiver know about. They then stay synchronized by repeating the pattern. A good example of an inexpensive consumer electronic device that does this is a wireless video baby monitor.

One of the most famous examples of FHSS is Bluetooth, which uses Adaptive-Frequency Hopping Spread Spectrum, a form of FHSS, to detect "bad channels" and avoid them. This can allow Bluetooth devices to dynamically avoid interference.

The primary advantage of FHSS is that it allows devices to avoid or cope with interference. Narrowband devices stay rooted in one frequency, and if they encounter interference, there's nothing they can do about it. FHSS constantly moves, limiting its exposure to interference. If an FHSS device does encounter interference, it will only be for a tiny moment, before the device moves on to the next channel.

Cellular Modulation Methods

Beyond all of the other modulation types that have been discussed here, there are two additional modulation methods that are often seen in the realm of cellular networks.

First is Time Division Multiple Access (TDMA), in which all wireless stations on the network share the same transmission frequency. To avoid stations transmitting at the same time and corrupting each other's signals, all of the stations essentially take turns transmitting on the channel, all controlled by a centralized authority. TDMA was used in legacy 2G GSM cellular networks, but it lives on in a variety of other technologies today, such as in some point-to-point wireless networks. Some point-to-point wireless bridges that use Wi-Fi can be placed in a dedicated TDMA mode to increase performance.

Instead of giving each device a time slot, *Frequency Division Multiple Access* (FDMA) provides separate sub-channels for each transmitting device.

TDMA and FDMA are illustrated in Figures 4.27 and 4.28 respectively.

The other notable modulation method for modulation in cellular networks is *Code Division Multiple Access*, or CDMA. CDMA uses a *Walsh Code* to convert user data into a series of *chips*. Chips from each user are then converted into simple waveforms, which are then merged into a *composite waveform* for transmission. When the composite waveform is received, the Walsh Code is then reapplied to recover the original data. CDMA is used in WCDMA, CDMA 2000, 1xEVDO, and HSDPA/HSUPA, all of which are 3G cellular technologies.

Additional Modulation Methods

Let's take a quick look at a handful of other modulation methods that could be considered "legacy" technologies, but are still in use today.

At first glance, Amplitude Modulation (AM) might just seem like Amplitude Shift Keying (ASK), but they are two different types of modulation. While ASK modulates 1's and 0's by changing the amplitude, Amplitude Modulation (AM) is instead used to transmit an analog signal.

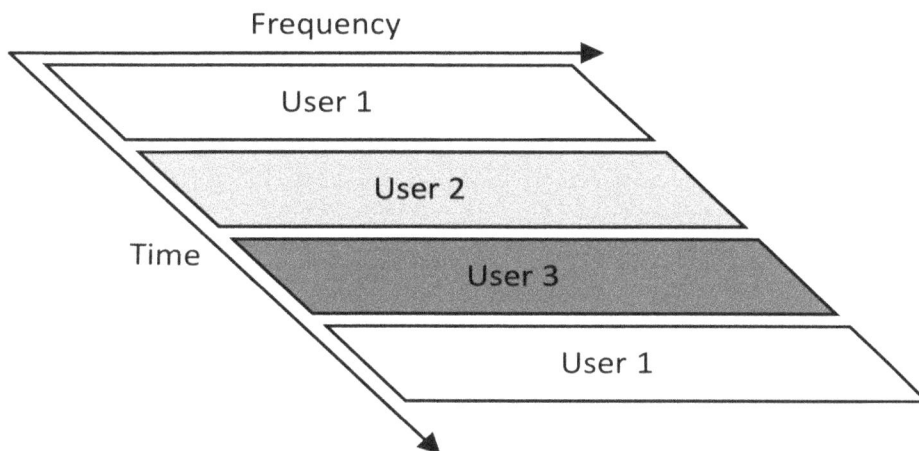

Figure 4.27: Time Division Multiple Access, showing how the channel is divided up by time to accommodate each wireless station

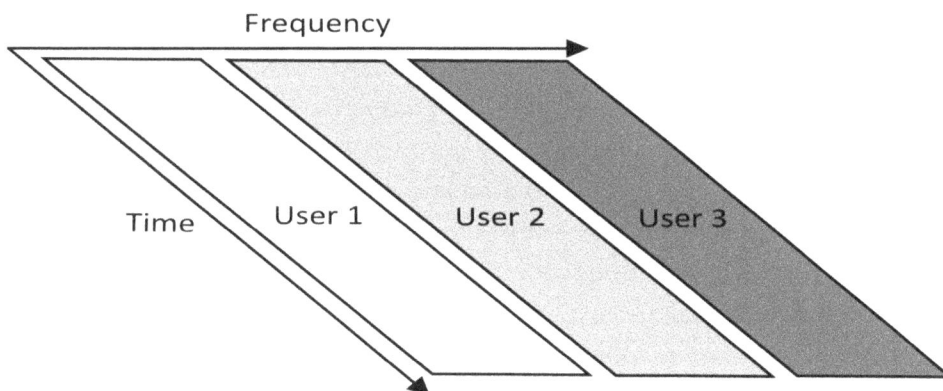

Figure 4.28: Frequency Division Multiple Access, showing how the channel is divided up into multiple frequencies to accommodate each wireless station

Today's most visible example of Amplitude Modulation feels a bit dated, but it's still readily accessible: AM radio. There are two sine waves involved. The first is the radio frequency *carrier wave*, which is from 535 to 1605 kHz, depending on the radio station. The second wave is the actual audio wave; the representation of human voice in waveform. The *carrier wave* modulates the audio wave by modifying the amplitude (height) of the wave, proportional to the audio wave. In Figure 4.29, you can see how the carrier wave "carries" the audio signal.

The advantage of AM is that it is very simple, but the disadvantage is that it is highly susceptible to any interference or atmospheric effects that modify amplitude. For example, if you've ever listened to AM radio near power transmission lines, you may notice a buzzing effect. This is interference from the transmission lines, increasing the amplitude of the AM carrier wave, which your car stereo then interprets as audio.

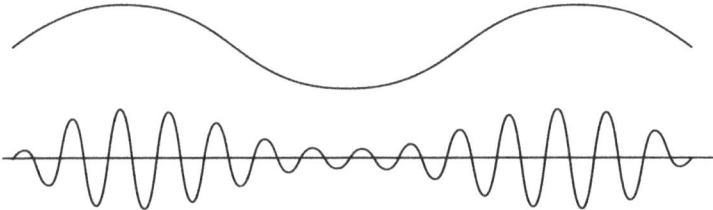

Figure 4.29: An audio waveform, and associated carrier wave, which is modulating the audio waveform

Frequency Modulation (FM) works very similarly to AM, except for a critical difference: instead of changing the amplitude as AM does, Frequency Modulation varies the wavelength to carry the analog signal, while the amplitude stays constant. In the case of FM radio, the analog signal is a waveform that represents music or audio programming. FM radio isn't influenced by atmospheric disturbances as easily as AM radio, because changes in amplitude don't significantly impact the frequency modulation and thus the quality of the received signal. Figure 4.30 shows an audio wave and the proportional variations in frequency.

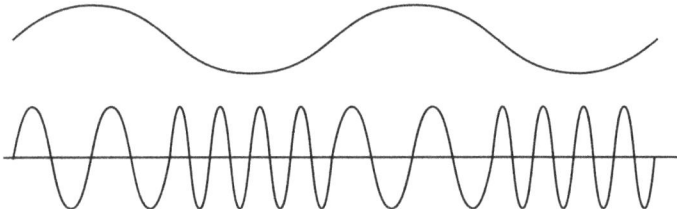

Figure 4.30: An audio waveform, and associated carrier wave with variations in frequency to modulate the audio

Perhaps the oldest type of modulation discussed here is *Continuous Wave* (CW), which could also be called *on/off carrier keying.* Essentially, a CW transmission keys a signal by abruptly beginning the carrier wave, maintaining it for a specific amount of time, and abruptly ending the carrier wave again. If this sounds a lot like a radiotelegraph (that is, a telegraph that uses radio instead of transmission wires), then you are correct! CW is the exact modulation scheme that was used by radiotelegraph systems to transmit Morse code.

CW is still a viable way to transmit data. Because it is so simple, it is very robust, allowing it to function in very adverse radio frequency conditions. That said, it is extremely slow, so today's use of CW is primarily by amateur radio enthusiasts.

Other Wireless Carriers

Recent news stories around light-based wireless communications systems like Li-Fi make communication with light seem futuristic, but the reality is that light-based communication has been with us for a very long time. In fact, light-based communication has been available in the home audio space since the mid-1980s, with the introduction of Toshiba's TOSLINK fiber-optic cables. While very rudimentary compared to the fiber optics that connect our modern networks, TOSLINK gave consumers the ability to transmit digital audio signals to and from audio receivers and amplifiers. One of the supposed advantages of this was to electrically isolate pieces of equipment to negate electrical noise. Check the back of your modern high-definition television, where you're more likely than not to find an optical TOSLINK port.

A decade later, another light-based technology emerged that you might remember: IrDA, which stands for Infrared Data Association. Infrared transceivers were built into all kinds of electronics throughout the late 1990s and early 2000s, including PDA's, laptops, and printers. IrDA gave away to newer wireless communication standards like Bluetooth and Wi-Fi, but infrared communication isn't gone; most modern TV remotes still use infrared.

Today, when you think "light-based communications," you probably think about Li-Fi, which is currently in the "technical evaluation" stage. Whether Li-Fi, as we imagine it (as a replacement for Wi-Fi), is yet to be seen. There are many advantages to transmitting data over visible spectrum instead of in radio spectrum, including visible spectrum's *isolation* from radio spectrum. For example, in environments with substantial limitations or restrictions around radio usage, Li-Fi could provide a completely new and safe way to transmit data. This is very similar to TOSLINK's advantage of being able to electrically isolate pieces of audio equipment. Li-Fi also has the exciting advantage of confinement to a specific room; Wi-Fi in the 2.4 and 5 GHz frequency bands can easily transit through walls and out of buildings, whereas light cannot. Even the problem of light leaving windows could be easily solved by applying special coatings that block specific wavelengths of light.

Li-Fi also offers huge potential for speed and throughput, with researchers reaching up to 224 gigabits per second in a lab setting, using *light-emitting diodes* (LED's) as a transmission source. Li-Fi certainly looks promising, but just like many exciting technologies that are in development: we'll have to wait and see.

After a lengthy discussion about radio frequency communications, and a quick look at the past and potential future of light-based communications, it makes sense to take a quick look at *acoustic communication*. We already use acoustic communication regularly: when we listen to the radio, play an audio clip from the internet, or even talk and listen. As we established earlier, all of these communications use waves in the audio spectrum. What if we could use acoustic communication not just for music or human voice, but for digital communications as well?

An environment where acoustic communication flourishes is underwater. Light-based communications can render a range of a couple of hundred meters, but that isn't far enough for some applications, and radio waves suffer from heavy, heavy attenuation from water. Acoustic communication, on the other hand, has the potential to enable long-distance communication. Whales use it, so why not humans? By utilizing pressure waves, it's possible to communicate underwater using many of the same modulation schemes that we discussed earlier, including *Phase Shift Keying* (PSK), Frequency Shift Keying (FSK), and even *Orthogonal Frequency Division Multiplexing* (OFDM)! This really underscores the similarities between acoustic (sound) waves and radio waves.

Common Frequency Bands

The following represents a number of common or interesting frequencies and frequency bands used by various technologies. It is by no means an exhaustive list; a complete list would most likely consume all of the pages of this book. Instead, this list is intended to illustrate the sheer number of various frequency band allocations that exist. Also keep in mind that this is for North America only, as other regulatory domains (such as *European Telecommunications Standards Institute*, or ETSI in Europe) will assign different allocations to different frequencies in some cases.

Frequency	Technology
1 MHz	Amatuer Radio (160 meter band)
144 MHz	Amatuer Radio (2 meter band)
440 MHz	Amatuer Radio (70 centimeter band)
535-1605 kHz	AM Radio
88-108 MHz	FM Radio
600 MHz	LTE (Cellular)
700 MHz	LTE (Cellular)
900-928 MHz	Unlicensed Band (Zwave, consumer electronics)
1227.60 MHz	GPS L1
1575.42 MHz	GPS L2
1700 MHz	LTE (Cellular)
1900 MHz	LTE (Cellular)
2100-2120 MHz	NASA Deep Space Network Uplink (S band)

2290-2300 MHz	NASA Deep Space Network Downlink (S band)
2.36-2.4 GHz	Wireless Body Area Network (WBAN)
2.4 GHz	Unlicensed Band (Wi-Fi, Bluetooth, ZigBee, consumer electronics)
2.5 GHz	LTE (Cellular)
3.55-3.7 GHz	Citizens Broadband Radio Service (CBRS)
4.9 GHz	Fixed and mobile services for public safety use only
5 GHz	Terminal Doppler Weather Radar, Unlicensed Band (Wi-Fi, consumer electronics)
6 GHz	Point-to-point communications, possible future Wi-Fi expansion
7145-7190 MHz	NASA Deep Space Network Uplink (X band)
8400-8450 MHz	NASA Deep Space Network Downlink (X band)
27-60 GHz	mm-Wave technologies (5G, WLAN, bridging)

Figure 4.31 shows the common terms used for the various frequency bands. There is some disagreement over the exact ranges of these bands, but the terms are very common in RF engineering. Originally, the band frequency ranges were chosen so that wavelengths had rounded numbers at the edges of the bands (The IEEE Wireless Dictionary, James P. Gibb, Wiley, 2005). Many wireless solutions span these bands or use frequency blocks in different bands depending on their implementation.

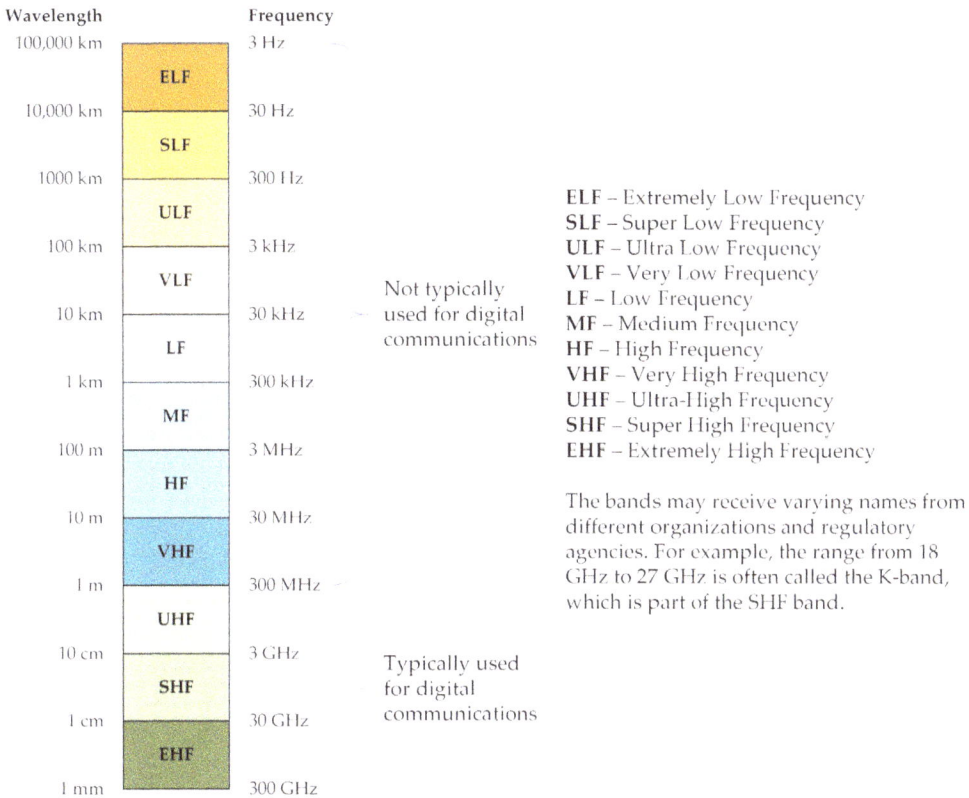

Wavelength | Frequency
ELF
SLF
ULF
VLF
LF
MF
HF
VHF
UHF
SHF
EHF

100,000 km — 3 Hz
10,000 km — 30 Hz
1000 km — 300 Hz
100 km — 3 kHz
10 km — 30 kHz
1 km — 300 kHz
100 m — 3 MHz
10 m — 30 MHz
1 m — 300 MHz
10 cm — 3 GHz
1 cm — 30 GHz
1 mm — 300 GHz

Not typically used for digital communications

Typically used for digital communications

ELF – Extremely Low Frequency
SLF – Super Low Frequency
ULF – Ultra Low Frequency
VLF – Very Low Frequency
LF – Low Frequency
MF – Medium Frequency
HF – High Frequency
VHF – Very High Frequency
UHF – Ultra-High Frequency
SHF – Super High Frequency
EHF – Extremely High Frequency

The bands may receive varying names from different organizations and regulatory agencies. For example, the range from 18 GHz to 27 GHz is often called the K-band, which is part of the SHF band.

Figure 4.31: RF Bands, Names, and Wavelength Estimations

In addition to the information in Figure 4.31, it is important to know that many RF engineers use shorthand terminology to reference various frequency ranges. The following list will help you in understanding what is meant by this colloquial terminology:

- 900 MHz - 900-928 MHz range

- 2.4 GHz - from 2.400 to 2.500 GHz

- 4.9 GHz

- 5 GHz - various frequency blocks in the 5 GHz to 6 GHz range

- 60 GHz - from 57 GHz to more than 60 GHz

153

Chapter Summary

In this chapter, we began by defining what a wave is, as well as identifying the key characteristics of a wave, including frequency, wavelength, amplitude, and phase. The frequency of a wave determines how quickly the wave completes a full cycle and is usually measured in hertz (1 cycle per second), kilohertz (1000 cycles per second), megahertz (one million cycles per second), and gigahertz (one billion cycles per second). A wavelength is the distance between two adjacent identical points on a wave, such as from one crest to the next crest. Higher frequencies mean shorter wavelengths, and lower frequencies mean longer wavelengths.

Amplitude refers to the height or power of a wave. Just like loud sounds can be heard from greater distances than quieter sounds, an high-amplitude radio signal can be detected and understood at greater distances than low-amplitude signals.

While frequency, wavelength, and amplitude are characteristics of single RF waves, the *phase* is a comparison between two waves. Two waves that arrive at the receiver in synchronization with each other are said to be *in-phase* and have the effect of amplifying the signal at the receiver. Two waves that arrive 90° out-of-phase with each other will be more difficult for other receivers to understand, as the sine wave is no longer a pure form and has experienced a certain amount of corruption. Two waves that arrive at a receiver 180° out-of-phase with each other will completely cancel each other out, nullifying the radio transmission.

Just like audio amplifiers increase the volume of an audio signal, radio amplifiers increase the amplitude of a radio signal. *Active Gain* refers to the process of amplifying a radio signal using an external power source, while *Passive Gain* uses no external power source, instead relying on more focused antennas to provide gain for the radio signal.

Attenuation occurs when a radio frequency signal is reduced in amplitude. This usually happens when RF passes through an object. Free Space Path Loss refers to the weakening of an RF wave, due to the broadening wavefront, and *Absorption* occurs when RF energy is dissipated inside an object. Just like light reflects off of a mirror, RF reflects off of smooth, non-absorptive surfaces, sharply changing the signal's direction. Refraction occurs when RF passes from one object density to another, changing the speed of the RF, and bending it into a slightly new direction. This phenomenon is observable by placing a spoon in a bowl of water, and observing how the spoon appears to be bent in the water. Scattering happens when RF strikes an uneven surface.

Watts, milliwatts, and microwatts are all absolute units of RF power. Decibels (dB) represent a logarithmic change in signal strength. Remember the rule of 3's and 10's:

- +3 dB (a gain of 3 dB) means double the power

- -3 dB (a loss of 3 dB) means half the power

- +10 dB (10 dB of gain) means 10x more power

- -10 dB (10 dB of loss) means 10x less power

- Gains and losses a cumulative (increasing by success additions or subtractions)

Decibels in relation to a milliwatt (dBm) show absolute power, and greatly simplifies the readability of signal strength in many cases. 1 mW = 0 dBm, and 100 mW = 20 dBm.

Noise is any signal that a receiver cannot decode or distinguish from other signals, and the *noise floor* refers to the ambient noise in the RF environment, whether it be natural background noise or noise from other radio devices. The definition of *signal-to-noise ratio* (SNR) is how much signal strength can be heard above the background noise.

Next, we discussed different ways of modulating signals on radio waves. The first was *Amplitude Shift Keying* (ASK), which moves the amplitude up or down to communicate 1's and 0's. Similarly, *Frequency Shift Keying* (FSK) changes the frequency of the wave to communicate 1's and 0's. Finally, *Phase Shift Keying* (PSK) changes the phase on-the-fly for the same purpose as above.

While *Binary Phase Shift Keying* (BPSK) can only transmit a 1 or a 0, *Quadrature Phase Shift Keying* (QPSK) uses FSK to shift the phase of the wave in four distinct phases. Depending on the selected phase, a target vector from a *constellation* of four *symbols* can be selected, and with QPSK, they each contain two bits instead of just one.

Quadrature Amplitude Modulation (QAM) couples both Phase Shift Keying and Amplitude Shift Keying together to add more points to the constellation, which in turn makes symbols bitter. 16-QAM has a constellation of 16 symbols with four bits each. 64-QAM offers 64 constellation points, and so on.

Orthogonal Frequency Division Multiplexing (OFDM) uses subcarriers to transmit a lot of data at once, and *Orthogonal Frequency Division Multiple Access* (OFDMA) can split the channel into *Resource Units* to transmit data to multiple receivers at the same time.

Frequency Hopping Spread Spectrum (FHSS) devices use a narrow bandwidth to transmit but hop all over their allocated space to avoid interfering with other devices. The hopping is usually very rapid, minimizing the chance of receiving interference. *Amplitude Modulation* (AM) is exactly what an AM radio uses; an analog signal is carried on amplitude variations in the carrier wave. Similarly, *Frequency Modulation* (FM) varies the frequency of the carrier wave to convey an analog signal.

Review Questions

1. Which modulation scheme uses phase to encode either a 1 or a 0?

    a. BPSK

    b. QPSK

    c. 16-QAM

    d. 1024-QAM

2. Which RF propagation behavior causes a signal to bounce off a surface, immediately changing the direction in which it is propagating?

    a. Absorption

    b. Refraction

    c. Scattering

    d. Reflection

3. Which modulation scheme carries an analog signal via changes in amplitude?

    a. Amplitude Shift Keying (ASK)

    b. Frequency Shift Keying (FSK)

    c. Amplitude Modulation (AM)

    d. Frequency Modulation (FM)

4. Which answer indicates that signal strength has been doubled?

    a. 10 dB of gain

    b. 10 dB of loss

    c. 3 dB of gain

    d. 3 dB of loss

5. What is the correct definition of a wavelength?

   a. The distance between two adjacent identical points on the wave

   b. The speed at which the wave propagates through space

   c. The speed at which the wave oscillates

   d. The distance between 0° and 90° on the wave

6. How many bits does a QPSK symbol contain?

   a. 1 bit per symbol

   b. 2 bits per symbol

   c. 4 bits per symbol

   d. 12 bits per symbol

7. In OFDM modulation, what is the frequency space divided into for data transmission?

   a. Frequency spaces

   b. Sub-channels

   c. Device channels

   d. Subcarriers

8. Which modulation scheme does not stay on one frequency; instead it uses a technique to avoid causing and receiving interference?

   a. FHSS

   b. OFDM

   c. QPSK

   d. CW

9. If two waves line up perfectly when they arrive at the receiver, what effect do they have?

    a. The waves cancel each other out

    b. The waves are more difficult for the receiver to understand

    c. The waves' amplitude is increased

    d. The receiver demodulates the signal

10. Which frequency has the longest wavelength?

    a. 5 GHz

    b. 3600 MHz

    c. 1400 MHz

    d. 80 Mhz

Review Answers

1. The correct answer is **A.** Binary Phase Shift Keying (BPSK) encodes a digital signal using a single phase shift to indicate a 1 or a 0.

2. The correct answer is **D.** Refraction causes RF to bend slightly, while scattering weakens the signal by causing it to scatter into many different directions. Reflection completely changes the direction that the RF propagates in.

3. The correct answer is **C.** Amplitude Shift Keying (ASK) encodes a digital signal of 1's and 0's, based on the amplitude of the wave. AM radio uses a carrier wave with variations in amplitude to modulate the analog signal.

4. The correct answer is **C.** 3 dB of gain indicates a doubling in signal strength. 10 dB of gain would indicate 10 times the previous signal strength.

5. The correct answer is **A.** The definition of a wavelength is the distance between two adjacent, identical points on the wave. The wave can be measured from swell or swell, trough to trough, or beginning to end.

6. The correct answer is **B.** Quadrature Phase-Shift Keying (QPSK) offers 4 constellation points, so each constellation point and thus each symbol offers 2 bits. 1 bit is BPSK, 4 bits is 16-QAM, and 12-bits is 4096-QAM.

7. The correct answer is **D.** OFDM divides the operating channel up into subcarriers, which all work together to move information, usually with QPSK or QAM.

8. The correct answer is **A.** FHSS stands for Frequency Hopping Spread Spectrum, which continuously hops in a pseudorandom channel pattern to avoid receiving interference.

9. The correct answer is **C.** When two waves arrive in-phase with each other, they have the effect of increasing the amplitude of the wave.

10. The correct answer is **D.** 80 MHz is the lowest frequency, and lower frequencies have longer wavelengths.

Chapter 5: Radio Frequency Hardware

Objectives Covered:

2.3   Explain the basic capabilities of components used in RF communications

RF hardware can be considered at different detail levels. We will begin this chapter by exploring these levels. Next, we will investigate the inner components of a wireless device, including the chips and circuits that provide for RF communications. We will then move to the link types created by various wireless devices and conclude the chapter with a general summary of RF device types (the highest layer of hardware abstraction).

Understanding the functionality of RF hardware and the various hardware types is essential for the CWSA exam; however, it is also vital when administering wireless networks. As a wireless solutions administrator, you will encounter scenarios where you must select appropriate equipment and replace equipment when the in-use hardware is no longer available. In such scenarios, it is essential that you understand the functionality of the wireless devices and implement new devices that meet the needs of the organization.

Hardware Levels

RF hardware can be considered from many detail levels, but this chapter will focus on three levels:

- **Circuit Board Level:** At this level, you explore the individual chips and circuits that make up a radio. Understanding the hardware that provides functionality for wireless communication assists the wireless solutions administrator in making effective decisions. The section titled *Basic RF Hardware Components (Circuit Board Level)* will address this level.
- **Use Category (Link Types):** At this level, you explore the wireless communications functionality provided by the RF hardware or system. Does it provide bridging functionality, mesh functionality, ad-hoc communications, or something else? The section titled *RF Link Types (Use Category)* will address this level.
- **RF Device Types:** At this level, you explore the various devices, as a whole unit, and the capabilities they provide and the features they offer. For example, a wireless sensor is a specific device type that may participate in a mesh or other wireless network. As another example, a Bluetooth device may be used in a location tracking system, and it may not connect to a network continually (on-demand). The section titled *RF Device Types* will address this level.

The remainder of this chapter will explore these three levels in detail. Understanding this information from a conceptual and practical perspective will allow you to grasp better any RF hardware you work with in the future.

Basic RF Hardware Components (Circuit Board Level)

The circuit board, in an electronic device, is the foundation on which circuits, transistors, chips, resistors, sensors, and other components can be placed, or to which they may be connected, to interact with each other. In mass manufactured devices, a printed circuit board (PCB) is used, and these can be seen inside of nearly every wireless device manufactured and sold today. Figure 5.1 shows an example of an assembled PCB used in a ZigBee access point for a wireless sensor network (WSN). The PCB brings together radio chipsets, filters, amplifiers, resistors, transistors, and various other components depending on the needs of the device. For wireless devices, antennas may be printed into the PCB, or they may be attached to it as needed.

Figure 5.1: A ZigBee Device PCB

Hobbyists or small production runs may use a component known as a breadboard to evaluate and test the functionality of a design in the prototype stage. Breadboards can be acquired and used to build basic radio communications devices within the lab. Breadboards are beneficial for prototyping because they are solderless and allow for quick assembly of a solution.

The following sections explain some of the most common components used to build radios, antennas, amplifiers, attenuators, and splitters, which are all used in different ways within wireless devices and wireless solutions.

Radios

RF-based wireless devices use radios to transmit and receive signals. A wireless link is used to transport information between the two nodes in the link. Information is encoded and transported across a carrier wave as a signal. A *transmitter* sends a signal to be received by a *receiver*. A radio that can both transmit and receive is called a *transceiver* (a combination of the words transmitter and receiver).

The transmitter is sometimes called the source, and the receiver is sometimes called the *destination* or the *sink*. In most wireless links, transceivers are used on both ends of the link because they send signals back-and-forth to each other. This behavior is not always the case, for example, a Bluetooth Low Energy (BLE) beacon may be transmitted and received by another device, but the receiving device may not reply in any way using BLE (though it may take action using another wireless radio within the same device).

A solution that transmits only in one direction (from the transmitter to the receiver) is known as a *simplex* system. When transmission can occur in both directions, it is known as a *duplex* system. A duplex system can be either half-duplex or full-duplex. A half-duplex system can either transmit or receive at any given moment, but it cannot do both concurrently. A full-duplex system can both transmit and receive concurrently. Both half-duplex and full-duplex wireless systems are categorized as transceivers.

Technically, the radio is the part that generates the RF signal, and the antenna is that part that "leaks" the signal into the environment. While the radio can process an incoming signal, it requires an antenna to "capture" that signal from the environment.

Three categories of RF signal radiators are commonly defined:

- **Incidental Radiators:** Electric or mechanical devices that generate RF energy and radiate it into the environment, but they are not designed to produce RF energy. According to the FCC, "*An incidental radiator (defined in Section 15.3 (n)) is an electrical device that is not designed to intentionally use, intentionally generate or intentionally emit radio frequency energy over 9 kHz. However, an incidental radiator may produce byproducts of radio emissions above 9 kHz and cause radio interference.*"

- **Unintentional Radiators:** Electric devices that generate electrical or radio frequency signals that are intended to be contained within the system or a conductive link (such as a wire-based cable) and not radiated into the environment, but they may emit some RF energy in spite of the intended design. According to the FCC, "*An unintentional radiator (defined in Section 15.3 (z)) is a device that by design uses digital logic, or electrical signals operating at radio frequencies for use within the product, or sends radio frequency signals by conduction to associated equipment via connecting wiring, but is not intended to emit RF energy wirelessly by radiation or induction. Today the majority of electronic-electrical products use digital logic, operating between 9 kHz to 3000 GHz and are regulated under 47 CFR Part 15 Subpart B.*"

- **Intentional Radiators:** Electric devices designed to generate radio frequency signals and transmit them into the environment. According to the FCC, "*An intentional radiator (defined in Section 15.3 (o)) is a device that intentionally generates and emits radio frequency energy by radiation or induction that may be operated without an individual license. Examples include: wireless garage door openers, wireless microphones, RF universal remote-control devices, cordless telephones, wireless alarm systems, Wi-Fi transmitters, and Bluetooth radio devices.*" The wireless solutions discussed in this book fit into the intentional radiator category.

Why is it important to know about these three categories of radiators? Because all three result in RF energy or signals in the environment at various amplitudes. When troubleshooting a wireless interference problem, the interferer is not always an intentional radiator. Often, the interferer is in the incidental or unintentional radiator categories, and the trained wireless solutions administrator must remember to look for such interferers as well.

Careful reading of the FCC regulations indicates that the intentional radiator is everything up to, but not including, the antenna(s) in their conceptualization. For example, CFR 15.203 Antenna Requirement states that *"An intentional radiator shall be designed to ensure that no antenna other than that furnished by the responsible party shall be used with the device."* While some resources include the antenna as part of the intentional radiator, for purposes of consistency, CWNP does not.

To understand the components in a logical radio transceiver, consider the block diagram in Figure 5.2. The figure is not intended to represent any specific wireless device, but to explain the parts and their purposes. The listed components will be explained in the remainder of this section.

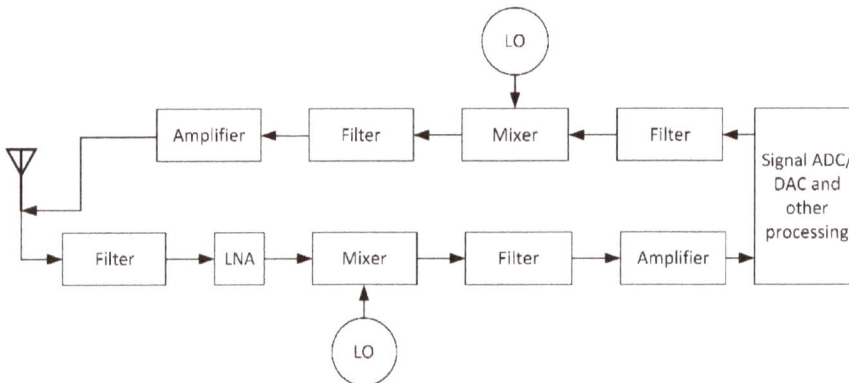

Figure 5.2: Block Transceiver Diagram Showing Logical Components

Important components used to build a radio system or transceiver include:

- Amplifiers
- Filters
- Oscillators and Mixers
- ADC/DAC Convertors
- Modulators/Demodulators

Radios are designed in various ways. The components discussed here may be integrated into chipsets or separated in PCBs or a mixture of both. Think of these components as actions required and not necessarily individual devices or chips in a wireless device.

Amplifiers include low-noise amplifiers (LNAs), Intermediate Frequency (IF) amplifiers and other amplifiers. The purpose of an amplifier is to increase the amplitude (power or strength) of the RF signal for transmission or during the reception.

Every component added in the radio chain has the potential (and usually does) to reduce the SNR at a receiver. The component itself adds noise as the RF energy passes through it. The difference between the SNR entering a component and the SNR exiting a component is called the *noise figure*. Noise figure is the same as *noise factor* except that noise factor is expressed as a ratio and noise figure is expressed in dB.

LNAs are used in the receive chain at or near the entry of the signal into the radio chain. An LNA is designed to amplify the signal for further processing while introducing as little noise as possible. Receivers can process very weak signals, in large part, due to the use of LNAs. An LNA may amplify the signal by 50 dB or more with a noise figure as low as 1 dB. Therefore, if a received signal is -90 dB with an SNR of 10 dB and it is amplified by 50 dB with a noise figure of 1 dB, the resulting signal will be -40 dB with an SNR of 9 dB. If the amplifier had a noise figure of 5 dB, the resulting SNR would only be 5 dB. As you can see, the noise figure in an LNA is crucial and impacts the overall receive sensitivity of the receiver significantly.

BEYOND THE EXAM: Why All the Details of Components and Functionality?
At this point, you are probably wondering why we are going into all the details of the components that make up a radio. Why not just say that a transceiver can transmit and receive and leave the rest out? The answer is simple: understanding

the components in the system and the impact they have on signal processing helps you to understand why one vendor's device may be superior to another vendor's device. Specification sheets may help if they list details like receiver sensitivity, but they may not.

Consider, for example, a wireless device that uses five internal primary components (amplifiers, filters, etc.). Imagine that the device has an average noise figure of 1.9 dB for each component. That's a total of 9.5 dB loss in SNR due to the cumulative noise figure. Further, another device has an average noise figure of 1.1 dB for each component. That's a total of 5.5 dB loss in SNR due to the cumulative noise figure. The difference in SNR between these two devices means that the latter device may be able to achieve higher data rates or even connect when the former device cannot. This concept is the foundation of receiver sensitivity.

So, why wouldn't every vendor use the better-quality components? The answer lies in total cost of development. For example, at the time of this writing, an LNA with 40 dB of gain and a 0.8 dB noise figure sold for $936.55 (this is an inline external LNA, hence far more expensive than the internal components used within devices). The same manufacturer was selling a 50 dB of gain LNA with a 1 dB noise figure for only $524.65. While the individual component prices are not as high for PCB implementations, the price variance added up over thousands of units adds up quickly. This reality (and the software factor) is why consumer-grade devices are often inferior to enterprise-grade devices, and it's also part of the reason for the significant price difference between them.

The result of this knowledge is an understanding that you must test equipment whenever possible before purchasing one hundred or ten thousand units. Image the impact on your networks and users if you purchase several thousand units that are significantly inferior to those that cost only a few dollars more. Lab testing, in your environment (not with high-end engineering gear), can help to validate that a device performs as required.

Filters are used to limit the signal to the desired frequencies. Several types of filters can be used, but we will focus on image rejection filters and bandpass filters here.

Before looking at filters specifically, it is important to understand the concept of an *intermediate frequency (IF)*. For simpler processing, many radio systems down-convert the received signal to an IF. The IF is then processed for actual demodulation. Using an IF provides several possible benefits:

- Selectivity is improved. When converting to an IF, the radio can improve selective filtering at the IF so that just the desired information is preserved.
- Simpler processing of high-frequency signals. It's easier to build solutions for lower frequencies, so converting the high-frequency signal to an IF allows for this processing.
- Using many different frequencies. Converting to an IF allows the receiver to receive many different frequencies, but process them all the same.

The phrases *image rejection* or *RF image rejection* refer to the undesired frequencies received by a wireless device that are the image of the desired frequencies. The image frequencies are basically any signals the same distance on either side of the IF used for processing in the radio. Filters perform image rejection so that these unwanted frequencies (from the mirror image of the wanted frequencies) do not introduce errors during down conversion to the IF.

Filters can also perform *selectivity*. In this case, such as the functionality of a bandpass filter, the filter allows the desired frequencies through. Bandpass filters help to prevent or diminish adjacent channel interference. Adjacent channel interference occurs when the next or previous channel is in use near the receiver, and the signal is sufficiently strong. Bandpass filters can help to reduce this interference as long as the adjacent channel energy is not too strong.

Oscillators generate electromagnetic waves and, when controlled, signals. In many wireless transceivers, the received signal is processed by a mixer with an oscillator (known as a local oscillator (LO)). The mixer "mixes" the LO RF with the received RF and converts to an IF. The IF is then used for actual processing of the received signal.

Analog-to-Digital Converters (ADC) and Digital-to-Analog Converters (DAC) convert received signals to baseband digital signals (ADC) on reception and baseband digital signals to RF signals (DAC) on transmission. Modulators and demodulators can be said to encompass the ADC/DAC and oscillation functions to generate RF signals at transmission and convert RF signals on reception.

Many more details could be provided related to these and other components used in RF and microwave engineering. However, the CWSA will not be building wireless devices. The CWSA will be implementing and supporting the devices. Understanding the high-level concepts presented here helps you to comprehend the functionality of a wireless device, in relation to RF signals, and better select quality hardware.

Antennas

Antennas are essential to RF and microwave communications. Without an antenna, unless a receiver is exceptionally close to a signal leak, communications cannot occur. Some wireless solutions use a single antenna and others use multiple antennas on each device. This section provides essential information regarding this important component in wireless communications.

The antenna is the radiating element in an RF system. It is the component that results in the propagation of RF waves through space. It is also the device that receives the RF signals from other propagating antennas. Different antennas have different coverage capabilities and different characteristics.

If you stand on the top of a tall building, you can see for a very great distance. You may even be able to see for many miles on a very clear day. If you can physically see something, it is said to be in your visual line of sight (visual LOS) or just LOS, for simplicity. Visual LOS is also called physical LOS. This *visual LOS* is the transmission path of the light waves from the object you are viewing (transmitter) to your eyes (receiver). Visual LOS is an apparently straight line from your perspective, but light waves are subject to similar behavior as RF waves, like refraction and reflection, and therefore the line may not actually be straight. Consider an object you are viewing in a mirror. The object is not directly in front of you, and yet it appears to be, showing that visual LOS is not necessarily a straight line between two objects.

Because RF is part of the same electromagnetic phenomenon as visible light, behaviors similar to visual LOS exist. However, RF LOS is more sensitive than visual LOS to interference near the path between the transmitter and the receiver, particularly when creating bridge links over some distance. You might say that more space is needed for the RF waves to be seen by each end of the connection. This extra space can be calculated and has a name: *the Fresnel Zone*. The Fresnel Zone is only important for long-distance point-to-point links, as indoor propagation patterns ensure the signals get through to the clients as long as the entire wireless solution is implemented based on proper design principles. Figure 5.3 illustrates the Fresnel Zone concept.

Before getting into specific antenna types, we need to explore a bit more math and some concepts related to outdoor links. The Fresnel zones (pronounced frah-nell) are named after the French physicist Augustin-Jean Fresnel and are a theoretically infinite number of ellipsoidal areas around the LOS in an RF link. The first Fresnel zone is the zone with the most significant impact on point-to-point long-distance links, in most scenarios. The Fresnel zones have been referenced as an ellipsoid-shaped area, an American football-shaped area, and even a Zeppelin-shaped area.

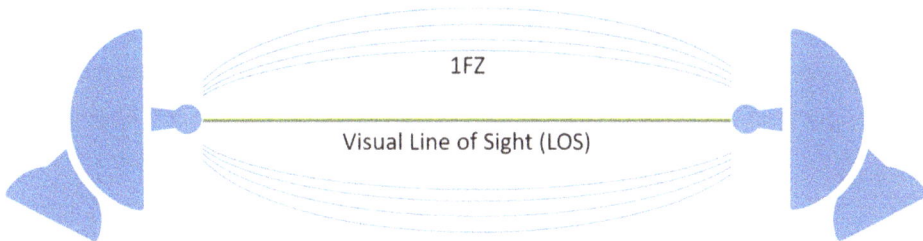

Figure 5.3: Fresnel Zone Illustrated

In this text, we will call Fresnel zone 1 1FZ from this point forward for simplification. Since 1FZ is an area surrounding the LOS, and this area cannot be largely blocked and still provide a functional link, it is important that you know how to calculate the size of 1FZ for your links. You'll also need to consider the impact of earth bulge on the link and 1FZ.

To calculate the radius of the 1FZ, use the following formula:

$$radius = 72.2 \times \sqrt{(D / (4 \times F))}$$

Where D is the distance of the link in miles and F is the frequency used for transmission in GHz and radius is reported in feet (72.2 is the constant for feet and 17.32 would be the constant for meters). For example, if you are creating a link that will span 1.5 miles and you are using 900 MHz radios, the formula would be used as follows:

$$72.2 \times \sqrt{(1.5 / (4 \times .9))} = 46.6 \text{ feet}$$

This formula provides you with the radius of the 1FZ, and doubling the result would give you the diameter, if you needed it to be calculated. However, it is important to realize that a blockage of the 1FZ of more than 40% can cause the link to become non-functional. To calculate the 60% radius, so that you can ensure it remains clear, use the following formula:

clearance radius = 43.3 x √ (D / (4 x F))

Where D is the distance of the link in miles and F is the frequency used for transmission in GHz and radius is reported in feet. Using the same example, we used to calculate the radius of the entire 1FZ, you will now see that the 60% clearance radius is only 27.96 feet. However, this leaves no room for error or change. For example, trees often grow into the 1FZ and cause greater blockage than they did at the time of link creation. For this reason, many wireless administrators choose to use a 20% blockage or 80% clearance guideline, and this is the recommended minimum clearance of the CWNP program as well. So how would you calculate this? Use the following formula:

recommended radius = 57.8 x √ (D / (4 x F))

Once you've processed this formula, you will see that the recommended minimum of 80% clearance (recommended maximum of 20% blockage) results in a 1FZ radius of 37.28 feet in our example.

As it is always better to be safe rather than sorry when creating point-to-point links, you will probably want to make it a habit to round your Fresnel zone calculations upward. For example, we would round the recommended radius to 38 feet in our example.

Remember that 80% clearance of the first Fresnel zone is recommended for most point-to-point links and 60% clearance is required. The additional 20% clearance allows for environmental changes without significantly impacting the link.

You might be wondering why we calculate the radius instead of the diameter. The reason is simple: we can determine where the visual LOS resides and then measure outward in all directions around that point to determine where the 1FZ required clearance area resides. Remember, the 1FZ does not reside in a downward direction only. It might seem that way since we are usually dealing with trees and other objects protruding up from the ground as interference and blockage objects. However, it is entirely possible that something could be hanging down from a very high position — such as a bridge — and encroach on the 1FZ from above the visual LOS. Additionally, buildings and other objects can cause blockages from the sides. For example, if you are attempting to create a point-to-point link that has visual LOS between two buildings on either side of the link in a downtown area, the two buildings may encroach on the 1FZ required clearance area resulting in insufficient signal for a consistent connection.

A lesser-known fact is that the reflections off of these obstacles often cause the problems in point-to-point links. Given that such links are often single-input/single-output (SISO) links, the reflected signals result in multipath, which may corrupt the received direct signal at the receiver.

Another factor that should be considered in 1FZ blockage is the Earth itself. As you know, the Earth — it turns out — is round. When any two objects are farther apart, there will be a higher likelihood that the Earth is between them. This scenario is demonstrated in Figure 5.4. Note the encroachment of the earth on the 1FZ over a significant distance.

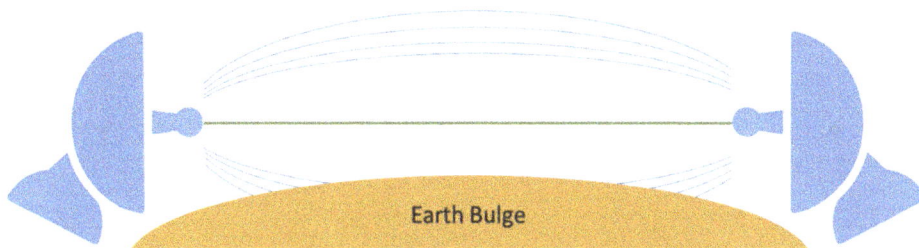

Earth Bulge

Figure 5.4: Earth Bulge Demonstrated

If you are creating point-to-point links over distances greater than 7 miles using wireless technologies, you will need to account for earth bulge in your antenna positioning formulas. You will not need to memorize the following formula for the CWSA examination, but you will need to know that earth bulge is a potential problem in outdoor wireless links over greater distances. The formula for calculating the extra height your antennas will need to compensate for earth bulge is:

$$Height = D^2 / 8$$

Where height is the height of earth bulge in feet, and D is the distance between antennas in miles. Therefore, if you are creating a 10-mile link, you would process the following formula:

$$100 / 8 = 12.5 \text{ feet}$$

Using our guideline of rounding up, we would raise the antenna height by 13 to 14 feet to accommodate for earth bulge.

To bring all the discussion of Fresnel zones together, it is important that you learn to deal with 1FZ obstructions. If the obstructions are coming up from the ground into the 1FZ and there are no obstructions anywhere above it, you can often solve the problem by simply raising the antennas involved in the communication link. For example, if there is a forest with maximum tree height of 23 feet that is between the two antennas, and there is a distance of 11 miles that must be spanned, we can calculate the needed height for the antennas, including earth bulge, with the following formula:

$$\text{minimum antenna height} = (57.8 \times \sqrt{(11 / (4 \times 2.4))}) + (121 / 8)$$

This calculation might seem complex, at first, but it is a simple combination of the recommended 1FZ clearance formula and the earth bulge formula. The result is rounded up to 77 feet. You will need to install very high towers and you will also need to monitor the forest, though it is unlikely that the trees would grow that much more into the 1FZ in a few years. Additionally, you will likely be required to acquire permits for the towers in most regulatory domains or lease/license space on existing towers.

If the obstructions are coming into the 1FZ from the sides — such as buildings intruding into the pathways, you will have to either calculate the 1FZ for a different frequency to see if you can get the clearance, or you will have to raise the antennas above the buildings. You may also be able to create a multi-hop link to "shoot" around the buildings if you can gain access rights to a third location that can be seen (RF LOS — including 1FZ) by both of your locations.

Notice that it was an option to calculate the 1FZ with a different frequency. Because the Fresnel zones are a factor of wavelengths (hence frequencies) and not a factor of antenna gain or beamwidth (covered later in this chapter), which is very important to differentiate, you can often implement a point-to-point link successfully using different frequencies. For example, the 77-foot antenna height to allow us to communicate over the top of the forest across 11 miles can be lowered to only 54 feet, if you are using devices in the 5 GHz range. However, the trade-off is in the distance. The 2.4 GHz signals are detected more easily than 5 GHz signals at a distance due to the receiving area of the antenna element and the length of the signal wave, but 5 GHz signals have a narrower 1FZ. The formula, when using the 5 GHz band changes to the following, assuming you use a center frequency of 5.745 GHz:

minimum antenna height = (57.8 x √ (11 / (4 x 5.745))) + (121 / 8)

An example of this is a link that travels only about a city block (0.1 miles). In the 2.4 GHz spectrum, the 1FZ radius would be approximately 6 feet. In the 5 GHz spectrum, the 1FZ would only be about 4 feet. Remember, this means 6 feet or 4 feet out from the center point in all directions. Therefore, a 5 GHz link traveling between two buildings for 0.2 miles would require a space between the buildings of about 8-9 feet, while the 2.4 GHz link would need a space between the buildings of about 12-13 feet. These factors are important considerations.

Different antennas have different *beamwidths*, and this beamwidth is the measurement of how broad or narrow the focus of the RF energy is as it propagates from the antenna along the main lobe. The main lobe is the primary RF energy coming from the antenna. It is the intended direction of propagation. Beamwidth is measured both vertically and horizontally, so don't let the term width confuse you into thinking it is a one-dimensional measurement. Specifically, the beamwidth is a measurement taken from the center of the RF signal to the points on the vertical and horizontal axes where the signal decreases by 3 dB or half power. In the end, there is a vertical and horizontal beamwidth measurement that is stated in degrees. Figure 5.5 shows both the concept of the beamwidth and how it is measured and Table 5.1 provides a table of common beamwidths for various antenna types (these antenna types are each covered in detail later in this chapter).

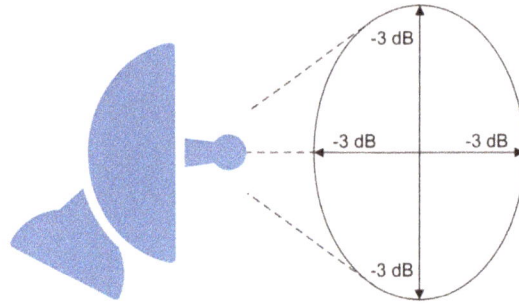

Figure 5.5: Beamwidth Concept and Measurement

Antenna Type	Horizontal Beamwidth	Vertical Beamwidth
Omnidirectional	360 degrees	7 to 80 degrees
Patch/panel	30 to 180 degrees	6 to 90 degrees
Yagi	30 to 78 degrees	14 to 64 degrees
Sector	60 to 180 degrees	7 to 17 degrees
Parabolic dish	4 to 25 degrees	4 to 21 degrees

Table 5.1: Common Beamwidths of Various Antenna Types

While beamwidth measurements give us an idea of the propagation pattern of an antenna, they are less than perfect at illustrating the actual areas that are covered by the antenna. For more useful visual representations, you will want to reference Azimuth and Elevation charts. However, when textual documentation of an antenna's characteristics is desired, the beamwidth is typically the best choice.

Where the beamwidth calculations provide a measurement of an antenna's directional power, Azimuth and Elevation charts, which are typically presented together, provide a visualization of the antenna's propagation patterns. Figure 5.6 shows an example of an Azimuth chart and Figure 5.7 shows an example of an Elevation chart.

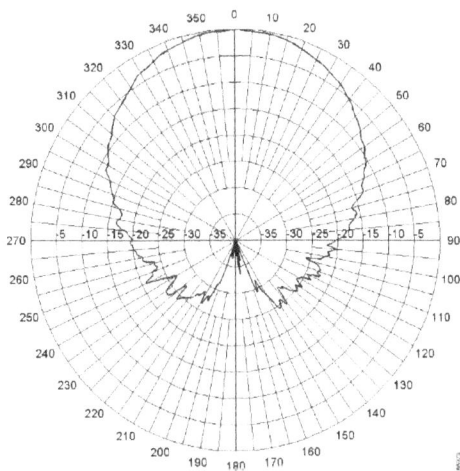

Figure 5.6: Antenna Azimuth Chart

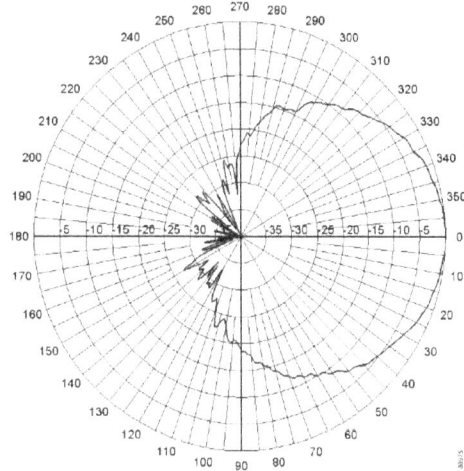

Figure 5.7: Antenna Elevation Chart

The difference between an Azimuth and Elevation chart is simple: The Azimuth chart shows a top-down view of the propagation path (to the left, in front, to the right and behind the antenna) and the Elevation chart shows a side view of the propagation path (above, in front, below and behind the antenna). Think of these charts in terms of a dipole antenna that is positioned vertically upright. If you are standing directly above it and looking down on it, you are seeing the perspective of an Azimuth chart. If you are beside it and looking at it from a horizontally level position, you are seeing the perspective of an Elevation chart.

The Azimuth chart in Figure 5.6 is reporting the different signal strength you can expect at different degrees from the antenna. For example, at 90 and 270 degrees (to the immediate left and right of the antenna's intended propagation direction), you will see a loss of approximately 20 dB. Directly behind the antenna, at 180 degrees, you will see a loss of approximately 35 to 50 dB. This is a sector antenna and is intended to propagate its energy in one direction but in a fairly wide path.

The Elevation chart in Figure 4.5 shows similar information in the vertical view for the same antenna. You will notice that the pattern of propagation is very similar to the Azimuth pattern. Like most Elevation charts, it is shown with the primary radiation direction to the right. Remember, this is intended to represent you looking at the antenna's propagation pattern from the side view. You can see that this antenna has very similar levels of loss along the same degree levels as the Azimuth chart.

The *isotropic radiator* is a fictional device or concept that cannot be developed using today's technology. Many say that it is not only impossible now, but because of the constraints of physics, it will always be impossible. While the future may be debatable, we know that you cannot currently create an antenna that propagates RF energy equally in all directions. This truth is due to the fact that the antenna must have some length (it must exist) and it must receive power from some source (it must be connected to something). These two constraints alone make it impossible to create an isotropic radiator at this time.

Even though we cannot create such a device, it is a useful theoretical concept in that we can use it as a basis for measurements. In fact, dBi — as was stated earlier in the book — is a measurement of the gain of an antenna in a particular direction over the power level that would exist in that direction if the RF energy were propagated by an isotropic radiator. In other words, dBi is a measurement of the difference between the power levels at a point in space generated by a real antenna versus the theoretical isotropic radiator. Since we can all agree on the behavior of an isotropic radiator, we can all use it as a basis for such power level measurements. Figure 5.8 illustrates the concept of the theoretical isotropic radiator.

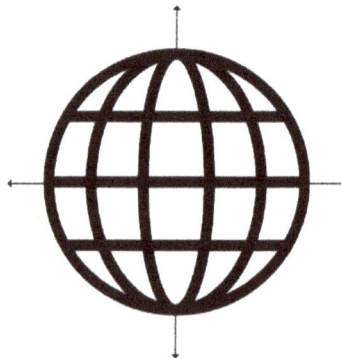

Figure 5.8: Theoretical Isotropic Radiator

The sun is often used as an analogy of an isotropic radiator. While this is an acceptable analogy, specific theories in physics — such as the hairy ball theorem, would even exclude the sun from being a true and complete isotropic radiator. However, it is one of the objects we've found to be closest to an isotropic radiator in that light indeed propagates from it in all directions. If we could analyze that light at the molecular level — or even the individual wave level, it is questionable as to whether the rays are truly radiated "equally" in all directions.

A factor that significantly impacts the performance of RF antennas is the polarization of the antennas. Antenna polarization refers to the physical orientation of the antenna in a horizontal or vertical position — typically. Technically, it's about how the antenna is designed to be used, but in practical application, it's about how you position it.

You'll remember from previous discussions in this book, that the electromagnetic wave is made up of electric and magnetic fields. The electric field forms what is known as the E-plane, and the magnetic field forms what is known as the H-plane. The E-plane is parallel to the radiating antenna element, and the H-plane is perpendicular to it. Simply stated, the E-plan runs alongside the antenna regardless of how you position it. If you put the dipole antenna upright, the E-plan is upright alongside it; however, if you tilt it, the E-plane is tilted as well.

The E-plane, or electric field, determines the polarization of the antenna since it is parallel to the antenna. Therefore, if the antenna is in a vertical position, it is said to be vertically polarized, if the antenna is in a horizontal position (the electric field and antenna a parallel to the Earth), it is said to be horizontally polarized.

A vertically polarized omnidirectional antenna propagates the signal horizontally, and a horizontally polarized omnidirectional antenna propagates the signal vertically, which is not what you typically desire unless you are creating a bridge link between floors in a tall building. If you configure a link like this (horizontally between floors and along walls), a best practice would be to place the antennas approximately 2 feet out from the wall to prevent the wall from creating 1FZ interference. The link would likely work anyway, but communications can be improved with this consideration. The spacing of 2 feet should keep the 1FZ 80% clear for up to 60-70 feet.

The impact of polarization is seen when antennas are not polarized in the same way. For example, if you have one device with the antennas positioned vertically (vertical polarization), and you have another device with the antenna down (horizontal polarization), your connectivity will be less stable and, at greater distances, may even be lost. However, in most cases, due to indoor reflections, the polarization of antennas does not have as great an impact indoors as it does with outdoor links. In outdoor links, the proper polarization of the antennas can make or break the connection.

Remember this: vertical polarization usually means that most of the signal is being propagated horizontally and horizontal polarization means that most of the signal is being propagated vertically, as previously stated. Therefore, the most popular polarization is vertical polarization because we are typically trying to send the signal along the direction of the Earth's surface, whether it is a five-meter link or a five-kilometer link.

Antenna diversity is a feature offered by many wireless devices, that allows the device to receive signals using two antennas and one receiver. In a traditional antenna diversity implementation, only one antenna is used at a time, so this should not be confused with Multiple-Input/Multiple-Output (MIMO) configurations.

The device supporting antenna diversity will look at the signal that comes into each antenna and choose the signal that is best, on a communication-by-communication basis. Again, remember that there is only one receiver that has two connections and two antennas.

An additional type of diversity is Multiple-Input/Multiple-Output (MIMO) diversity. MIMO systems use more than one antenna in several different ways, but they can also support diversity. For example, a device may be a 2x3 device, which means that it can transmit on two antennas, but receive on three. Such a configuration would allow for diversity selection during frame reception providing MIMO receive diversity. Such a solution may also support maximal ration combining (or maximum ratio combining, depending on whose whitepaper you're reading) as discussed in the following section.

Wireless devices with multiple antennas can use several special techniques with the multiple antennas provided. These techniques include:

- Spatial Multiplexing — MIMO
- Transmit Beam Forming

- Maximal Ratio Combining

Spatial Multiplexing (*SM*) uses advanced algorithms to create separate data streams for each transmitting antenna with MIMO hardware. It requires multiple radio chains, which are effectively radios linked to transmitting (Tx) and receiving (Rx) antennas.

If a spatial multiplexing link is to function at the highest possible data rate, the following factors must be true:

- The receiving device must have an equal number of radio chains to the transmitting device.

- The signal strength must be strong enough to avoid changing to lower data rates.

- The link must be operating with spatial multiplexing enabled.

In many systems, the "center" of the network, such as a cell tower, may have many more antennas than the devices connecting to it possess. This configuration allows for multiple user transmissions concurrently and even multiple beamformed transmissions concurrently. An example of this would be massive MIMO in some LTE and 5G deployments allowing for more multi-user MIMO (MU-MIMO) communications to happen concurrently. With a 64x64 massive MIMO configuration (or larger), many devices can communicate with the same receiver concurrently.

Transmit Beamforming (*TxBF*) is a specialized antenna technology that allows the signal to be focused on a specific destination. To use TxBF, the characteristics of the signal received at the remote wireless node must be known. Special communications called *channel sounding* occur between the devices to discover this information. TxBF uses multiple antennas and adjusts those antennas to simulate a sector array of antennas.

To better understand TxBF, you must first understand the phenomenon of multipath. Multipath occurs when the transmitted signal reflects, refracts, diffracts, and scatters as it travels. The result is often that more than one copy of the signal arrives at the receiver. If two copies arrive at the receiver in phase with each other, upfade occurs, and the signal strength is increased. If the signals arrive out of phase, the signal can be downfaded (resulting in a loss), corrupted, or canceled out entirely.

TxBF devices work with each other to determine how to calibrate the transmissions so that multiple signals arrive in phase. Any time a device moves, the TxBF transmission must be recalculated. The result is that TxBF is best for nomadic roaming or in non-congested networks that will not be significantly impacted by extra transmissions for TxBF calibration operations. With nomadic roaming, the clients move but remain stationary most of the time.

Maximal Ratio Combining (*MRC*) uses antenna diversity to increase the strength of the received signal through combination algorithms. Traditional antenna diversity uses only one antenna to receive a transmission, even if both antennas receive the signal fine because only one radio exists. With MIMO devices, MRC can combine the signals of two antennas to increase signal strength at greater distances. The result is an increase in received signal quality, which can result in higher data rates and throughput.

Now that you understand several antenna systems concepts, in this section, we will cover the basic types of antennas that are available to you, including their RF propagation patterns and their intended use. Keep in mind that modern wireless solutions do not always use antennas that fit neatly into one of the three primary categories, but understanding these concepts will help you understand the actual antenna propagation patterns within your solution.

Three primary categories of antennas are used today:

- Omnidirectional
- Semi-directional
- Highly directional

Also, variations on the implementation and management of these antenna types exist, which results in the sectorized and phased array antennas among others. These antenna types will also be addressed in this section. Finally, we'll review the MIMO (Multiple-Input/Multiple-Output) antenna systems that are used by many wireless solutions today.

Omnidirectional antennas, the most popular type being the dipole antenna in early days, and being internal device antennas today, are antennas with a 360-degree horizontal propagation pattern. In other words, they propagate most of their energy outward in a 360-degree pattern shaped much like a doughnut — though a very thick one, in low-gain omnidirectional antennas. The omnidirectional antenna provides coverage at an angle upwards, downwards and directly out horizontally, as is shown in the Elevation chart in Figure 5.9.

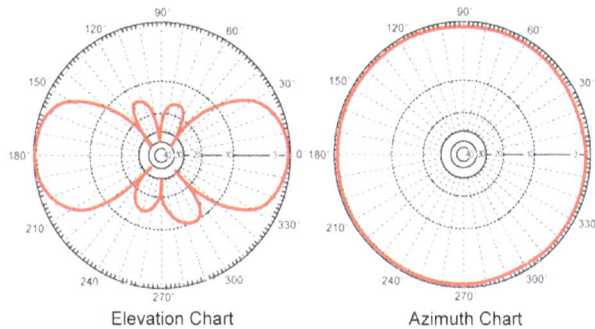

Elevation Chart   Azimuth Chart

Figure 5.9: Omnidirectional Elevation and Azimuth Charts

Inspecting the Elevation chart in Figure 5.9 reveals that an omnidirectional antenna propagates most of its energy to the right and left of the antenna (from a side view) and very little energy directly above the antenna. At the same time, the Azimuth chart shows a fairly even distribution around the antenna (from a top-down view). This pattern is the common propagation characteristic of omnidirectional antennas. Figure 5.10 shows a typical omnidirectional antenna of the dipole design.

Figure 5.10: Omnidirectional Antenna

The omnidirectional antenna is most commonly used indoors to provide coverage throughout an entire space; however, they have become more and more popular in outdoor usage for either hotspots or private access outdoor networks. Omnidirectional antennas may be mounted on poles, masts, towers, ceilings or desktops and floors. They provide coverage on a horizontal plane with some coverage vertically and outward from the antenna. They may provide some coverage to floors above and below where they are mounted in some indoor installations. Many wireless devices with internal antennas have an omnidirectional propagation pattern. Always consult the vendor specifications to determine the pattern implemented by a given device, when available. Figure 5.11 shows the LORD MicroStrain WSDA-2000 gateway used to build an 802.15.4 wireless sensor network (WSN). Note the omnidirectional antenna in use.

Figure 5.11: LORD MicroStrain WSDA-2000 WSN Gateway with Omnidirectional an Antenna

Because all antennas use passive gain — they focus the RF energy — it is important to consider the impact of this passive gain on any antenna that you implement. In the case of omnidirectional antennas, the result is that devices directly above or below the omnidirectional antenna may have a very weak signal or even be unable to detect the signal. This behavior is due to the primary signal being focused outwardly on a horizontal plane (vertical polarization).

You can use antennas that have higher dBi gain such as 12 or 15 dBi omnidirectional antennas; however, you must keep the impact of these higher gain antennas in mind.

As an example, consider the two Elevation charts side-by-side in Figure 5.12. The one on the left is from a 4 dBi omnidirectional antenna and the one on the right is from a 15 dBi omnidirectional antenna. You can see the flattening of the signal. It is very plausible that a higher gain antenna, such as the one on the right, could cause devices on the floors above and below the antenna to lose their connection. Ultimately, when using omnidirectional antennas, choosing between a higher gain and a lower gain is choosing between reaching device farther away horizontally (higher gain) or reaching devices farther up or down vertically (lower gain). In most situations, you'll place separate antennas (or devices with attached or internal antennas) on each floor of a multi-floor installation to get the coverage you need.

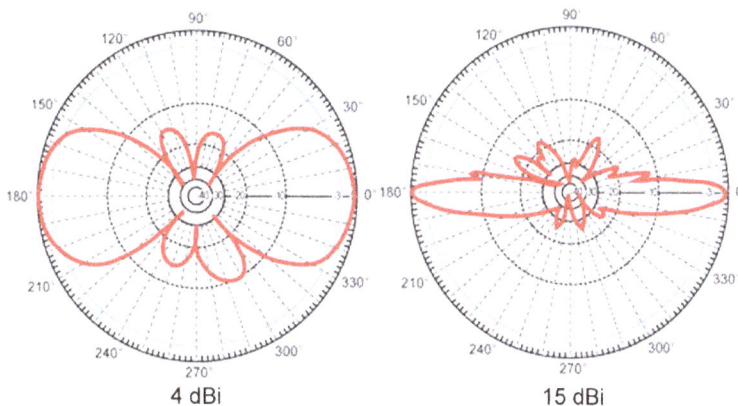

4 dBi   15 dBi

Figure 5.12: 4 dBi vs. 15 dBi Antennas

Semi-directional antennas are antennas that focus most of their energy in a particular direction. Examples include patch, panel, and Yagi antennas. (Yagi is pronounced yah-gee.) Patch and panel antennas come in flat enclosures and can be easily mounted on walls. Yagi antennas look a lot like TV antennas — a long rod with tines sticking out; however, the Yagi antennas are usually enclosed in a plastic casing that hides this appearance. Patch and panel antennas usually focus their energy in a horizontal arc of 180 degrees or less, where Yagi antennas usually have a 90 degree or less coverage pattern. Some Yagi antennas can be categories as highly directional antennas as well. Figure 5.13 shows examples of patch, panel and Yagi antennas.

The Azimuth and Elevation charts for Yagi antennas often look the same. They often have the same coverage pattern from the top-down view (horizontal coverage) as they do from the side view (vertical coverage). Figure 5.14 shows an example coverage pattern of a 9 dBi Yagi antenna. Panel antennas usually have a similar pattern to Yagi antennas except that the "fish-like design" appears quite a bit fatter or thicker.

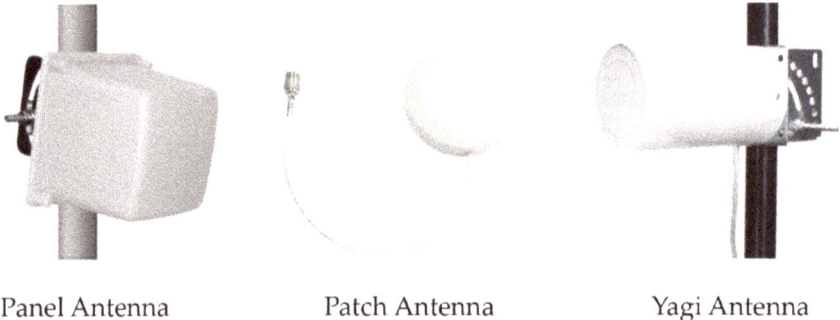

Panel Antenna   Patch Antenna   Yagi Antenna

Figure 5.13: Semi-Directional Antennas

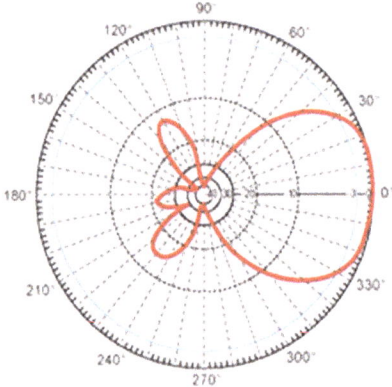

Figure 5.14: 9 dBi Yagi Antenna Pattern (Elevation Chart)

Semi-directional antennas are useful for providing RF coverage down long hallways or corridors when using Yagi-style antennas. They are also useful when providing RF coverage in "one" direction using patch or panel antennas. The patch and panel antennas will have some level of energy propagated behind their intended direction. This energy is known as the rear lobe. However, most of the energy will be directed inward. For this reason, patch and panel antennas are usually mounted on outside walls facing inward when they are intended to provide coverage inside an area only. Additionally, they can be used on the outside of a building to create an "external-only" coverage area.

Creatively using Yagi, patch and panel antennas can prevent the use of large numbers of omnidirectional antennas for many situations. For example, a single patch antenna placed on a wall facing inward may provide all the coverage needed, when two omnidirectional antennas would otherwise be needed. The energy coming from the patch antenna is forced directionally inward instead of being forced in all horizontal directions equally. The RF energy is going where it is needed instead of losing a third to half of it outside the walls of your facility. It is also assumed MIMO is not required when Yagi antennas are used. MIMO patch and panel antennas are available, but MIMO Yagi is not a typical installation.

An example of a patch antenna device is shown in Figure 5.15. This device is the LORD MicroStrain G-Link 200 and is a WNS accelerometer sensor used to track movement, impact, and other factors related to motion. It connects to a gateway like the earlier referenced WSDA-2000. It also uses the 802.15.4 protocol for Physical and Data Link layer communications.

Figure 5.15: LORD MicroStrain G-Link-200 with an Internal Patch Antenna

A common misconception that enters at this point is the fear that using a Yagi, patch or panel antenna will get the signal to the remote device, but that it will not get the signal from the remote device to the local device (or the Yagi, patch or panel antenna). Stated another way, it is often assumed that you must use semi-directional antennas at the remote device if you use a semi-directional antenna at the local device; however, this is not the case.

I usually explain this by saying this, "When you place the megaphone over the antenna's mouth, it is smart enough to move it over its ear to listen." What I mean by this statement is simple: the very quality of the antenna that increases its gain in a particular direction, also allows it to "hear" better (have receive gain) from that same direction. Therefore, as Joseph Bardwell says, "If you can hear me, then I can hear you." For more information, see the website:

www.connect802.com/non-cad-wireless-network-design#Law-of-Reciprocity

Highly-directional antennas are antennas that transmit with a very narrow beam. These types of antennas often look like the satellite dish that is so popular with people who do not have access to wired cable television or do not desire to use it. They are generally called parabolic dish or grid antennas. The parabolic dish is the one that looks like a satellite dish, and the grid antenna looks like an antenna with a curved grill grate behind it. Figure 5.16 shows examples of each antenna type.

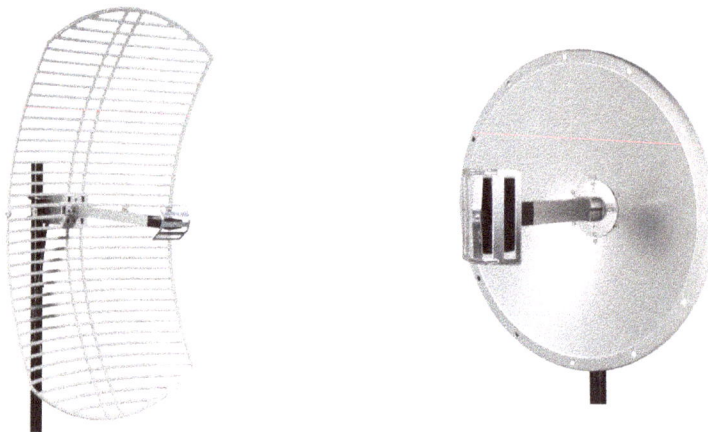

Figure 5.16: Parabolic Dish and Grid Antennas

Due to the high directionality of these antennas, they are mostly used for point-to-point or point-to-multi-point (PtMP) links. PtMP links will usually use an omnidirectional or semi-directional antenna at the center and multiple highly or semi-directional antennas at the remote sites. They can transmit at distances of 35 miles or more and usually require detailed aiming procedures that include a lot of trial and error. By positioning one antenna according to visual LOS and then making small movements at the other antenna, accurate alignment can usually be achieved.

The grid antenna provides the added benefit of allowing air to pass through the back panels so that the antenna does not shift as much as the parabolic dish in high wind load scenarios.

A *sectorized antenna* (or sector antenna) is a high-gain antenna that works back-to-back with other sectorized antennas. They are often mounted around a pole or mast and can provide coverage in indoor environments, such as warehouses, or outdoor environments, such as university campuses or hotspots. Figure 5.17 shows an example of sectorized antennas mounted on a pole.

A *phased array antenna* is a special antenna system that is actually comprised of multiple antennas connected to a single processor. The antennas are used to transmit different phases that result in a directed beam of RF energy aimed at client devices.

When mounting antennas, always abide by vendor recommendations. If the vendor provides a mounting kit, that's usually the best solution to use. However, if you have to create your own kit — because the vendor doesn't provide one, keep the following tips in mind:

- Be careful not to bend or twist antennas in an effort to "get them into" a specific location. Treat the antennas with love and care.

- Make sure the antennas are firmly mounted so that they are not continually moving and changing the resulting coverage patterns.

- Mount antennas according to their intended use. For example, antennas designed to be mounted on walls, of course, work best when mounted on walls.

- Pole or mast mount antennas usually come with a wrapping type of mounting kit. The wrappers may be simple zip ties or similar to them.

- Ceiling mount antennas are usually omnidirectional and should be mounted in the center of the targeted coverage area.

- Wall mount antennas are usually semi-directional and should be mounted on the perimeter of the targeted coverage area so that they propagate inward.

Figure 5.17: Sectorized Antennas

As usual, if you have to climb a ladder or hang from a rafter to mount the antenna, please make sure you abide by safety best practices and regulations for your region. In many areas, this recommendation means that we abide by OSHA specifications. Whether OSHA has any influence on your area or not, please be careful, and DON'T break a leg — in this case.

Finally, it is essential to note that the RF signal radiated out of the antenna is known as Equivalent Isotropic Radiated Power (EIRP), also called Effective Isotropic Radiated Power, or Effective Radiated Power (ERP).

The difference between EIRP and ERP is simple. ERP is the radiated power in the intended direction of propagation by the antenna under test as compared to the output power of a half-wave dipole antenna. That is, how much power would a half-wave dipole antenna have to generate to equal the power of the antenna under test? The answer to that question is ERP.

EIRP is the radiated power in the intended direction of propagation by the antenna under test as compared to the output power of a theoretical isotropic radiator. That is, how much power would an isotropic antenna have to generate to equal the power of the antenna under test? The answer to that question is EIRP.

$EIRP(dBm) = AntGain(dBi) + P_{in} (dBm)$

$EIRP(dBm) – ERP(dBm) + 2.15$

$ERP(dBm) = AntGain(dBd) + P_{in} (dBm)$

$ERP(dBm) = EIRP(dBm) – 2.15$

What is the real difference? A half-wave dipole antenna already has a gain of 2.15 dB over an isotropic radiator. Therefore, ERP is always 2.15 dB less than EIRP when measuring the same antenna. ERP is related to the dBd measurement and EIRP is related to the dBi measurement. dBd and dBi are used to specify antenna gain and they are relative as the actual output power depends on the input power to the antenna. ERP and EIRP are absolute as they are a measurement of the output power from the antenna for a fixed input power to the antenna.

RF Cabling, Connectors and Components

RF cables are used to connect the transceiver to the antenna (and possibly other in-series devices). Cables have different levels of loss, and this should be considered when selecting the cabling for your system. Keep the following factors in mind when selecting RF cables for your implementation:

- Different cables have different levels of loss, so not all cables are the same.

- Make sure the impedance of the cable matches the rest of your system.

- Be sure to select the cable that is rated for the frequency you will be using.

- Check with the vendor to discover the loss incurred per foot or per 100 feet before selecting the cable.

- Higher frequencies mean more significant loss in the same cable.

- Either master the art of building cables or hire a professional to cut the cables and install the connectors, so you do not unnecessarily introduce extra loss.

RF connectors come in many shapes and sizes. The following types are common:

- N-TYPE
- SMA
- BNC
- TNC

In addition, there are common variations of these types, such as reverse polarity and reverse threading. These different types exist in an effort to comply with FCC and other regulations for components used in a wireless system. While dongles and pigtails exist, if they are used to convert from one type to another for transmission, they may constitute a breach of regulatory agency regulations.

These connectors are found on the ends of cables, the back of wireless devices, and the ends of antennas (in the case of dipole or rubber ducky antennas). Figures 5.18-5.21 shows examples of common connector types.

RF splitters are installed in series between the transceiver and the antennas. The splitter receives a single input and has two or more outputs. They may be used with sectorized antennas. RF splitters should be avoided unless absolutely necessary as they create insertion loss.

Figure 5.18: N-TYPE Connector

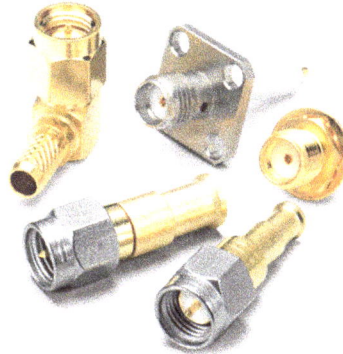

Figure 5.19: SMA Connectors

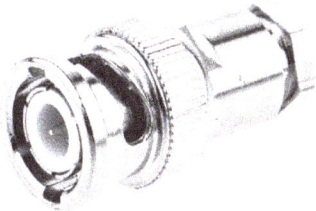

Figure 5.20: BNC Connector

Figure 5.21: TNC Connector

Much like wireless devices have amplifiers and even attenuators internally, these devices can be used in-line between a wireless device and an antenna. Amplifiers increase the signal strength before it reaches the antenna and attenuators decrease it. Amplifiers are used to increase the range of bridge links in many cases. Attenuators are used to keep the ERP or EIRP within regulatory limits when the output power of the wireless radiator cannot be adjusted low enough.

When using amplifiers, it is important to ensure that the input signal is sufficiently low so that the amplifier does not cause signal compression. Alternatively, you can select an amplifier with circuitry to prevent signal compression.

Signal compression occurs when the input signal is high enough that the "top" of the signal gets "flattened" (compressed) and the quality of the signal is reduced such that it may not be receivable on the other end of the link. Consider what happens when audio speakers are turned up too high, and the result is distortion and unintelligible sound. A similar phenomenon occurs when amplifiers over-saturate the RF signal.

When an amplifier works appropriately, based on proper implementation, it amplifies the signal linearly. The output signal simply looks like a bigger (stronger) version of the input signal. However, when saturation occurs, the signal is compressed, and the output signal no longer looks like the input signal. Both phase and amplitude may be impacted, though amplitude tends to be impacted the most. Given that amplitude is part of Quadrature Amplitude Modulation (QAM) modulation schemes, as well as many others, saturation can result in signals that cannot be properly demodulated even though they may appear strong from a pure signal strength perspective – they are distorted.

Figure 5.22 illustrates the impact of compression, resulting from amplifier saturation, on an amplitude modulated signal. Note that the higher amplitude waveform comes out of the amplifier apparently unchanged, but it's actually compressed. The lower amplitude waveform comes out at a higher amplitude. The resulting signal has lost the variance between the peak and low amplitude levels, which can prevent proper demodulation at the receiver.

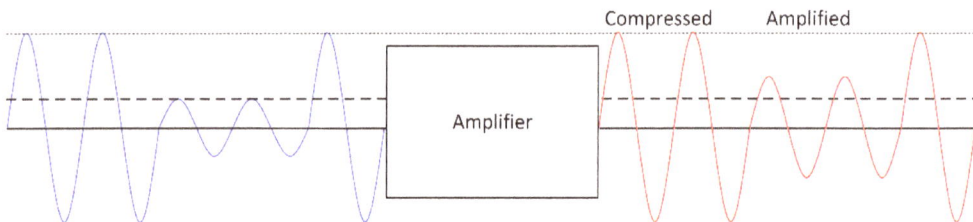

Figure 5.22: Compression from Amplifier Saturation Illustrated

RF Link Types (Use Category)

Wireless links can be created between two devices, among several devices, and on-demand. This section explains the various link types with examples of wireless solutions that utilize them.

Point-to-Point (PtP)

A PtP link is one that exists exactly between two devices. Technically, an ad-hoc network may be a PtP link, if only two devices are involved in the network, but the later section on ad-hoc networks explains why it best fits in its own category. In most cases, PtP links refer to bridge links. A *bridge link* is used to connect two otherwise disconnected networks. Bridge links can be created using wired or wireless solutions.

Wireless bridges may use standard protocols, like IEEE 802.11, or proprietary vendor protocols. To establish the link, both ends must support the same protocol. If one end of the PtP link is using 802.11ac and the other end is using 802.11af, they will still be unable to establish a link, because these two Physical Layers of 802.11 use very different frequencies and modulations in communications. The point is simply that both ends of the link must be speaking the same language (protocol) and using compatible hardware (radios).

Figure 5.23 shows a typical scenario where a PtP link would be beneficial. In the image, we have two buildings on either side of a forest. The buildings are high enough that antennas can be placed on top of each building and a bridge link can be established between them.

Figure 5.23: A Typical Bridge (PtP) Link Scenario

The actual wireless solution used to implement a PtP link, like the one in Figure 5.23 can vary significantly. Selecting the right solutions is a factor of link distance, required speed, compatibility with existing networks, licensing, and expertise of the installer and administrator.

Distance is a critical factor. The 802.11a 5 GHz bridges from EnGenius, still available at the time of writing as many PtP links require only 30-50 Mbps data rates, have a range of up to 1 mile or 1.6 kilometers. They include IP55 rating enclosures for use outdoors in the weather. Figure 5.24 shows these bridges.

Figure 5.24: EnGenius 802.11a 5 GHz Bridges with One Mile Range

Ingress Protection (IP), also called *International Protection,* ratings categorize enclosures for their protective abilities. Two-digit numbers are used. The first digit indicates protection against objects (fingers, dust, tools, etc.). The second digit indicates protection against fluids at varying pressure levels. IP55 is protected against most dust and low-pressure water, including heavy rainstorms. IP68 is completely protected from dust and can be submerged in water for an extended period.

The next example is an Avalan Wireless AW900xTR, which is a 900 MHz wireless bridging solution that offers only 1.536 Mbps data rate with 935 Kbps throughput. As you can see, this device would likely only be acceptable when very low-speed communications are required. However, it can create links up to 15 miles or 24 kilometers with the right antennas and proper clearance of the signal path. The AW900xTR also supports local connections when an omnidirectional antenna is used instead, which can suffice for remote connections to some monitoring devices or other low data rate devices. This device is shown in Figure 5.25.

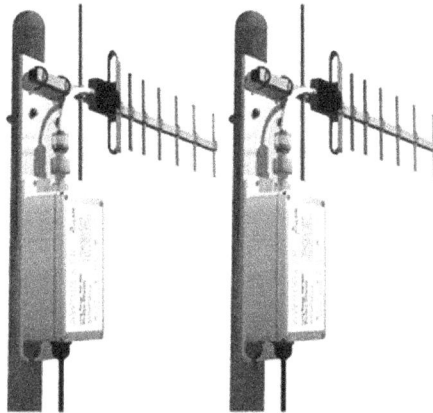

Figure 5.25: Avalan Wireless AW900xTR 900 MHz Bridge

Devices are available offering higher data rates with an acceptable range as well. The vital thing to keep in mind is that two specifications are typically provided (among others) for wireless bridges: range and throughput (or data rate). In most cases, specification sheets list the maximum range and the maximum throughput only. The reality is that you may get one but not the other. If you need the maximum throughput, you're not likely to achieve the maximum range. If you need the maximum range, you're not likely to achieve the maximum throughput. Some bridges will rate shift in steps, and others have only two rates: a high rate (like 150 Mbps) and a low rate (like 11 Mbps). Carefully select the bridges that meet your needs.

The *data rate* of a wireless link is the speed at which bits can be sent across the link. The *throughput* of a wireless link is the speed at which useful data can be transmitted. Do not confuse the two. One is a reference to how fast YOUR data gets through (throughput), and the other is how fast wireless bits are transmitted (data rate).

It is also essential to ensure that the bridge you select is compatible with your network. Given that the vast majority of wireless bridges, regardless of operating frequencies and protocols in use, have Ethernet connectivity, they will typically work with most networks. However, some may have 100 Mbps Ethernet (Fast Ethernet) connections and others have 1000 Mbps Ethernet (Gigabit) connections. If you are building a 200 Mbps bridge link, you do not want a 100 Mbps Ethernet connection.

Licensing is not an issue as long as you select a bridge solution that uses unlicensed frequencies in your regulatory domain. 900 MHz, 2.4 GHz, 5 GHz, and 60 GHz are all common unlicensed frequencies used today. However, you must be aware of your local regulations and use only devices authorized for operation in that regulatory domain. If you choose to use licensed frequencies, the process is far more arduous, and you are typically better off using a managed network operator (MNO) that handles all of the licensing for you.

Point-to-Multi-Point (PtMP)

PtMP links include a central device to which other devices connect. The typical wireless LAN (WLAN) with an access point (AP) and clients is an example of a PtMP implementation. PtMP bridge links can also be implemented if the wireless solution allows for it. The previously referenced AW900xTR solution can implement PtMP configurations.

The MetroLinq 2.5G 60 GHz base station is an example of a PtMP bridging solution. Shown in Figure 5.26, it has a 120-degree beamwidth allowing for significant coverage. At the same time, it uses 16 narrower beam steps of 10 degrees each reducing noise in the communications with remote clients (which are remote bridges in PtMP bridging configuration).

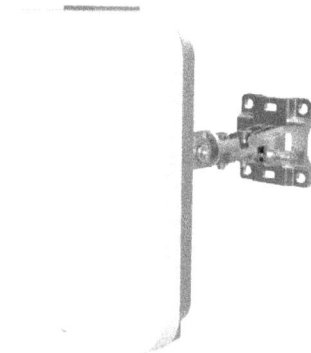

Figure 5.26: MetroLinq 2.5G 60 GHz Base Station for PtMP Links

To assist in the alignment of the base station with remote nodes (which MetroLinq calls clients – even though they are bridges), a mobile iOS application is provided (shown in Figure 5.27). With this application, you can view local signal strength and SNR, remote signal strength and SNR, and combined signal strength (the average of the two). The bridges can be considered "aligned" when the signal strength is best. Given that this solution is a 60 GHz solution, range is limited to less than 1 kilometer in most cases, though the vendor indicates that some have successfully established consistent links with 80% efficiency at 2.5 kilometers.

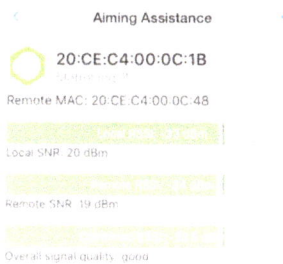

Figure 5.27: MetroLinq Bridge Alignment App

An additional feature of the MetroLinq bridging solution is failover to 5 GHz. This feature will likely be seen more frequently for 60 GHz devices. 60 GHz is susceptible to significantly reduced SNR during rainstorms and 5 GHz is much less susceptible to it. Therefore, to ensure the link can continue to operate, even though it will drop from a maximum data rate of 4.62 Gbps to 866 Mbps, the link can switch over to the 5 GHz band.

When implementing a PtMP solution, it is important to consider airtime management. For example, 802.11 uses a contention algorithm to minimize interference among clients. In outdoor bridge PtMP deployments, some algorithm or scheduling solution should be implemented to prevent collisions at the central point of the various bridge links.

Ad-Hoc

The term *ad-hoc* is a general use term meaning, "when necessary or needed," or "created for a particular purpose as necessary." In other words, it is a temporary and dynamic action or process that is implemented when required. In wireless networks, an ad-hoc (also spelled ad hoc) wireless network is a group of wireless devices that dynamically create a network without requiring an existing network infrastructure to function. Many wireless links are categorized as ad-hoc network links, including:

- Bluetooth and other peripheral wireless device connections
- 802.11 direct wireless links between client devices without an access point
- Vehicular ad-hoc networks (VANETs) used to communicate with other vehicles and roadside equipment
- WSNs sometimes use self-forming and self-healing ad-hoc networks

Ad-hoc networks may be subcategorized as stationary ad-hoc networks and mobile ad-hoc networks (MANETs). Stationary ad-hoc networks are those formed between devices, and the devices do not move at all or move very short distances while participating in the network. The network may be formed and continue to exist for hours or even days. MANETs, as they are usually conceptualized, involve more dynamics with devices coming and going from the network, moving significant distances while participating in the network, or the network completely dissolves very shortly after creation.

An example of a stationary ad-hoc network would be two mobile phones connected by either Bluetooth or Wi-Fi, while the users stay in the same room sharing photos or playing a collaborative game. Another example would be two laptops forming an ad-hoc network and being used for several hours as the two users collaborate. Typically, in a stationary ad-hoc network, all of the participating devices can communicate directly with the other devices. There is no need for a dynamically built routing solution.

The IEEE 802.15.4 standard defines protocols that can be used to implement a peer-to-peer network like the one we're calling a stationary ad-hoc network. All of the devices can communicate directly with all of the other devices in the network. 802.15.4 provides all that is needed to implement a stationary ad-hoc network; however, it does not provide sufficient capabilities alone to implement a full-scale routed ad-hoc network where devices communicate with remote devices using intermediaries. MANETs often require these extra layers to be defined.

An example of a MANET would be a WSN where the some or all of the sensors are mobile, and the sensors form an ad-hoc (though it is sometimes a mesh network instead) network among them for routing (or hopping) to other destinations in the network. Imagine a WSN in a large warehouse. If forklifts have accelerometer sensors mounted on them and these sensors are part of the ad-hoc network, you can quickly realize that the network requires an ad-hoc nature so that all of the sensors continually have a route through the network. The forklift sensors will require new routes as they move and they may become a hop in another sensor's route for a short time until they move again. This functionality is the very definition of a MANET. For industrial, agricultural, military, emergency response, vehicular, healthcare, and many other ad-hoc implementations, the MANET model is beneficial.

> The IEEE 802.15.4 standard defines protocols that are often used in ad-hoc networks; however, it does not define how multi-hop communications might occur when the Physical and Data Link layers of 802.15.4 are used. Such capabilities are often proprietary or based on another complementary standard, such as ZigBee.

Mesh

Two essential differentiators exist between an ad-hoc network and a mesh network. First mesh network nodes typically have more than one radio today. With this implementation, the mesh nodes can communicate with multiple other nodes concurrently. Second, one, or more, mesh node(s) will have a connection to the rest of the network and possibly the Internet. Typically speaking, the primary difference is that a network can be ad-hoc or MANET without connecting to any other network, but a mesh network is usually connected to other networks.

This primary point of difference is where much confusion enters as you begin to evaluate vendor equipment. Some vendors implement a mesh network but call it an ad-hoc network. Others implement an ad-hoc network but call it a mesh network. It is most important for the wireless solutions administrator to understand the vendor's architecture and ensure it meets the requirements of the desired solution. Whether the vendor calls it by the right name or not, you must be able to select and implement the right technology for your needs.

Mesh networks are made up of mesh nodes. Some of the nodes are pure mesh nodes in that they only provide connections between other mesh nodes. Other mesh nodes may connect to the mesh and provide client access. Still, other mesh nodes may provide connections to other mesh networks or wired networks. Mesh nodes providing connections to other networks are referred to as mesh gateways, mesh portals, mesh sinks, or mesh coordinators, depending on the vendor terminology and the mesh protocol implemented.

Examples of mesh networks include ZigBee, LoRa, 802.11 mesh, and proprietary solutions.

On-Demand

On-demand wireless link types are those that are created, data is transmitted, and the link is torn down (ended). Many wireless networks are defined as on-demand, and the meaning is simply that they are there when users want to use them. For the CWSA exam, we are referencing devices that enter a sleep state such that they may not connect for several minutes, days, or hours, until the actual need to transmit data arises. Such behavior can be seen with some wireless sensor devices. For example, a motion-tracking device that is battery powered needs to be operational with those batteries for as long as possible. Therefore, powering the wireless module down completely, until motion is detected, can be a useful power saving mechanism. When motion is detected, the sensor can power on the wireless module, connect to the network, transmit the required data, and the power down the wireless module again.

In other scenarios, wireless sensors may store data locally and only transmit the data at regular intervals, such as every five minutes, every hour, or even once each day. Environmental sensors may fit into this category if real-time monitoring is not required. For example, environmental research projects may require only that the data is captured for analysis once each week or month. Given that the data is analyzed weekly or monthly, a single transmission of the data each day will suffice. Using this on-demand model, battery life can be extended to years for many sensors.

RF Device Types

The final topic of this chapter is the RF device types. The purpose of this section is to familiarize you with common terminology related to wireless devices. We will explore nodes, infrastructure devices, mesh devices, and client devices.

Nodes

In networking terminology, any device that connects to the network and participates in any way is a node. Network nodes include servers, clients, access points, mesh points, sensors, and other devices.

Another common term for a node is endpoint, but this term is not correctly inclusive and is not synonymous with node. For example, a router is a node on the network, but it is not an endpoint. Nodes include endpoints, but they also include other devices that participate in the network, including routing, switching, bridging, repeating, and other devices. Generally speaking, if the device is active on the network it is a node.

The remaining device types are particular types of nodes.

Infrastructure Devices

Traditionally, networking professionals think of infrastructure devices as routers, switches, and bridges. However, we are now far past the wired networking age, and many wireless devices can also be considered infrastructure devices. Such devices are not clients on the network, but instead, they provide network access for the clients or they coordinate or control network operations.

In a wireless network, the following devices may be considered infrastructure devices as they help to build the wireless infrastructure:

- Access points
- ZigBee coordinators
- Mesh portals
- Mesh gateways
- Wireless sensor network sinks
- Wireless sensor network gateways
- LTE/5G gateways

Mesh Devices

Two basic kinds of mesh nodes or devices exist:

- **Mesh Points/Stations/Nodes:** These devices participate in the mesh network and communicate only with other mesh nodes, including other mesh points, portals, and gateways. Mesh points may require multiple hops to reach the edge of the mesh. Mesh points may allow for direct client access, or they may participate in the forwarding of communications through the mesh.

- **Mesh Gates/Gateways/Sinks/Portals:** These devices connect to other non-mesh networks, such as Ethernet, LTE, 5G, and more. They are the egress point from the mesh to the rest of the networking world and the ingress point to the mesh from the same. These devices may also connect to separate mesh networks so that the two networks may interact.

Figure 5.28 shows the mesh points and mesh portals in a typical mesh network. The three mesh points provide the internal mesh connections and two of them, in this case, connect to the portal. In this scenario, clients can connect to the mesh points. Notice that only two of the mesh points can directly communicate with the mesh portal due to signal strength issues. Therefore, the devices on the far left must pass through two mesh points before reaching the mesh portal on the right.

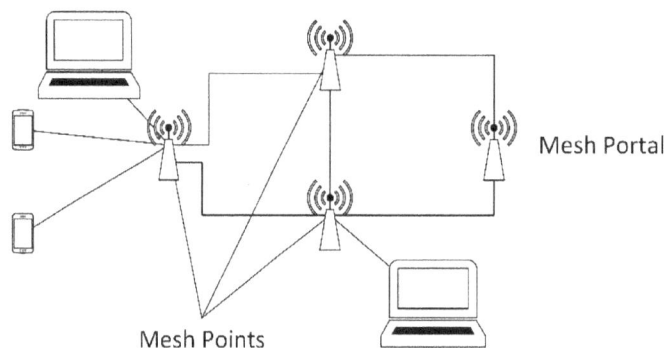

Figure 5.28: Mesh Network Devices Illustrated

Client Devices

Client devices are those that use the resources of the wireless network but do not help to build the wireless network. Therefore, laptops, mobile phones, tablets, sensors, barcode scanners, door locks, garage door openers, ceiling fans, and more can all be client devices. When deploying any wireless solution, the administrator must remember that the network is being built to support the use of the network. When this includes client access, the client devices should be evaluated so that the network can be designed for optimal performance.

Several factors should be discovered, when possible, about the client devices, including:

204

- Supported protocols: 802.11, 802.15.4, ZigBee, Bluetooth, etc.
- Supported speeds: data rates.
- Supported security: WPA/WPA2/WPA3, AES encryption, authentication, proprietary solutions.
- Data requirements: the amount of data (Kbps), type of data (real-time, burst, low-rate, high-rate), frequency of data transfer.
- Mobility requirements: some wireless clients have little to no mobility and others are highly mobile moving at rapid speeds.
- Location: where will the devices be used? Some are used within a facility, others without, and others at the perimeter.

In some cases, more information will be required, but these six items form the foundation of what must be known about all clients to support them well.

Chapter Summary

In this chapter, you learned about RF hardware, including the components used within wireless devices and the various device types and their uses. You also explored antennas and RF link types. This information will help you to better understand the chapters still to come as they explain specific use cases of wireless technologies, including cellular, short-range, low-rate, sensor, and IoT networks.

Review Questions

1. What is another term for a radio that can send and receive signals?
    a. Demodulator
    b. Modulator
    c. Transceiver
    d. Radiator

2. What is an example of an incidental radiator?
    a. Wired networking device
    b. Motor
    c. Wireless sensor
    d. Bluetooth beacon generator

3. What is represented as a ratio and indicates the noise added by an RF component?
    a. Noise figure
    b. Noise factor
    c. SNR
    d. SINR

4. When a radio down-converts a received signal to a different frequency for further processing, what is this different frequency called?
    a. NF
    b. SNR
    c. IF
    d. Image

5. What percentage of the Fresnel Zone must remain clear for a successful link?
    a. 60%
    b. 80%
    c. 90%
    d. 100%

6. What kind of chart shows the top down view of an antenna propagation pattern as it propagates outward along the Earth's surface?
    a. Azimuth
    b. Elevation
    c. E-Plan
    d. Polar view

7. What happens to an RF signal when an amplifier is over-saturated?
    a. Expansion
    b. Equalization
    c. Nullification
    d. Compression

8. What is an example of an ad-hoc wireless network?
    a. Laptops connecting to an 802.11 access point
    b. Cellular phones connecting to a service provider
    c. Automobiles communicating with roadside equipment
    d. LoRaWAN devices interconnecting

9. What is more important in bridge links than in indoor client wireless access?
    a. Antenna alignment
    b. Proper output power settings
    c. Proper security settings
    d. Proper data rate configuration

10. What does the first digit in an IP rating indicate?
    a. Protection level against fluids
    b. Protection level against objects
    c. Protection level against explosives
    d. Protection level against children

Review Answers

1. The correct answer is **C**. A transceiver is a wireless device that can both transmit and receive.

2. The correct answer is **B**. An incidental radiator is not designed to transmit signals internally or externally, but it radiates them incidentally, like a motor.

3. The correct answer is **B**. The noise factor is a ratio and the noise figure is a decibel value.

4. The correct answer is **C**. The intermediate frequency (IF) is the frequency to which signals are down-converted.

5. The correct answer is **A**. The Fresnel Zone must remain 60% clear and it is recommended that it remain 80% clear.

6. The correct answer is **A**. The Azimuth chart or H-Plane chart or Horizontal Chart shows the top-down view.

7. The correct answer is **D**. Compression occurs when an amplifier is over-saturated.

8. The correct answer is **C**. Automobiles communicating with other automobiles or roadside equipment fit into the ad-hoc category.

9. The correct answer is **A**. Antenna alignment is more important in bridge links than in indoor wireless access because it cannot depend on reflections to allow the signal to reach the destination.

10. The correct answer is **B**. The first digit of an IP rating addresses objects and the second digit addresses fluids.

Chapter 6: Cellular Networks

Objectives Covered:

2.2   Describe the fundamentals of modulation techniques used in wireless communications

3.7   Plan for technical requirements of the wireless solution

3.8   Understand the basic features and capabilities of common wireless solutions and plan for their implementation

The goal of this chapter is not to teach you how to engineer a cellular wireless network. That is beyond the scope of the CWSA exam. However, the goal is to provide you with the information you require to select and use the right kind of cellular network, when required, for your wireless projects. You will review the history of cellular networks and understand the modern architectures used. This knowledge includes the modulation methods (at a high level), frequency bands, and commonly supported end devices. Finally, you will explore various applications of cellular networks in industry, business and government.

We will begin with an overview of cellular networks in general and then explore both LTE and 5G specifically.

Cellular Networks

A cellular network is a network that is cellular in nature, meaning that it consists of cells across which mobile devices may roam to maintain connectivity. In this case, the term cell refers to the coverage area of radios on towers. In fact, the tower is often referred to as a cell tower. Wireless telephones have been around since the 1970s, but in real production since 1983. This section will briefly introduce you to the history of cellular communications and then provide an abstract overview of the architecture of cellular networks. Later sections will cover the important facts a CWSA needs to know related to LTE/4G and 5G networks.

History

The first mass-market handheld mobile phone was the Motorola DynaTAC 8000x available starting in 1983. The series of phones continued in production, with new models, until 1994. A prototype of this phone was used to place a call as early as 1973. Before the release of handheld phones starting in 1983, all mobile phones were larger and required significant power sources. Therefore, they were either permanently used in automobiles or carried about in large cases.

During the 80s and 90s, the mobile phone market continued to grow. By the end of the 90s, millions of people were using mobile phones all over the world. According to Pew Research in 2019, more than 5 billion people have mobile phones, and over half of them are smartphones. In advanced economies, the median ownership of a mobile phone among adults is 94%, with Canada having the highest percentage of adults claiming no mobile phone ownership at 25%.

The first generation of handheld mobile phones were known as 1G (G for generation). This network was an analog system and was used only for voice conversations. It had none of the unique features provided by digital networks that we're used to today.

2G cellular introduced digital communications to the networks for the first time. Because of this digital technology, mobile phones could be manufactured in smaller form factors, and new capabilities were introduced, such as text messaging. The most popular 2G system was the Global System for Mobile Communications (GSM). Most of the world standardized on GSM and the United States commonly used Code-Division Multiple Access (CDMA). It was also with the release of 2G that telephone communications became encrypted. With 1G systems, scanners could be used to listen in on cellular conversations. The data rates of 2G networks are between 48 kbits/s and 384 kbits/s. EDGE, a stop-gap solution between 2G and 3G often called 2.75G, provided data rates up to 473 kbits/sec. Interestingly, these solutions would be sufficient for many low-rate IoT device implementations; however, continuing to run them alongside modern cellular physical and MAC layers wouldn't make much sense. Modern low-rate solutions are available for LTE and 5G.

Most 2G networks will be shut down by either the end of 2019 or in 2020. For example, Verizon plans to disable 2G support at the end of 2019, and T-Mobile plans to disable it sometime in 2020. Existing 2G devices must be replaced with newer technologies, though few of them still exist.

3G cellular is known as the Universal Mobile Telecommunication System (UMTS) and altered the technology used on the air interface (radios) though the core network (behind the towers) remained mostly the same. Wideband CDMA (WCDMA) was the most commonly implemented solution in 3G. Data rates were between 144 kbits/s and several Mbits/s.

4G and 5G are discussed in more detail than these older solutions in later sections of this chapter. The vast majority of cellular connections today will be based on 3G and 4G; however, in some remote areas, 3G-only coverage does still occur. If you are planning to implement devices requiring cellular connectivity, the first task is to determine the service providers covering the target space and the technologies offered. Then, and only then, you can select the appropriate radio interfaces.

Architectures

As a CWSA, your role will not involve implementing cell tower radios or implementing and configuring the cellular core network. Therefore, your in-depth understanding of it is not required. Given that your primary goal is to select an appropriate cellular provider and technology when required, you need only understand the general way in which the wireless network side of the cellular system is built.

The first thing to know is that cellular networks use a cell-based coverage plan. The plan is, in part, the result of early desires for large geographic coverage in mobile networks. In 1947, the FCC chose to constrain the frequency bandwidth available for mobile telephone networks. The limitations they imposed would allow fewer than 30 concurrent telephone conversations in a service area. This limited the profit potential for telephone service providers, such as AT&T at the time, and so little was done to further development even though technologies existed to implement it at this early time.

In 1968, the FCC changed its position and stated that "if the technology to build a better mobile service works, we will increase the frequency allocation, freeing the airwaves for more mobile phones." With this exciting step, Bell Labs and AT&T presented a mobile telephone system that would use low-powered broadcast towers covering small cells (in this case, a few square miles). Multiple towers would be implemented to provide large-scale coverage and, with the low-power output in each cell, extensive frequency reuse. Based on this suggestion, the FCC open or frequency space and the rest is history.

In different regions of the world, the term *cell* can mean different things in relation to cellular networks. In North America, it typically means all sectors covered by a tower (the 360-degree area around the tower). In other regions, it is often synonymous with *sector*, where the term cell refers to one of the three areas covered by a cell site (the location of the tower). 4G and 5G networks have changed this somewhat, but it is useful to know about this terminology.

Traditional three-sectored cell sites use three radios/antenna system to provide coverage to three sectors from the tower (cell site). Figure 6.1 illustrates this in the typical hexagon pattern used for planning of cellular networks (not that coverage is every anything like a hexagon).

Figure 6.1: Three-Sector Cellular Architecture

With proper planning, the cellular network can cover vast areas with effective frequency reuse and multiple cell towers, which might look something like that shown in Figure 6.2. Assume the varied colors represent frequencies used within the sector. This architecture allows for client devices to connect to the sector they are in without receiving significant interference from other same-frequency sectors.

Figure 6.2: Frequency Reuse in Cellular Architecture

Communications across the network function through base station transceivers communicating with a local base station controller at the cell site. The base station controllers connect back to a mobile switching center (through either wired or wireless connections), which is connected to the PSTN and possible private networks and the

Internet. Figure 6.3 illustrates this high-level architecture of cellular networks.

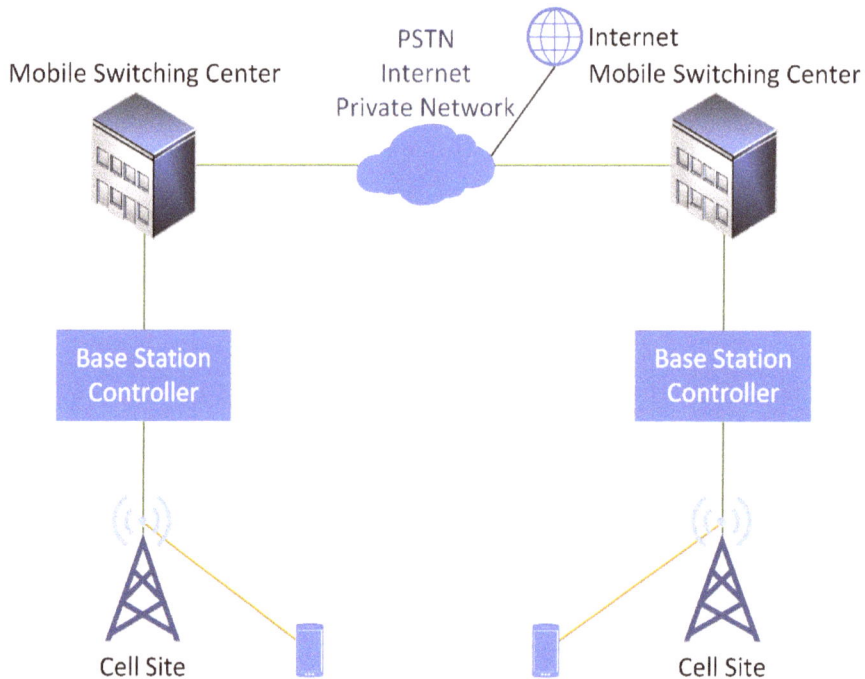

Figure 6.3: Cellular Network Communications Architecture

Each cell site can service multiple carriers. For example, Figure 6.4 shows a monopole cell tower with three service providers. Each provider implements multiple sectors from the tower to provide adequate coverage around the cell site location. Cell towers may be as short as 70-80 feet or well above 100 feet in height. The surrounding terrain dictates the height requirements of the towers for proper area coverage.

At the base of the towers are the base station controllers. These are usually small metal or block enclosures that house the local radio and switching equipment that communicates with the antennas on the towers. Additionally, the base station controllers will have some kind of connection back to a central or regional mobile switching center.

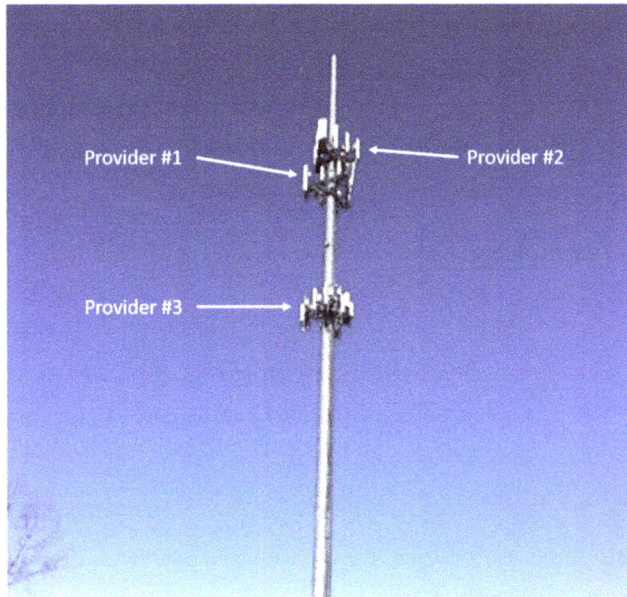

Figure 6.4: A Typical Cell Site Tower

Services Provided

While all historical cellular networks have provided telephone conversation services, modern networks (starting with 2G) added the ability to do much more on the network.

The first major added service was the short messaging service (SMS), which allowed for text messaging. In the decades since SMS was introduced, it has all but replaced pager services for most industries. However, pagers are still used in some markets, like health care. A study released in late 2018 revealed that nearly 80% of hospitals still use pagers. The primary motivation for continued use seems to be greater reliability, particularly within buildings and rooms with special wall materials, like X-ray rooms. In fact, more than 10% of all pagers still in use in 2016 were used in the healthcare industry. However, this too is changing. For example, UK National Health Service announced that 130,000 pagers still in use in February 2019 will be phased out. This reduction in pager usage is almost entirely based on the ability to use SMS and other texting services and in-building paging solutions with smart badges and other technologies.

The next added service was locationing. Locationing can be performed using multilateration with cell towers or using internal GPS components in the mobile device. Cellular-based locationing depends on the mobile device being seen by multiple towers. Based on the signal strength seen by the towers, a general location can be achieved. While not as accurate as GPS-based locationing, it can provide useful enhancements to the mobile experience, such as way-finding and integration with mapping utilities.

The final added service and one that changed the face of mobile devices forever was Internet access. By interconnecting cellular networks with the Internet, service providers allowed mobile devices to communicate with the Internet and use applications on those mobile devices that synchronized with web applications or accessed the Internet in real-time. This added service is essential in modern IoT cellular deployments that depend on cloud services or use the Internet as the connection to remote enterprise data centers.

Now that you have a basic understanding of cellular networks and the services that they provide, you can explore the two primary services offered today: LTE (4G) and 5G.

LTE

The Long Term Evolution (LTE) network is the next step before 5G and is also known as the 4G network. It is essential to know that all 4G devices are not the same just as all 5G devices are not the same. The original 4G was established in Release 10 from the 3GPP organization. Between 4G and 5G are Release 11, 12, 13, and 14, which provide enhancements to 4G networks. This reality is why you see things like LTE Advanced, LTE-M, and other enhancements.

In the world of Wi-Fi, you have amendments that add capabilities to the IEEE 802.11 standard. Therefore, you can say that an 802.11 device is not necessarily the same as all other 802.11 devices. For example, one may support 802.11n, and another supports the newer 802.11ac. However, other than speed of communications, all 802.11ac devices are indeed 802.11ac devices. They may implement different capabilities of 802.11ac, but there is only one 802.11ac amendment.

The same is not valid in cellular. Because 3G, 4G, and 5G are marketing terms to make things simpler for users. They hide even more than Wi-Fi marketing terms hide, like Wi-Fi 6 (which is a newer marketing term for 802.11ax). Quite literally, a 4G device that uses LTE-M is entirely different than a 4G device that uses LTE Advanced. For this reason, the CWSA must be very careful when selecting equipment to operate on 4G and 5G networks.

4G networks can provide up to 100 Mbps speeds, though the average speeds of client devices are typically much closer to 20 or 30 Mbps. Generally speaking, upload speeds are between 5 and 10 Mbps and download speeds are between 20 and 30 Mbps on today's networks. These speeds are sufficient for most medium-rate and even high-rate communication devices in industrial, medical, retail, transportation, and other systems.

LTE Advanced systems can accomplish peak rates in the hundreds of megabits per second, though most implementations achieve lower rates due to client radio limitations and network saturation.

Keeping in mind that 4G is not 4G is not 4G, LTE-M (LTE-Machine Type Communications) is a special LTE network with speeds ranging from 1 to 10 Mbps on the downlink and 1 to 5 Mbps on the uplink. It is designed for use with IoT-type deployments. LTE-M is defined in Release 12 and Release 13 of the 4G standards. Network latency is estimated to be between 10 and 15 milliseconds.

Narrowband IoT (NB-IoT) is also designed for such communications and was defined in Release 13 of the 4G standards. It has speeds ranging from 20 Kbps to 250 Kbps for uplinks and fixed at 250 Kbps for downlinks. Network latency is estimated to be between 1.6 and 10 seconds.

As you can see, LTE-M offers lower latency and higher speeds than NB-IoT; however, the applications of NB-IoT do not depend on high speeds or low-latency in most cases. For this reason, most predict that NB-IoT will be more popular in the coming years than LTE-M.

Both NB-IoT and LTE-M are generally placed in the Low-Power WAN (LPWAN) category. Chapter 7 will discuss NB-IoT and LTE-M in more detail.

Frequency bands used

More than fifty different frequency bands may be used by LTE/4G deployments depending on the region in which it is deployed. Table 6.1 provides a breakdown of the Frequency Division Duplexing (FDD) bands, and Table 6.2 provides a breakdown of the Time Division Duplexing (TDD) bands.

Table 6.1: FDD LTE Frequency Band Allocations

LTE BAND NUMBER	UPLINK (MHZ)	DOWNLINK (MHZ)
1	1920 – 1980	2110 – 2170
2	1850 – 1910	1930 – 1990
3	1710 – 1785	1805 -1880
4	1710 – 1755	2110 – 2155
5	824 – 849	869 – 894
6	830 – 840	875 – 885
7	2500 – 2570	2620 – 2690
8	880 – 915	925 – 960
9	1749.9 – 1784.9	1844.9 – 1879.9
10	1710 – 1770	2110 – 2170
11	1427.9 – 1452.9	1475.9 – 1500.9
12	698 – 716	728 – 746
13	777 – 787	746 – 756
14	788 – 798	758 – 768
15	1900 – 1920	2600 – 2620
16	2010 – 2025	2585 – 2600
17	704 – 716	734 – 746
18	815 – 830	860 – 875
19	830 – 845	875 – 890
20	832 – 862	791 – 821

21	1447.9 – 1462.9	1495.5 – 1510.9
22	3410 – 3500	3510 – 3600
23	2000 – 2020	2180 – 2200
24	1625.5 – 1660.5	1525 – 1559
25	1850 – 1915	1930 – 1995
26	814 – 849	859 – 894
27	807 – 824	852 – 869
28	703 – 748	758 – 803
29	n/a (DL only)	717 – 728
30	2305 – 2315	2350 – 2360
31	452.5 – 457.5	462.5 – 467.5
32	n/a (DL only)	1452-1496
65	1920-2110	2110-2200
66	1710-1780	2110-2200
67	n/a (DL only)	738-758
68	698-728	753-783
69	n/a (DL only)	2570-2620
70	1695-1710	1995-2020
252	n/a (DL only) unlicensed	5150-5250 Note: "offload" use only
255	n/a (DL only) unlicensed	5725-5850 Note: "offload" use only

Table 6.1: FDD LTE Frequency Band Allocations

LTE BAND NUMBER	ALLOCATION (MHZ)	WIDTH OF BAND (MHZ)
33	1900 – 1920	20
34	2010 – 2025	15
35	1850 – 1910	60
36	1930 – 1990	60
37	1910 – 1930	20
38	2570 – 2620	50
39	1880 – 1920	40
40	2300 – 2400	100
41	2496 – 2690	194
42	3400 – 3600	200
43	3600 – 3800	200
44	703 – 803	100
45	1447 – 1467	20
46	5150 – 5925 (unlicensed) Note: "offload" use only	775
47	5855 – 5925 (unlicensed) Note: "offload" use only	70
48	3550 – 3700	150

Table 6.2: TDD LTE Frequency Band Allocations

As a wireless solutions administrator, you should know the bands available in your regulatory domain. For the CWSA exam, you will not be required to know this information as the exam is universal. Memorizing all of the bands would serve no purpose for a CWSA practicing in a particular region.

It is important to know that frequency bands change during the life of a cellular technology. Local regulatory agencies may open more bandwidth or they may close it. Keeping up with the available frequency bands is, at minimum, an annual task for the wireless solutions administrator.

Modulation methods

The general modulation used in LTE/4G networks is Orthogonal Frequency Division Multiplexing (OFDM). Orthogonal Frequency Division Multiple Access (OFDMA) is used in the downlink communications, and Single Carrier Frequency Division Multiple Access (SC-FDMA) is used in the uplink communications.

OFDM (and therefore OFDMA, which is a multiple access implementation of OFDM) uses subcarriers within the allocated channel bandwidth to transmit data. Each subcarrier is separated by 15 kHz. The subcarriers are arranged into resource block of 180 kHz (12 15 kHz subcarriers). Actual channels are 1.4, 3, 5, 10, 15, or 20 MHz wide. Wider channels, of course, support more subcarriers and higher data rates. However, wider channels require the use of more power and more expensive processing components in the radios. Therefore, mobile devices are required to receive the full 20 MHz transmissions, but they are not required to send them.

Additionally, the mobile devices transmit using SC-FDMA (a form of OFDM), which also depends on lesser quality hardware and consumes less battery power. Table 6.3 outlines the downlink OFDM characteristics of the different LTE/4G channel bandwidths. The difference between the Bandwidth and Maximum Occupied Bandwidth values in the table is a result of the required 10% guard band in the modulation scheme.

Bandwidth (MHz)	Resource Blocks (Max)	Maximum Occupied Bandwidth (MHz)	Subcarriers (Max)
1.4	6	1.08	72
3	15	2.7	180
5	25	4.5	300
10	50	9.0	600
15	75	13.5	900
20	100	18.0	1200

Table 6.3: LTE/4G OFDM Channel Characteristics

The channelization of LTE/4G communications are documented in the 3GPP Technical Specification (TS) 36.211 and 36.213. They maintain the same numbers but are updated in releases. For example, TS 36.211 Release 8 (first LTE) is 87 pages in length, and TS 36.211 Release 15 (first 5G) is 238 pages in length.

It is useful to think of OFDM as a modulation scheme, because, in actuality, each subcarrier is modulated with one of three actual modulations:

- QPSK
- 16-QAM
- 64-QAM

QPSK provides 2 bits per symbol (waveform). 16-QAM provides 4 bits per symbol. 64-QAM provides 6 bits per symbol. The time slot (period to measure symbols) in LTE/4G is .5 milliseconds. Therefore, two time slots (also called symbol periods or slot times) exist in each millisecond. Each time slot consists of 7 symbols. The result is 14 symbols per millisecond or 14,000 symbols per second. With 14,000 symbols per second, the following list shows the bit rate on the RF medium for each modulation type and a single subcarrier:

- QPSK = 28,000 bits per second
- 16-QAM = 56,000 bits per second
- 64-QAM = 84,000 bits per second

Referring back to Table 6.3, the 20 MHz channel has a maximum of 1200 subcarriers. Therefore, with 64-QAM, the maximum bit rate is 100,800,000 bits per second or 100.8 Mbps (bits per second divided by one million). Now, you can see the upper bounds of the original LTE/4G data rates.

You will not be required to perform these modulation calculations for the CWSA exam. They are provided to help you understand the relationship between modulation methods, channel bandwidth, and data rates.

The preceding mathematical examples assumed a single spatial stream (one antenna used on both ends). If multiple spatial streams were used, simply multiplying the resulting data rate by the number of spatial streams gives you the new rate. For example, if three spatial streams are used, the data rate goes up to approximately 300 Mbps.

Devices

From the perspective of the CWSA, the client devices are the primary consideration. Most LTE/4G devices are mobile phones; however, many routers can be used to uplink to the LTE/4G network as well. They may connect a wired network to the LTE network or they may connect an 802.11 wireless network to the LTE network. Many other local networks may be supported as well. For example, Monnit, a wireless sensor network company, offers an LTE cellular gateway so that their sensors can reach the Internet and their cloud service through the cellular network. Figure 6.5 shows this device. The gateway is equipped with an internal battery providing 24 hours of operation even in the event of power failure. Therefore, if all wireless sensors use battery power (or some other form of energy harvesting) and the batteries are charged in the gateway, the wireless sensor network can continue to report events even when local power has failed.

Many Wireless Body Area Network (WBAN) devices will connect with the user's mobile phone through Bluetooth to gain access to the cellular network. Therefore, these devices also become proxied clients of the cellular network. Additionally, some home health monitoring devices are temporarily assigned to patients, and they come equipped with LTE/4G connectivity for transmission of health data back to the hospital or doctor.

Figure 6.5: Monnit 4G LTE Cellular Gateway for a Wireless Sensor Network

Finally, a note on LTE-A or LTE-Advanced. You may have heard of 5G Evolution, which means "not 5G yet, but evolving towards it." It is really LTE-A, which offers downlink speeds of up to 1000 Mbps (1 Gbps) and uplink speeds of up to 500 Mbps. To reach these speeds all capabilities of LTE-A must be in play. Key enhancements allowing the higher data rates include the addition of 256-QAM and carrier aggregation (allowing up to 100 MHz effective channels).

The LTE Release 10 marketing name was LTE-Advanced. The LTE Release 13 marketing name was LTE-Advanced Pro. Still LTE, but one is advanced, and the other is a professional at being advanced. My apologies for the geek humor.

5G

The fifth-generation (5G) of cellular networks is now upon us. 5G is still based on the foundation of LTE/4G and uses OFDM as the primary modulation scheme. Several areas of enhancement are seen in 5G networks:

- Support for frequency bands above 6 GHz

- Coexistence with LTE Evolution (the name for the reality of every changing LTE technical specifications)
- Ultra-low latency of at or under 1 millisecond
- Higher data rates

The supported frequencies will be addressed in the next section.

One key feature of the 5G networks as they roll out is support for coexistence with LTE as it has evolved. This coexistence is known as 5G non-standalone (NSA). 5G NSA is mostly used for standard mobile phone access scenarios. This scenario, technically called enhanced mobile broadband (eMBB), uses an LTE radio with alongside a 5G new radio (NR) with the supporting infrastructure being LTE infrastructure. It is a transitional solution and the one tested in early 5G deployments.

An additional feature of 5G is ultra-reliable low-latency communications (URLLC) and has applications in gaming and Industrial IoT where low-latency communications with actuators are essential and other low-latency, high availability scenarios exist.

Finally, 5G brings the promise of higher data rates through enhanced modulation and the use of wider channels above 6 GHz in the mmWave area.

Frequency bands used

When considering the frequency bands available for use in 5G, like 4G or another other wireless technology, it is essential to know that it will vary by regulatory domain. However, the 3GPP standards group has specified in TS 38.101-1 and TS 38.101-2 two frequency ranges as FR1 and FR2.

- FR1 – 410 MHz to 7125 MHz
- FR2 – 24.250 GHz to 52.600 GHz

FR1 is mostly within the frequency ranges used by LTE/4G. FR2 introduces the mmWave frequency ranges. Table 6.4 provides a listing of the 3GPP defined 5G NR bands for FR1. Table 6.5 provides a listing of the bands for FR2.

In Tables 6.4 and 6.5, a few terms need to be defined:

- BS – base station
- UE – user equipment (clients)
- Ful_low – the low-bounding frequency for uplinks

- Ful_high – the high-bounding frequency for uplinks
- Fdl_low – the low-bounding frequency for downlinks
- Fdl_high – the high-bounding frequency for downlinks
- FDD – Frequency Division Duplexing
- TDD – Time Division Duplexing
- SUL – Supplement uplink

Table 6.4: FR1 Bands for 5G NR

NR operating band	Uplink (UL) operating band BS receive / UE transmit $F_{UL\_low} - F_{UL\_high}$	Downlink (DL) operating band BS transmit / UE receive $F_{DL\_low} - F_{DL\_high}$	Duplex Mode
n1	1920 MHz – 1980 MHz	2110 MHz – 2170 MHz	FDD
n2	1850 MHz – 1910 MHz	1930 MHz – 1990 MHz	FDD
n3	1710 MHz – 1785 MHz	1805 MHz – 1880 MHz	FDD
n5	824 MHz – 849 MHz	869 MHz – 894 MHz	FDD
n7	2500 MHz – 2570 MHz	2620 MHz – 2690 MHz	FDD
n8	880 MHz – 915 MHz	925 MHz – 960 MHz	FDD
n12	699 MHz – 716 MHz	729 MHz – 746 MHz	FDD
n14	788 MHz – 798 MHz	758 MHz – 768 MHz	FDD
n18	815 MHz – 830 MHz	860 MHz – 875 MHz	FDD
n20	832 MHz – 862 MHz	791 MHz – 821 MHz	FDD
n25	1850 MHz – 1915 MHz	1930 MHz – 1995 MHz	FDD
n28	703 MHz – 748 MHz	758 MHz – 803 MHz	FDD
n30	2305 MHz – 2315 MHz	2350 MHz – 2360 MHz	FDD
n34	2010 MHz – 2025 MHz	2010 MHz – 2025 MHz	TDD
n38	2570 MHz – 2620 MHz	2570 MHz – 2620 MHz	TDD
n39	1880 MHz – 1920 MHz	1880 MHz – 1920 MHz	TDD
n40	2300 MHz – 2400 MHz	2300 MHz – 2400 MHz	TDD
n41	2496 MHz – 2690 MHz	2496 MHz – 2690 MHz	TDD
n48	3550 MHz – 3700 MHz	3550 MHz – 3700 MHz	TDD
n50	1432 MHz – 1517 MHz	1432 MHz – 1517 MHz	TDD
n51	1427 MHz – 1432 MHz	1427 MHz – 1432 MHz	TDD
n65	1920 MHz – 2010 MHz	2110 MHz – 2200 MHz	FDD

n66	1710 MHz – 1780 MHz	2110 MHz – 2200 MHz	FDD
n70	1695 MHz – 1710 MHz	1995 MHz – 2020 MHz	FDD
n71	663 MHz – 698 MHz	617 MHz – 652 MHz	FDD
n74	1427 MHz – 1470 MHz	1475 MHz – 1518 MHz	FDD
n75	N/A	1432 MHz – 1517 MHz	SDL
n76	N/A	1427 MHz – 1432 MHz	SDL
n77	3300 MHz – 4200 MHz	3300 MHz – 4200 MHz	TDD
n78	3300 MHz – 3800 MHz	3300 MHz – 3800 MHz	TDD
n79	4400 MHz – 5000 MHz	4400 MHz – 5000 MHz	TDD
n80	1710 MHz – 1785 MHz	N/A	SUL
n81	880 MHz – 915 MHz	N/A	SUL
n82	832 MHz – 862 MHz	N/A	SUL
n83	703 MHz – 748 MHz	N/A	SUL
n84	1920 MHz – 1980 MHz	N/A	SUL
n86	1710 MHz – 1780 MHz	N/A	SUL
n90	2496 MHz – 2690 MHz	2496 MHz – 2690 MHz	TDD

Table 6.4: FR1 Bands for 5G NR

These defined frequency bands are subject to change. The tables are updated as of August 2019. In fact, from April 2019 to July 2019, several new bands were added to the specification in FR1. Keep in mind that these bands are specified for use with 5G New Radio, but that does not mean they are legal for use in your regulatory domain.

Always verify which bands are allowed in your regulatory domain before you begin to utilize and radio technology. Of course, with cellular networks, the CWSA will not be implementing the cell sites, but, instead, using the network. For this reason, you will not be able to use client devices that do not support the frequencies in use within your provided cellular networks. Breach of frequency use regulations is less common for this reason as compared with unlicensed frequency bands.

Phase one of 5G rollouts focuses on the use of existing bands consumed by LTE/4G. Phase two will begin to explore the mmWave bands listed in Table 6.5. These bands offer the highest data rate improvements, but they also introduce the most significant range problems. The higher frequency ranges have the appearance of faster attenuation and improved transmit beamforming and antenna design will have to supplement for this. Additionally, at around 60 GHz, oxygen fade becomes a significant signal attenuation factor.

Operating Band	Uplink (UL) operating band BS receive UE transmit $F_{UL\_low} - F_{UL\_high}$		Downlink (DL) operating band BS transmit UE receive $F_{DL\_low} - F_{DL\_high}$		Duplex Mode
n257	26500 MHz	– 29500 MHz	26500 MHz	– 29500 MHz	TDD
n258	24250 MHz	– 27500 MHz	24250 MHz	– 27500 MHz	TDD
n260	37000 MHz	– 40000 MHz	37000 MHz	– 40000 MHz	TDD
n261	27500 MHz	– 28350 MHz	27500 MHz	– 28350 MHz	TDD

Table 6.5: FR2 Bands for 5G NR

Modulation methods

5G still supports OFDM modulation and the use of QPSK, 16-QAM, and 64-QAM. However, it adds support for BPSK and 256-QAM as well. BPSK represents 1 bit per symbols, and 256-QAM represents 8 bits per symbol (as opposed to the 6 bits per symbol of 64-QAM). The introduction of 256-QAM increases potential data rates. However, using carrier aggregation (do not take carrier to mean service provider, it is a reference to use of multiple frequency ranges) total channel bandwidth can reach 100 MHz even in FR1.

The ultimate goal of 5G networks is a maximum downlink speed of 20 Gbps and a maximum uplink speed of 10 Gbps. At the cell edge, the goal is 100 Mbps downlink speeds and 50 Mbps uplink speeds. In early 5G deployments, it has already been clear that the cell edges are significantly inferior in speeds to the better signals available closer to the cell site. However, even those inferior speeds tend to be far better than those at the LTE/4G cell edge.

Use Cases

From the perspective of the CWSA, the use cases for LTE/4G and 5G networks fall into two categories:

- Service provider network use cases
- Private network use cases

This section will provide examples of both use case categories

Service Provider Network Use Cases

When the end devices simply need Internet access, service provider use cases are a simple factor of data usage costs and ensuring proper coverage where it's needed. For example, it is entirely possible that one cellular provider has significantly better coverage in your facility than another. Choosing the provider with better coverage can result in increased performance for your end devices. Of course, this also assumes that the service provider has sufficient infrastructure in place for your needs.

Use cases for service provider access include:

- IoT
- Standard mobile phone access
- Laptops and tablets
- Backup Internet connections through LTE/5G routers
- Mobile vehicular access and fleet tracking

Private Network Use Cases

The concept of private LTE and private 5G has grown in significant interest and actual implementation in recent years. Conceptually, it is the simple task (though not so simple in frequency management planning) of implementing your own cells within your facility or coverage area. You may connect your LTE or 5G cells to a service provider's network for backhaul or connect them to your own network. Either way, you are moving beyond the use of service provider network towers and implementing your own completely private LTE or 5G network.

Private LTE and 5G networks are most commonly implemented in unlicensed frequency bands. Today, those bands are mostly 1.9 GHz, 2.4 GHz, 3.5 GHz, and 5 GHz. The 2.4 GHz and 5 GHz bands are well-known for their use in Wi-Fi networks. The 1.9 GHz and 3.5 GHz band are lesser-known bands but may be used as well.

The CBRS Alliance (www.cbrsalliance.org) is focused on promoting the use of LTE and 5G in the 3.5 GHz Citizens Broadband Radio Service (CBRS) band. In Table 6.4, this is known as band 48. The range is from 3550 MHz to 3700 MHz for a total of 150 MHz bandwidth. This band was first approved for unlicensed use in 2015 by the FCC and certain regulations are in place to protect incumbent technologies that existed in this band. OnGo is the CBRS Alliance standard for LTE and 5G in the CBRS band.

Of course, when using private LTE or 5G, it is vital to ensure that your user equipment (UE) devices can communicate with the network. They must use a module or be manufacture to support connections on the frequency band selected.

Chapter Summary

In this chapter, you learned about the basic functionality of cellular networks and the specifics of both LTE and 5G. There is much more you can learn about both LTE and 5G and all of their variations, but this information is sufficient to help you select an appropriate service provider, implement the proper private solution, and prepare for the CWSA exam.

Review Questions

1. What major feature was added with the introduction of 2G cellular networks?

   a. Digital communications

   b. OFDM

   c. OFDMA

   d. MIMO

2. What is the difference between service provider LTE and private LTE?

   a. Service provider LTE does not use encryption and private LTE does

   b. Service provider LTE uses unlicensed frequency bands and private LTE uses licensed frequency bands

   c. Service provider networks are built by the service provider and private networks may be built by the implementing organization

   d. Private LTE does not allow phone calls; it only allows data transmissions

3. What was the maximum data rate of the first LTE systems?

   a. 300 Mbps

   b. 200 Mbps

   c. 1000 Mbps

   d. 100 Mbps

4. What does LTE-A stand for?

   a. LTE-Advanced

   b. LTE-AES

   c. LTE-Aggregated

   d. LTE-Awesome

5. For what is the CBRS band used related to cellular networks?

   a. Private LTE or 5G

   b. Licensed LTE

   c. Licensed 5G

   d. Unencrypted LTE

6. What band is used by CBRS as defined by the 3GPP?

   a. 45

   b. 48

   c. 90

   d. 86

7. What is an example of the mmWave frequency bands according to the 3GPP numbering system?

   a. 100

   b. 160

   c. 200

   d. 261

8. What new QAM level is introduced in 5G NR?

   a. 64-QAM

   b. 16-QAM

   c. 2048-QAM

   d. 256-QAM

9. What does a device need to support to properly participate in a private 5G network?

    a. 802.11

    b. 802.15.4

    c. The frequencies used

    d. AES-512 encryption

10. What is best used to connect a wireless sensor network directly to the cellular network?

    a. An LTE or 5G gateway

    b. A mobile phone

    c. A wired switch

    d. A wired firewall

Review Answers

1. The correct answer is **A**. 2G networks were the first to offer digital communications introducing features like SMS texting.

2. The correct answer is **C**. Service provider LTE networks are always constructed by the service provider and/or their contractors. Private networks may be built by the implementing organization.

3. The correct answer is **D**. While later enhancements to LTE increased the possible maximum data rate, Release 8 from 3GPP provided up to 100 Mbps.

4. The correct answer is **A**. LTE-A is LTE-Advanced. The next iteration was LTE-Advanced Pro, which is also often called Gigabit LTE.

5. The correct answer is **A**. The CBRS band is used for private LTE/5G.

6. The correct answer is **B**. The 3GPP numbering system placed the CBRS band in band 48.

7. The correct answer is **D**. The mmWave bands are numbered 257, 258, 260 and 260 according to 3GPP.

8. The correct answer is **D**. 5G networks add the 256-QAM modulation to the specification.

9. The correct answer is **C**. Just because a device is an LTE or 5G device does not mean it supports the private LTE/5G bands.

10. The correct answer is **A**. An LTE or 5G gateway can be used to connect a WSN to the cellular network.

Chapter 7: Short-Range, Low-Rate, and Low-Power Networks

Objectives Covered:

2.2   Describe the fundamentals of modulation techniques used in wireless communications

3.7   Plan for the technical requirements of the wireless solution

3.8   Understand the basic features and capabilities of common wireless solutions and plan for their implementation

This chapter is not about a single technology that is a short-range, low-rate, and low-power solution. Instead, it explains the various technologies that fit into one or more of these categories. In later chapters, wireless sensor networks and IoT are discussed. These network types often employ the solutions addressed here. Before we discover specific network types that fit into one or more of these categories, we will explore the factors that impact speed, range and power consumption in RF communications.

RF and Speed

Several factors impact the speed of wireless transmissions. This speed is known as the *data rate*. The primary factors influencing speed are modulation, coding, channel bandwidth, SNR/SINR, and spatial streams. In this section, each one will be explained with sufficient detail to allow you to understand data rates regardless of the wireless solution in question.

Modulation and Data Rates

The modulation used within a system significantly impacts the data rates available. In digital wireless communications, the concept of a symbol is fundamental. A symbol is simplified as a waveform that represents bits. Modulation schemes support the following common levels for symbol-to-bits mapping:

- Two waveforms: 1 symbol = 1 bit (0 or 1)
- Four waveforms: 1 symbol = 2 bits (00, 01, 10, 11)
- Sixteen waveforms: 1 symbol = 4 bits
- Sixty-four waveforms: 1 symbol = 6 bits
- Two hundred fifty-six waveforms: 1 symbol = 8 bits

The first key to accomplishing higher data rates through various modulation methods is increasing the number of individual waveforms that can be mapped to a set of bits. The most basic modulation method supports only two waveforms, for example, one waveform with a particular amplitude and the next with a significantly different amplitude. Two amplitude positions will provide two waveforms, and we can represent a 1 or a 0 with these two waveforms.

If we can use four amplitude positions, we can represent two bits in each waveform. While no production modulation method uses the exact model represented, consider Figure 7.1 concerning this concept. Amplitude A could represent 00, Amplitude B could represent 01, Amplitude C could represent 10, and Amplitude D could represent

11. The result is that, with four amplitude levels, we can represent four different values. Because we are representing binary information in modern network, this means we are using a two-bit modulation method.

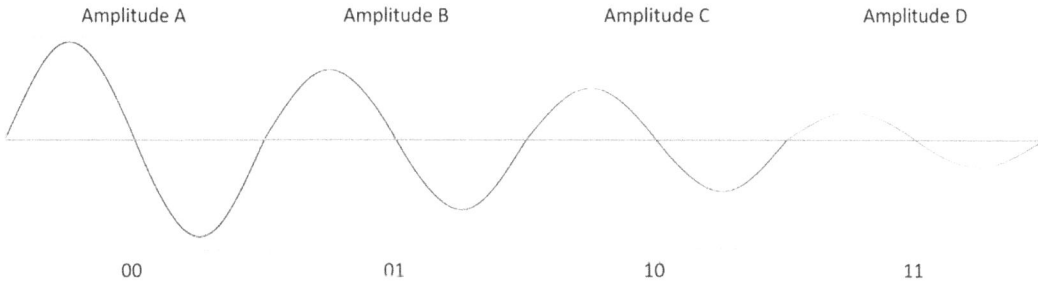

Figure 7.1: Theoretical Amplitude Modulation Example

With the targets defined, we can generate amplitude B when we need a 01, amplitude D when we need a 11, and so forth. Of course, actual modulation methods are not this simple, but this example illustrates the possible ways a symbol (waveform) can represent data. In the perfect environment (extremely low noise), we could carry on this example to 16 amplitude levels, 64 amplitude levels, 256 amplitude levels, or 1024 amplitude levels or more. Because we don't have perfect environments, we can't use amplitude in this extreme manner, but conceptually, it would allow for more and more bits to be represented in a single waveform.

Instead, we couple amplitude with phase to get ever more bits per symbol. For example, with two amplitude levels and two phases, we have four waveforms that can represent the same four binary values (00, 01, 10, and 11) represented in Figure 7.1. As noise rises, it is easier to detect the phase shifts than the amplitude shifts.

The final piece to the modulation puzzle is something called the time slot, slot time, or symbol period. It is the minimum window of time defined in the modulation scheme to look for symbols, and a particular number of symbols (remember, waveforms) will be seen in this window. For example, with a time slot of .5 milliseconds and 7 symbols per time slide, the result would be 14 symbols in 1 millisecond or 14,000 symbols per second. If each symbol can represent only 1 bit, the resulting rate is 14 kbps. If each symbol can represent 2 bits, the resulting rate is 28 kbps. As a final example, if each symbol can represent 8 bits, the resulting rate is 112 kbps.

These rates are based on a single carrier in a transmission. Advanced modulation schemes, like OFDM, use multiple carriers in a transmission (called subcarriers). If 300 subcarriers are used with each symbol representing 8 bits and a rate of 14,000 symbols per second, the result is a data rate of 33,600 kbps or 33.6 Mbps.

When selecting a wireless solution, the modulation method is very important. It is not just important from the perspective of achieving very high data rates, but it is also important for achieving stability and range. More complex modulation schemes require better signal quality for demodulation. Less complex modulation schemes can work in noisier environments. Therefore, you are not "going slower" just to go slower, but because it is needed for either range or functionality in an environment with a high noise floor. Low-rate wireless can be beneficial in the following situations:

- Noisy environments: environments with limited SNR require the use of less complex, and therefore slower, modulation methods.
- Long-range communications: longer distances result in weaker signals at the receivers and may require slower modulation methods.
- Limited frequency bandwidth available: narrower channels have fewer carriers and require slower data rates even with more complex modulation methods.

Coding and Data Rates

Another factor in the speed of wireless links is a concept called coding. *Coding* is the process used to select the bits that represent bits in transmission. That is, if I desire to send the value 1, I may actually transmit 1010. If I desire to send a 0, I may actually transmit 0101. Why do this? The answer is found in stability. For example, if I transmit a 1 for a 1 and the transmission is interfered upon, I cannot properly demodulate the data. However, if I transmit a 1010 pattern to represent a 1, at the receiver, I know that a demodulated 1 - - 0 is equal to a 1 even though I couldn't demodulate the second and third values. I also know that a - - - 0 is a 0 even though I couldn't demodulate the first, second or third values.

In wireless modulation methods, coding is defined by a coding rate, which is a ratio of information bits to transmitted bits. For example, a coding rate of 1/4 indicates that I am sending one information bit for every four transmitted bits. A coding rate of 3/4 indicates that I am sending three information bits for every four transmitted bits. The latter is better for speed, and the former is better for resiliency.

Like modulation itself, noisy environments and long-range links can benefit from coding rates that result in increased resiliency.

Channel Bandwidth and Data Rates

The next factor in wireless link speeds is the channel bandwidth. Generally speaking, wider channels result in higher speeds, and narrower channels result in lower speeds. Of course, this assumes that the same modulation and coding rate is used in the wider or narrower channels.

Many low-rate networks use channel widths of less than 5 MHz. High-rate networks typically use channel widths of 20 MHz or more. Remember that OFDM modulation methods use subcarriers measured in kHz within the defined channel bandwidth.

SNR/SINR and Data Rates

Signal-to-Noise Ratio (SNR) and Signal-to-Interference and Noise Ratio (SINR) impact speeds in a link as well. While oversimplifying the situation, links with better SNR can transmit faster than links with lower SNR. At more advanced levels of knowledge, it is good to know the details behind this, but, for now, this will suffice.

Wider channels often require higher SNR to achieve maximum data rates. This fact is important and impacts decisions in different network types. For example, in 802.11 networks, you can use 20, 40, 80 or even 160 MHz channels. The wider channels require increased SNR to maintain high-rate links.

Spatial Streams and Data Rates

Spatial Multiplexing (*SM*) uses advanced algorithms to create separate data streams for each transmitting antenna. It requires multiple radio chains, which are effectively radios linked to transmitting (Tx) and receiving (Rx) antennas. Spatial multiplexing is the use of multiple, concurrent spatial streams in transmission.

Considering what we've covered so far, including modulation, coding, channel bandwidth, and SNR/SINR, the added element of spatial streams is a multiplying factor. Generally speaking, when you have multiple spatial streams available, you calculate the data rate by multiplying a single stream data rate times the number of spatial streams. If a single stream data rate is 24 Mbps, then three streams would provide 72 Mbps.

It is important to remember that the use of multiple spatial streams may increase the required SNR for maximum data rates in some wireless solutions.

Accomplishing Desired Speeds

Now, that you understand the individual concepts that impact wireless link speeds, you can select the appropriate gear for a specific scenario. Consider the following example specifications for a link:

- 200 meters
- 100 kbps
- End Device: Wireless sensor
- Sink Device: Wireless sensor gateway

Upon investigation, you find the possible solution listed in Table 7.1. Because a wireless sensor is in question, low-rate communications are acceptable. The range requirement of 200 meters is easily met by the 1200-foot range of the devices shown (200 meters is approximately 660 feet). Of course, the selection of a solution is about more than simply meeting the technical requirements of a single link. The CWSA must consider other devices that may be required to communicate on the network and ensure that the selected solution can support the other devices as well. If that is true, the solution shown in Table 7.1 meets the specifications and may be a good choice for your scenario.

RF and Range

The *range* of a wireless connection is a reference to the physical distance over which a connection may be maintained. Three primary factors impact the maximum capable range of a wireless link and one impacts the desired range. The capable range is the distance at which a link can be maintained regardless of performance. The desired range is the distance at which a link can be maintained while achieving performance goals. The first three factors are output power, frequency bands, and antennas. The fourth factor that impacts the desired range is link performance.

Possible Solution	Image	Characteristics
Alta Ethernet Gateway 4 (Monnit)		Range: 1200 feet Modulation: FHSS Data Rate: >100 kbps
Alta Wireless Motion Detector (Monnit)		Range:1200 feet Modulation: FHSS Data Rate: >100 kbps

Table 7.1: Solution for Specifications

Output Power

The output power is the actual amplitude of the RF signal generated by the radio before entering the antenna. You can adjust the output power of many devices; however, some devices provide no control over this attribute. When no control is provided, you must know the fixed output power setting and ensure the receiver on the other end matches it as closely as possible.

Frequency Bands

Higher frequencies are more challenging to receive at greater distances than lower frequencies. This reality is why so many long-range communication systems use low frequencies like 433 MHz and 900 MHz.

Antennas

Higher gain antennas can receive weaker signals and transmit signals farther. The highest gain levels typically come from directional antennas. To increase range in a wireless link, consider replacing the antennas with high gain antennas. However, be sure to understand your local regulations related to antenna replacement.

Desired Performance and Link Range

Finally, the range of an RF link must be considered in relation to performance requirements. In many cases, it is simply impossible to achieve high data rates with very long-distance links. You will either have to accept lower data rates or implement hops along the path. For example, instead of creating one 50-kilometer link, you might create one 15-kilometer link to another 20-kilometer link to another 15-kilometer link. Such configurations may result in significantly improved performance.

RF and Power

Power must be considered from two perspectives: the output power of the RF signal (in mW, W, or dBm) and the power consumed by the device. We are really discussing low-power wireless and wireless device power management.

Low-Power Wireless

The phrase *low-power wireless* is usually a reference to the output power of the wireless solution. Low-power solutions include:

- Bluetooth Low Energy (BLE) (100 meters/500 meters with version 5)
- Near Field Communication (NFC) (10 centimeters)
- ZigBee (100 meters)
- ANT – a proprietary protocol (30 meters)
- Wi-Fi (distance based on output power and antennas)

Output power in low-power wireless solutions is typically in the 1 to 100 mW range with ultra-low-power devices in the 1 to 10 mW range in many cases. For those systems supporting variable output power, increased power on both ends of the link results in increased link range. Decreased power results in decreased range.

Power Management and Wireless

The other perspective of RF and power is power consumption. When selecting a wireless technology, it is important to consider the power consumption requirements and this is particularly true when devices will be powered by a battery. Technologies like NFC and BLE are among the lowest power-consuming solutions. Technologies like Wi-Fi and Cellular consume much more power.

In a wireless communications device, several components result in power consumption:

- Radio components
- CPU/ASIC
- Local storage
- Wired ports
- Additional interfaces in use

When a device has more of these components, it consumes more power. This fact is why MIMO 802.11 APs require more power than SISO APs. MIMO APs have more radio chains resulting in more consumed power. This power issue is also why most cellular phones with integrated Wi-Fi do not implement the maximum number of spatial streams that could be supported by the standard and available chipsets. Power management is very important for battery-powered devices.

Now that you understand some of the fundamental factors related to speed, range, and power, you can begin to explore various wireless solutions that may fit into one or more of the categories of short-range, low-rate, and low-power networks.

802.11

The first network type we will explore is the 802.11 Wi-Fi network. This network fits into the short-range and low-rate categories, depending on the Physical Layer (PHY) implemented. For example, when using the older Direct Sequence Spread Spectrum PHY (802.11-Prime or 802.11-1997), the maximum data rate is 2 Mbps, which is far lower than the possible 1+ Gbps supported in 802.11ac and 802.11ax. This section will summarize the various PHYs supported in 802.11 networks.

The IEEE 802.11 Standard

IEEE standards are managed by working groups. For example, there is an 802.3 working group and an 802.11 working group. The working group oversees the creation and maintenance of the standard. When the initial standard is created, several drafts are generated, and feedback is received and incorporated as needed in the drafting process. When the final draft is ratified (approved by a vote of active members), it becomes a standard.

After a standard exists (i.e., it has been ratified), it must be maintained. Figure 7.1 illustrates the lifecycle for standards used by the IEEE. Draft amendments are created and may go through several drafts. When a draft amendment is ratified, it becomes part of the standard. A ratified amendment may add features to the standard or it may add completely new ways of communicating on the network (known as physical layers [PHYs]). For example, 802.11n was an amendment that added the High Throughput (HT) PHY to the standard.

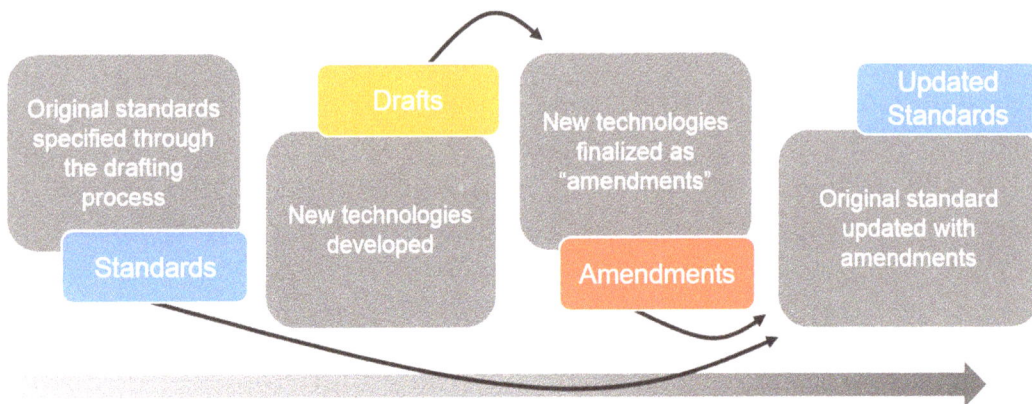

Figure 7.1: The IEEE Standards Lifecycle

The phrase "802.11 as amended" refers to the most recent revision document (currently 802.11-2016) plus any ratified amendments released after the revision. Therefore, 802.11ah modified 802.11-2016 to add the Sub-1 GHz (S1G) PHY to the standard, and the standard is actually 802.11-2016 plus the changes introduced in the 802.11ah amendment.

802.11 wireless devices can operate in one of four primary frequency bands:

- Sub-1 GHz (S1G)
- 2.4 GHz
- 5 GHz
- 60 GHz

Traditional WLAN devices operate in either the 2.4 GHz or 5 GHz frequency bands. These devices include laptops, tablets, mobile phones, wireless APs and many other specialty devices (e.g., video cameras, door locks, mobile VoIP handsets, push-to-talk devices, wireless video systems). 802.11b, 802.11g, 802.11n, and 802.11ax devices can all operate using the 2.4 GHz frequency band and are backward compatible with each other. 802.11a, 802.11n, 802.11ac and 802.11ax devices can all operate using the 5 GHz frequency band and are compatible with each other. 2.4 GHz-only devices cannot communicate with 5 GHz-only devices. Therefore, a device operating as a 2.4 GHz-only 802.11n device cannot communicate with a device operating as a 5 GHz 802.11n device, because they communicate with different frequencies.

The frequency ranges specifically used for 2.4 GHz and 5 GHz are as follows:

- 2.4 GHz uses the range from 2400 MHz (2.4 GHz) to 2500 MHz (2.5 GHz), and the actual usage range for 802.11 channels is from 2.401 GHz to 2.495 GHz
- 5 GHz uses the range from 5000 MHz (5 GHz) to 5835 MHz (5.835 GHz), and the actual usage range for 802.11 channels is from 5.170 GHz to 5.835 GHz, though not all areas within this range are used.

Within these ranges, specific portions are defined as channels of 22 MHz for the oldest wireless devices and 20 MHz or some factor of 20 MHz (e.g., 40 MHz, 80 MHz, 160 MHz) for newer devices.

The 2.4 GHz and 5 GHz bands are used differently depending on a region's radio authorities and their adoption of a set of rules within the regulatory domain. It is important to know which frequencies are available in the regulatory domain where the wireless LAN (WLAN) will be installed. As a CWSA, you may want to pay attention to this to avoid ordering mishaps that could prevent your customer from taking full advantage of the products you are offering.

The S1G bands are used for both 802.11ah devices (also called Wi-Fi HaLOW) and 802.11af devices (TV whitespaces). The specific frequency ranges used vary based on regulatory domains for 802.11ah and available frequency space for 802.11af. More details about these bands will be provided in the next section of this chapter (802.11 PHYs).

The 60-GHz band, with respect to the 802.11 standard, is used only for the Directional Multi-Gigabit (DMG) PHY and is not covered in detail in this book or the CWSA exam. However, it is important to know that it is mostly used with video devices and other short-range wireless devices that need very high throughput rates but require short-distance links. The DMG PHY was first defined in 802.11ad and is now part of the 802.11-2016 standard. It is basically an implementation of 802.11ac (VHT) in the 60-GHz band.

802.11 PHYs

The 802.11-2016 standard defines several different physical layers (PHYs) that provide different data rates, channel widths, and operational frequency bands. Additionally, the 802.11ah and 802.11ax amendments define added PHYs, and more PHYs will be added in the future. The CWSA exam requires that you are aware of the basic features of the PHYs defined in 802.11-2016, 802.11ax and 802.11ah, including the following:

- Data rates (the speed of Wi-Fi frame transmissions supported by the PHY)
- Bands used (the frequency bands supported by the PHY)
- Supported technologies (for example, the devices that may use a particular PHY)

Remember that data rates and throughput are two very different things in wireless networks. The data rate is the rate at which data is transmitted on the wireless medium (RF channel). Throughput is the rate at which higher-layer data (usually TCP or UDP) is transmitted and is often 20-60% less than the wireless link data rate.

The channel width and modulation used significantly impact the actual data rates that are available. Each PHY supports specific data rates based on the combination of channel width, modulation, coding, and a few other features. The data rates are not in some way arbitrarily variable (for example, going from 11 Mbps to 10.9 Mbps to 10.8 Mbps, and so forth), but they are specific data rates supported based on combinations of these factors (for example, going from 11 Mbps to 5.5 Mbps to 2 Mbps, and so forth).

The following subsections provide a brief overview of each PHY with all of the information that a CWSA should understand. This information is sufficient for the exam and can be used to explain the technologies to purchasing decision-makers and to make good purchasing decisions. The sections are named based on the PHY names used in the 802.11 standard itself.

DSSS (Direct Sequence Spread Spectrum)

The oldest PHY, still supported by modern 802.11 devices, is the Direct Sequence Spread Spectrum (DSSS). DSSS uses a 22-MHz wide channel and operates only in the 2.4 GHz band. Each channel is assigned based on a channel center frequency (e.g., 2.412 GHz for channel 1) and uses 11 MHz on either side of the center frequency. Therefore, channel 1 would use the range from 2.401 to 2.423 for the 22 MHz channel.

As with all PHYs introduced before 802.11n (HT), DSSS only supports one spatial stream. This means that the transceiver (transmitter/receiver) sends one stream of data and receives one stream of data at a time. More recent PHYs will support multiple streams for transmission and reception, greatly increasing the available data rates.

Remembering the supported data rates for DSSS is simple. Only two data rates are supported: 1 Mbps or 2 Mbps. By today's standards, this PHY is very slow. However, the DSSS PHY is supported by all 802.11 devices that operate in the 2.4 GHz band, including the newest 802.11n devices.

In summary, the DSSS PHY supports data rates of 1 or 2 Mbps. It operates in the 2.4 GHz band and only supports a single spatial stream. All DSSS transmissions use a 22-MHz channel width. Figure 7.2 provides a summary of this information.

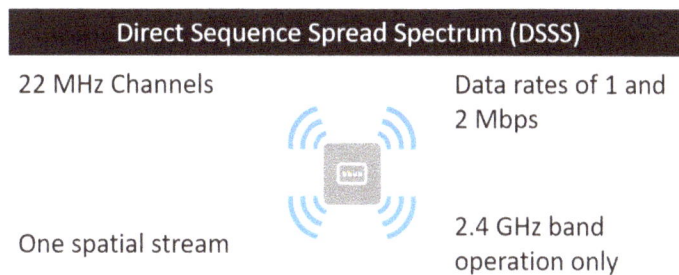

Direct Sequence Spread Spectrum (DSSS)

22 MHz Channels

Data rates of 1 and 2 Mbps

One spatial stream

2.4 GHz band operation only

Figure 7.2: DSSS PHY Summary

HR/DSSS (High Rate/Direct Sequence Spread Spectrum)

The High Rate/Direct Sequence Spread Spectrum (HR/DSSS) PHY was released with the 802.11b amendment in 1999. It introduced more advanced modulation techniques, allowing for data rates of 5.5 and 11 Mbps while still supporting the DSSS data rates of 1 and 2 Mbps. HR/DSSS uses the same 22-MHz wide channels as DSSS and only supports a single spatial stream. Like DSSS, HR/DSSS only operates in the 2.4 GHz frequency band.

All newer PHYs operating in the 2.4 GHz frequency band are designed to be backward compatible with earlier PHYs in the same band. This statement means that an HR/DSSS device can communicate with a device that only supports DSSS. Additionally, though the details of the HT PHY have not yet been explored, an HT PHY device can communicate with a DSSS PHY device as long as they both operate in the same 2.4 GHz frequency band. Figure 7.3 provides a summary of the HR/DSSS PHY.

LAN administrators may disable backward compatibility by disallowing lower data rates. However, this is a configuration constraint and not a radio or device constraint. With all data rates enabled, newer 2.4 GHz 802.11 devices can communicate with all older devices.

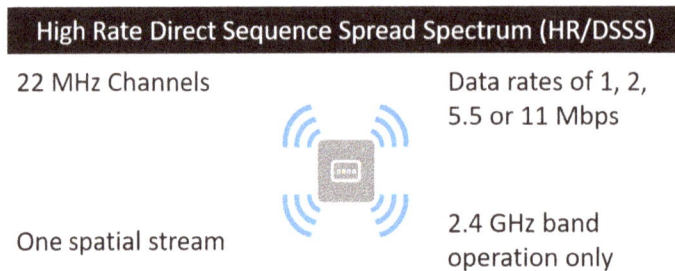

Figure 7.3: HR/DSSS PHY Summary

OFDM (Orthogonal Frequency Division Multiplexing)

The Orthogonal Frequency Division Multiplexing (OFDM) PHY was the first to support 5 GHz band operations. This PHY was made available through the 802.11a amendment in 1999. In addition to 5 GHz band support, OFDM was the first to use 20-MHz channels instead of 22-MHz channels. All modern PHYs that are based on OFDM use some factor of 20-MHz channels. For example, they may use 20-MHz (802.11a/n/ac) or 40-MHz (802.11n/ac) channels (as well as 80 and 160 MHz in 802.11ac). 802.11a was the first PHY amendment to use OFDM modulation, and the PHY is named after the modulation. All PHYs, introduced since 802.11a, also use OFDM modulation, but they have a different PHY name to clearly differentiate them from 802.11a OFDM.

OFDM still uses one spatial stream, but with enhanced modulation, it supports data rates of 6, 9, 12, 18, 24, 36, 48 and 54 Mbps. Notice that it does not support 1, 2, 5.5 or 11 Mbps. OFDM operates in the 5 GHz frequency band and has no need to be backward compatible with DSSS or HR/DSSS. Figure 7.4 provides a summary of the OFDM PHY.

Do not be confused by the difference in the PHY name of OFDM versus OFDM modulation. OFDM modulation is used in many different 802.11 PHYs (OFDM, ERP,

HT, VHT, HE, and TVHT), but the only PHY named OFDM is the 802.11a amendment PHY that operates in 5 GHz.

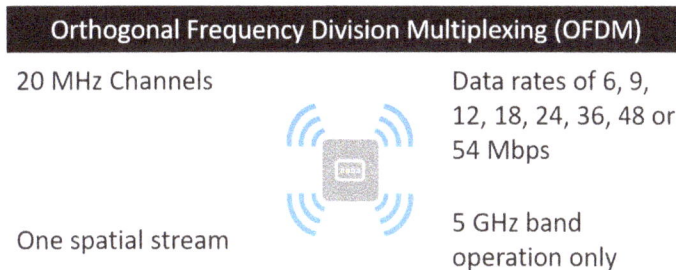

Orthogonal Frequency Division Multiplexing (OFDM)

20 MHz Channels

Data rates of 6, 9, 12, 18, 24, 36, 48 or 54 Mbps

One spatial stream

5 GHz band operation only

Figure 7.4: OFDM PHY Summary

ERP (Extended Rate PHY)

The Extended Rate PHY (ERP) was introduced to extend OFDM modulation down into the 2.4 GHz band. The implemented PHY features in 802.11 devices use the same OFDM modulation used in 5 GHz 802.11a devices and use 20-MHz channels. There are some slight differences in the way that the PHY was implemented, but for the CWSA, it is sufficient to know that it provides the same basic functionality as OFDM provided in the 5 GHz frequency band.

Operating at 2.4 GHz, all ERP devices (which are also called 802.11g devices based on the amendment that defined ERP) support backward compatibility with HR/DSSS and DSSS PHY devices. This fact is another characteristic that differentiates it from the OFDM PHY (802.11a) at 5 GHz. OFDM was the first PHY at 5 GHz and required no backward compatibility. To accomplish backward compatibility, ERP (802.11g) devices still support the DSSS data rates of 1 and 2 Mbps and the HR/DSSS data rates of 5.5 and 11 Mbps. Also, they support the same data rates as OFDM (802.11a) of 6, 9, 12, 18, 24, 36, 48 and 54 Mbps. To be clear, the ERP PHY supports only the data rates of 6, 9, 12, 18, 24, 36, 48 and 54 Mbps, but all devices implementing the ERP PHY also implement the DSSS and HR/DSSS PHYs so that the 1, 2, 5.5 and 11 Mbps data rates are also supported.

Figure 7.5 provides a summary of information for the ERP PHY and the important characteristics to know as a CWSA professional. Note that the data rates listed are those supported specifically by the ERP PHY, but remember that ERP or 802.11g devices support backward compatibility by also implementing the DSSS and HR/DSSS PHYs and the data rates they support.

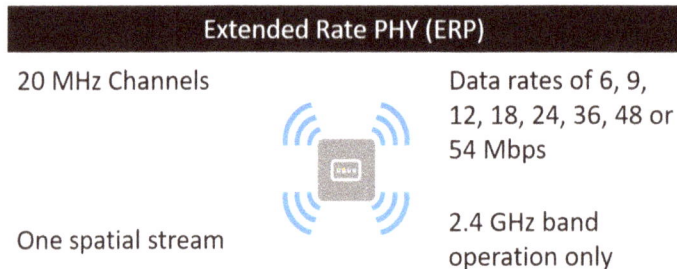

Extended Rate PHY (ERP)

20 MHz Channels

Data rates of 6, 9, 12, 18, 24, 36, 48 or 54 Mbps

One spatial stream

2.4 GHz band operation only

Figure 7.5: ERP PHY Summary

HT (High Throughput)

The High Throughput (HT) PHY was introduced in the 802.11n amendment and offers several advantages over older PHYs. HT provides wider channels by combining two 20-MHz sections into a 40-MHz channel. Therefore, HT provides either 20-MHz or 40-MHz channels. An AP that offers a 40-MHz channel can still service 20-MHz clients on its primary channel.

The primary channel will be one of the defined channel numbers, such as 1, 6, 11, 36 or 44. Then, the secondary channel, which provides a total of 40 MHz, will be 20 MHz above or below the primary channel. When a device connects to an AP that offers a 40-MHz channel and the connecting device supports only a 20-MHz channel, it will communicate with the AP using the primary channel. The 40-MHz client devices can use the entire 40-MHz channel.

Wider channels result in higher data rates even with no additional features. However, the HT PHY also introduces the capability to use multiple spatial streams through Multiple Input/Multiple Output (MIMO). MIMO takes advantage of RF propagation behaviors to send multiple concurrent streams of data from the transmitter to the receiver. The HT PHY supports up to four spatial streams; however, most devices support from one to three spatial streams when using the HT PHY.

Another unique feature of the HT PHY is that it operates at either 2.4 GHz or 5 GHz. More channels are available in the 5 GHz band, so it is the preferred band, but many devices operate only at 2.4 GHz (even some of the newest devices being sold), so this band continues to be supported in nearly all implementations.

Finally, the HT PHY offers many more data rate possibilities than earlier PHYs. The actual available data rates will depend on the channel width (20 MHz vs. 40 MHz), the number of spatial streams and the modulation and coding used. The maximum data rate achievable with the HT PHY, assuming a 40-MHz channel and the highest modulation and coding rate, is 600 Mbps. Most HT or 802.11n devices support maximum data rates of 150, 300 or 450 Mbps because the devices support from one to three spatial streams, but the standard allows for up to 600 Mbps.

It is beneficial to know that 2.4 GHz devices will support maximum data rates of 72.2, 144.4 and 216.7 Mbps (sometimes these numbers are rounded to 72, 144 and 217 Mbps) because they will only use 20-MHz channel widths in a proper implementation. While 2.4 GHz devices could be configured to support the higher data rates offered by 40-MHz channels, they should not be. When using 40-MHz, 2.4 GHz channels in a multi-AP deployment, the degradation in performance due to channel overlap is not worth the gains offered by 40-MHz channels. However, in 5 GHz standard deployments, 40-MHz channels can be beneficial, depending on the type of network being deployed. The next chapter will provide more details on this issue. Figure 7.6 shows a summary of information for the HT PHY.

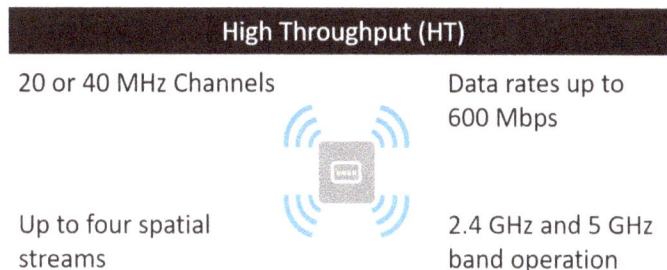

High Throughput (HT)

20 or 40 MHz Channels

Data rates up to 600 Mbps

Up to four spatial streams

2.4 GHz and 5 GHz band operation

Figure 7.6: HT PHY Summary

VHT (Very High Throughput)

The Very High Throughput (VHT) PHY moves 802.11 networks even further than the HT PHY. The VHT PHY now supports additional channel widths of 80 MHz and 160

MHz. The base channel width is still 20 MHz, but two, four or eight 20-MHz portions may be used to form the wider channels.

The wider channels achieve higher data rates, but VHT (802.11ac) also adds support for more spatial streams. A VHT device can use up to eight spatial streams. The first devices that were released supported three spatial streams, but devices that support four spatial streams are now on the market. Whether eight spatial streams will be used is yet to be seen, simply because the general trend in client devices is to stay with fewer spatial streams, which reduces battery consumption. It is very important to know that the VHT PHY works only in the 5 GHz frequency band. There is no support for VHT in the 2.4 GHz band, unlike the HT PHY. The primary reason for this decision to limit VHT to the 5 GHz band was simply the lack of frequency space for wider channels at 2.4 GHz.

Finally, VHT devices can achieve a maximum data rate of 6933.3 Mbps; however, this data rate would require eight spatial streams. Because 802.11ac devices implement no more than four spatial streams today, the real-world peak data rate is 3466.7 Mbps. To achieve this data rate of 3466.7 Mbps, the AP and client must both support four spatial streams and use a 160-MHz channel. Given the reality that few 802.11ac APs will be implemented with channels configured with a bandwidth of more than 40 MHz, it is more likely that you will see maximum data rates of 800 Mbps for four spatial streams on a 40-MHz channel.

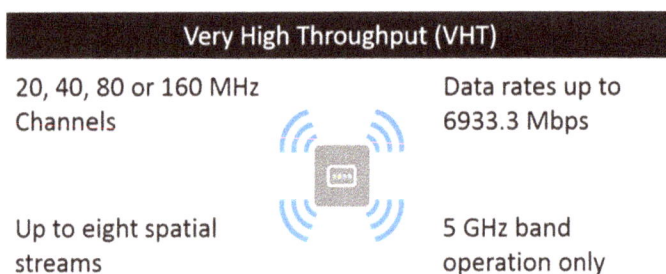

Very High Throughput (VHT)

20, 40, 80 or 160 MHz Channels

Data rates up to 6933.3 Mbps

Up to eight spatial streams

5 GHz band operation only

Figure 7.7: VHT PHY Summary

Always remember that the data rate available for a link is constrained by the least capable component of that link. For example, if an AP is configured with a 40-MHz channel and supports four spatial streams, a four spatial stream client supporting a 40-MHz channel could potentially connect with a data rate of 3466.7 Mbps. However, a single stream 40-MHz client will connect to the same AP with a maximum data rate of 200 Mbps. As you can see, the real world is often very different from marketing literature. Figure 7.7 provides a summary of this key information related to the VHT PHY.

HE (High Efficiency)

The *High Efficiency (HE)* PHY, formerly known as the High Efficiency Wireless PHY, is the latest addition to the 802.11 family of PHYs as of 2019. Introduced in the 802.11ax amendment, the HE PHY continues to support 20, 40, 80, and 160 MHz channels; however, it adds a very important feature: Orthogonal Frequency Division Multiple Access (OFDMA) modulation. OFDMA, already used heavily in the cellular communications world, adds the ability to break a channel apart into resource units and use these smaller portions of the channel to communicate with multiple clients at the same time. This transmission concurrency is a major part of the reason 802.11ax is called the High Efficiency PHY.

At the same time, HE added support for increased modulation rates with 1024-QAM. VHT supported a maximum modulation rate of 256 QAM. The result is that HE has a maximum data rate on eight spatial streams of approximately 9.6 Gbps where VHT had a maximum data rate of approximately 6.9 Gbps. The marketing professionals usually round VHT up to 7 Gbps and HE up to 10 Gbps, but the standard is clear that these rates are not supported. Figure 7.8 provides a summary of the HE PHY, including the fact that it works in both the 2.4 and 5 GHz bands.

TVHT (Television High Throughput)

The Television High Throughput (TVHT) PHY is not included on the CWSA exam beyond a general awareness of its target use and frequencies used, but information related to its capabilities is provided here for completeness. This PHY is designed to take advanced of unused frequencies in the bands often used for television and other broadcasts. Because it is designed to use such spaces, it supports very narrow channel widths of 6, 7 or 8 MHz, depending on the regulatory domain in which it operates. Additionally, the channel widths of 6, 7 or 8 MHz (called Basic Channel Units BCUs) can operate as 1, 2 or 4 BCUs. Therefore, with two 7-MHz BCUs, the total frequency space available for the transmissions would be 14 MHz.

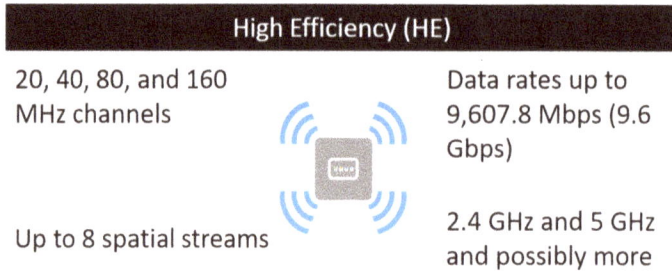

Figure 7.8: HE PHY Summary

The maximum data rate supported with the use of four 8-MHz BCUs (32 MHz of frequency space) is 568.9 Mbps. This is accomplished with four spatial streams. Like 802.11n devices, TVHT devices can use one to four spatial streams. Finally, TVHT operates in the frequency range from 50 MHz to 790 MHz and uses frequency space as allocated by the operating regulatory domain. Figure 7.9 provides a summary of this information related to TVHT.

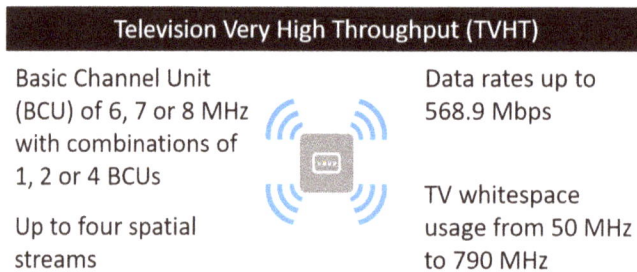

Figure 7.9: TVHT PHY Summary

S1G (Sub-1 GHZ)

The Sub-1 GHz (S1G) PHY was designed with long-range, low-data-rate communications in mind and is defined in the 802.11ah amendment. It is ideal for Internet of Things (IoT), industrial automation and monitoring networks. The S1G PHY operates on 1-, 2-, 4-, 8- or 16-MHz channels, and it appears likely that more devices will use the 1-, 2- and 4-MHz channels since their likely use cases do not warrant higher data rates.

The maximum data rate supported on the S1G PHY is 346.6667 Mbps. This rate is based on a 16-MHz channel and four spatial streams. Given the desire for extended battery life and the low need for high data rates, (e.g., devices supporting the S1G PHY), we are likely to continue to see many single-stream devices. Such devices, operating on a likely maximum of 4- or 8-MHz channels, will achieve a maximum data rate of 8666.7 Kbps for a 2-MHz, single-stream device or 20,000 Kbps (20 Mbps) for a 4-MHz, single-stream device. The actual frequencies used will vary greatly by regulatory domain but will all be less than 1 GHz. Some will operate in the 700-MHz range, others in the 900-MHz range and still others in between. For the CWSA, it is important to remember that all S1G PHY devices operate at frequencies below 1 GHz. Figure 7.10 provides a summary of this key information related to the S1G PHY.

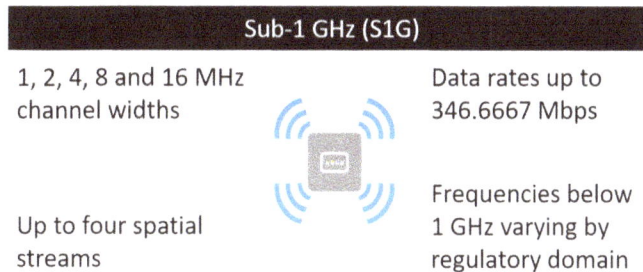

Sub-1 GHz (S1G)

1, 2, 4, 8 and 16 MHz channel widths

Data rates up to 346.6667 Mbps

Up to four spatial streams

Frequencies below 1 GHz varying by regulatory domain

Figure 7.10: S1G PHY Summary

802.15.4

The next technology to explore is the IEEE 802.15.4 standard for Low-Rate Wireless Networks. This standard defines a low-rate wireless personal area network (LR-WPAN and several other well-known protocols are based upon it. Figure 7.11 shows the common protocols that base their functionality on the 802.15.4 MAC and PHY layers. ZigBee and 6LoWPAN are the most well-known among those listed, by Wireless HART, MiWi, and ISA 100.11a are all supported and in production environments as well. Later sections of this chapter will address ZigBee and 6LoWPAN specifically. In this section, we will explore the basic functionality provided by the 802.15.4 specification.

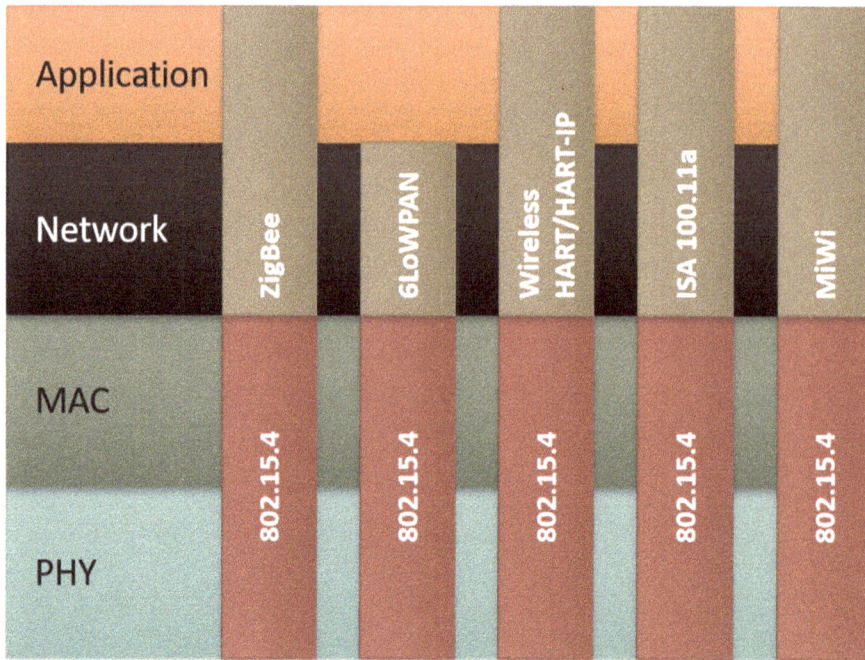

Figure 7.11: IEEE 802.15.4-Based Technologies

802.15.4 PHYs

The 802.15.4 standard specifies 18 different PHYs as of 802.15.4-2015. Some of these PHYs were created to support specific network types. For example, three PHYs specifically support Smart Utility Networks (SUNs) and two PHYs support Low-Energy, Critical Infrastructure Monitoring (LECIM) networks. An additional two PHYs are designed to support the Rail Communications and Control (RCC) networks. Table 7.2 provides an overview of the 18 PHYs specified in the standard.

Table 7.2: 802.15.4 PHYs

PHY Name	Description
O-QPSK	A DSSS PHY operating in the 780, 868, 915, 2380, and 2450 MHz bands.
BPSK	A DSSS PHY operating in the 868 and 915 MHz bands.
ASK	A parallel sequence spread spectrum (PSSS) PHY using amplitude shift keying in the 868 and 915 MHz bands.

CSS	A chirp spread spectrum (CSS) PHY operating in the 2450 MHz band.
HRP UWB	A burst position modulation (BPM) and BPSK modulation ultra-wideband (UWB) PHY operating in the sub-1 GHz and 3 to 10 GHz bands.
MPSK	An M-ary phase shift keying (MPSK) modulation PHY operating in the 780 MHz band.
GFSK	A gaussian frequency shift keying (GFSK) PHY in the 920 MHz band.
MSK	A minimum shift keying (MSK) modulation PHY.
LRP USB	A low rate pulse (LRP) UWB PHY.
SUN FSK	An FSK modulation PHY in support of SUN applications
SUN OFDM	An OFDM modulation PHY in support of SUN applications.
SUN O-QPSK	The O-QPSK PHY with modifications to support SUN applications.
LECIM DSSS	A DSSS PHY in support of LECIM applications.
LECIM FSK	An FSK PHY in support of LECIM applications.
TVWS-FSK	An FSK PHY operating in the Television Whitespace (TVWS) bands.
TVWS-OFDM	An OFDM PHY operating in the TVWS bands.
TVWS-NB-OFDM	A narrow band OFDM PHY operating in the TVWS bands.
RCC LMR	A land mobile radio (LMR) for use in RCC applications.
RCC DSSS BPSK	A DSSS BPSK PHY for use in RCC applications.

Table 7.2: 802.15.4 PHYs

802.15.4 devices can operate in many different frequency bands. Some PHYs work across any of the supported bands and others are specified for use in specific bands. For example, the CSS PHY can operate only in the 2450 MHz band, while the MSK PHY can operate in any frequency band allowed by local regulations. Table 7.3 provides an overview of the frequency bands supported by 802.15.4 devices.

Band Reference Name	Band Frequency Range (MHz)
169 MHz	169.400-169.475
433 MHz	433.05-434.79
450 MHz	450-470
470 MHz	470-510
780 MHz	779-787
863 MHz	863-870
868 MHz	868-868.6
896 MHz	896-901
901 MHz	901-902
915 MHz	902-928
917 MHz	917-923.5
920 MHz	920-928
928 MHz	928-960
1427 MHz	1427-1518
2380 MHz	2360-2400
2450 MHz	2400-2483.5
HRP UWB sub-gigahertz	250-750
HRP UWB low band	3244-4742
HRP UWB high band	5944-10,234
LRP UWB	6289.6-9185.6

Table 7.3: Supported Frequency Bands in 802.15.4

The most commonly used bands are the 868 MHz, 915 MHz and 2400 MHz bands, in large part because those bands may be used by ZigBee devices. Therefore, the supported modulation methods are DSSS and O-QPSK in those ZigBee solutions as well. The data rate of O-QPSK operating in 2.4 GHz is 250 kbps. The data rate of DSSS operating in 868 MHz is 20 kbps, and it is 40 kbps when operating in915 MHz.

802.15.4 Architectures

An 802.15.4 network forms either a star topology network or a peer-to-peer topology network. Figure 7.12 shows these two possible network topologies as defined in the standard.

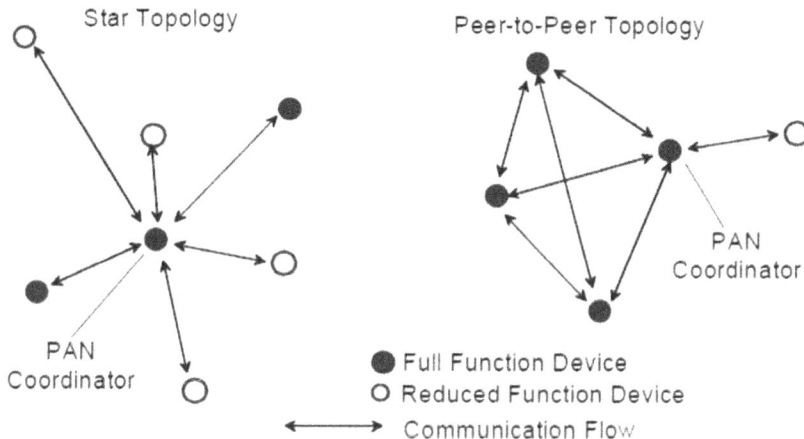

Figure 7.12: Star and Peer-to-Peer Topologies (Source: 802.15.4-2015)

As you can see, two basic device types participate in an 802.15.4 network: Full Function Device (FFD) and Reduced Function Device (RFD). An FFD can become a Personal Area Network (PAN) Coordinator and an RFD cannot. RFD units use the network for communications, but do not control the network in any way. Should the PAN Coordinator go off of the network, another FFD unit can take over PAN Coordinator operations. The PAN Coordinator starts the network and, in a star topology, all communications go through it. In a star topology, the PAN coordinator would usually be main powered (plugged into power or powered by Power over Ethernet). Devices not acting as the PAN coordinator may be battery-powered or mains-powered as well.

In a peer-to-peer topology, devices can communicate directly with each other. Generally, the first device communicating on the channel becomes the PAN coordinator, but if this device should leave the network, another FFD may be elected to the role.

Cluster tree networks are also supported. Such networks are built by interconnections among multiple PANs such that communications can be routed from a device in one PAN to a device in another PAN. This structure allows for coverage of much larger areas. Figure 7.13 shows an example of a cluster tree network.

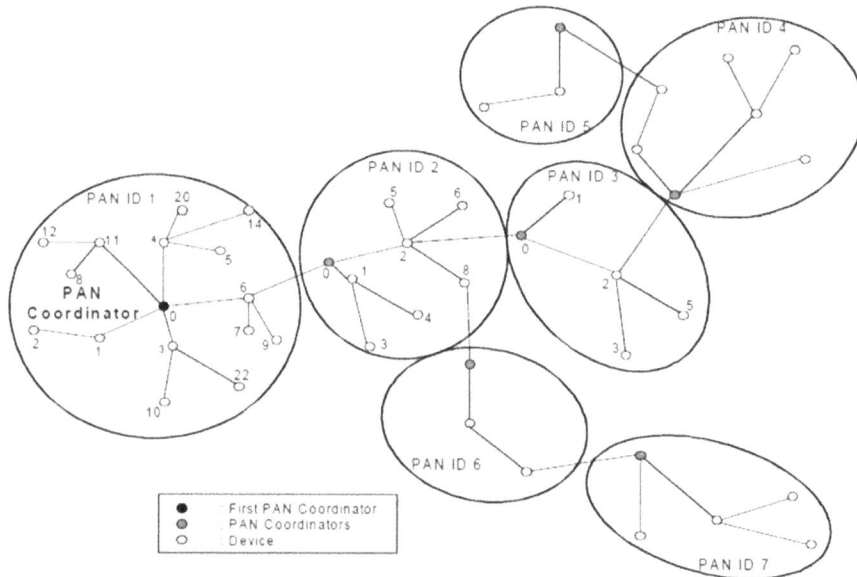

Figure 7.13: An 802.15.4 Cluster Tree Network

The cluster tree is built when the PAN coordinator in the first cluster (PAN) instructs another FFD to become the PAN coordinator of a new cluster adjacent to the existing one. This pattern can repeat until several clusters exist. Devices may then join their nearest cluster.

Very few devices exist on the market that are simply called 802.15.4 devices. In most cases, they either implement higher layer standard protocols like ZigBee or 6LoWPAN or proprietary protocols like MiWi. However, at the radio level, these systems are using 802.15.4 communications.

Bluetooth

Early on, Bluetooth was a peripheral connectivity solution. You could connect mice, keyboards, audio headsets, headphones, speakers and other such devices. Enhancements have been made to BLE over the years to the point where it can function as an IoT connectivity solution and provide links that, in some cases, have spanned 2 kilometers.

Several Bluetooth enhancements have provided modern applications of this technology, including:

- **Bluetooth Low Energy (BLE):** First introduced in 2010 with Bluetooth 4, this was the foundation that eventually led to beacons and other technologies. Bluetooth 5 enhanced BLE to provide twice the speed and four times the range. Before Bluetooth 5, the maximum speed was 1 Mbps, but version 5 added a 2 Mbps PHY as well. The new coded PHY was also added with Bluetooth 5 providing four times the range. You cannot get the extended range at the same time as the highest data rates, but both options are available in an either/or implementation option.
- **Beacons:** BLE beacons have been in Bluetooth for nearly a decade and are the functional foundation of iBeacons and other beaconing technologies used mostly for locationing. The best location systems use trilateration to locate the target, which means that beacons are read from multiple Bluetooth sensors to calculate the location. This locationing method could accomplish accuracy to between 1 and 2 meters. Accuracy levels may still not be to the level desired and so the Bluetooth Special Interest Group (SIG) has added Direction Finding to Bluetooth 5.1.
- **Direction Finding:** New in Bluetooth 5.1, direction finding will slowly make its way into production devices and our networks. Direction Finding uses Angle of Arrival (AoA) and Angle of Departure (AoD) for greater accuracy in location tracking and even movement tracking. In some environments, Bluetooth 5.1 devices can be located to within 10 centimeters with 86% accuracy.
- **Mesh:** Bluetooth mesh implements a many-to-many network that can scale from tens to thousands of devices communicating with one another. Mesh capabilities were introduced in Bluetooth 4.0 and are part of the core specification, so most of the newer Bluetooth devices can support it (given that Bluetooth 4 was finalized in 2010). However, to supported it, some devices may require upgraded firmware or software stacks. Bluetooth mesh is a networking technology layered over Bluetooth core communications. It uses BLE for communications and this is why field devices may be upgraded to Bluetooth mesh if they have sufficient processing power and memory.

Bluetooth effectively supports three topologies today: point-to-point, broadcast, and mesh. A point-to-point topology is the oldest solution in Bluetooth and is what is used by peripherals for connectivity. It can be used for data transfer as well, with speeds up to 3 Mbps. The broadcast topology is part of BLE and provides the beaconing and advertisement features for applications that provide information to user devices and notifications as well. The mesh topology is the modern game-changer in Bluetooth and may well introduce new opportunities for entrance into the IoT market.

Table 7.4 summarizes important information about different implementations of Bluetooth.

	Bluetooth 2.1	Bluetooth 4.0 (LE)	Bluetooth 5.0 (LE)
Range	100 meters	100 meters	400 meters
Max range (free space)	100 meters	100 meters	1000 meters
Data Rate	1-3 Mbps	1 Mbps	2 Mbps
Application Throughput	0.7-2.1 Mbps	Up to 305 kbps	Up to 1.36 Mbps
Topologies	Point-to-Point	Point-to-Point, Broadcast	Point-to-Point, Broadcast, Mesh

Table 7.4: Bluetooth Capabilities

LoRa (Long Range)/LoRaWAN

LoRa and LoRaWAN are not necessarily synonymous. LoRa is the base on which LoRaWAN is built. LoRa provides the machine-to-machine communications and LoRaWAN builds the network at large. LoRaWAN fits into the Low-Power WAN (LPWAN) category (like NB-IoT and LTE-M). The LoRa Alliance promotes and evolves the LoRaWAN open standard.

A LoRaWAN network is comprised of gateways that communicate using IP connections with the rest of the non-LoRa network and that communicate using single-hop LoRa or FSK communications with end-devices. All communications between end-devices use LoRa (with CSS modulation) or FSK on the radio channel until they reach a gateway, which may require multiple single-hops, where the communication is encapsulated as IP to the server or cloud managing the devices.

LoRaWAN functionality is defined in classes. All LoRaWAN devices must be able to perform Class A functionality at a minimum. This provides a baseline whereby all LoRaWAN devices will be able to perform at least minimal communications with all other LoRaWAN devices. The difference between the defined classes as of the 1.0.3 specification from 2018 are:

- Class A: All devices support bi-directional communications such that an uplink transmission from an end-device is always followed by to short downlink receive windows. Class A devices cannot receive communications from the network (downlink) except for the time during the receive windows immediately after an uplink. Downlink communications from the server will have to wait for the next uplink. These devices consume the least power.
- Class B: These devices provide more receive windows. The Class A receive process is still supported, but Class B devices open additional receive windows at scheduled times. A time synchronization beacon is sent from the gateway to provide the scheduling for Class B devices. These devices consume moderate power.
- Class C: These devices have open receive windows. They are unable to receive when transmitting, but other than that time, they can receive at any time. These devices consume the most power.

These device classes are illustrated and summarized in Figure 7.14.

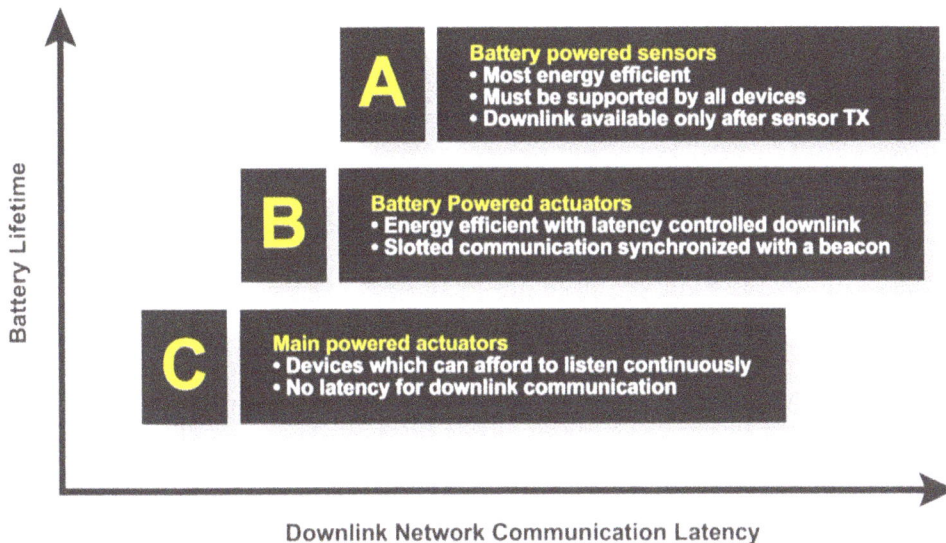

Figure 7.14: LoRa Device Classes (Source: LoRa Alliance)

The LoRA single-hop radio modulation results in data rates from 0.3 kbps to 50 kbps. Fitting LoRa nicely into the low-rate wireless category. LoRaWAN is commonly used for IoT devices, such as sensors, that require only the transmission and reception of small amounts of information and those transmissions may be infrequent in many scenarios.

The radio side of a LoRaWAN (LoRa) operated in different frequency bands depending on the regulatory domain. These bands include 430 MHz, 433 MHz, 868 MHz, and 915 MHz in common use. Figure 7.15 illustrates the LoRaWAN architecture with example use cases (on the left).

You may note that the image in Figure 7.15 references a 3G backhaul. Keep in mind that LoRaWAN communications come into the LoRaWAN gateways at a maximum of 50 kbps. Even with several dozen such communications incoming, they can be forwarded across even a 3G backhaul without problems. Of course, today, if a cellular backhaul is used, it is more likely to be LTE/4G or 5G.

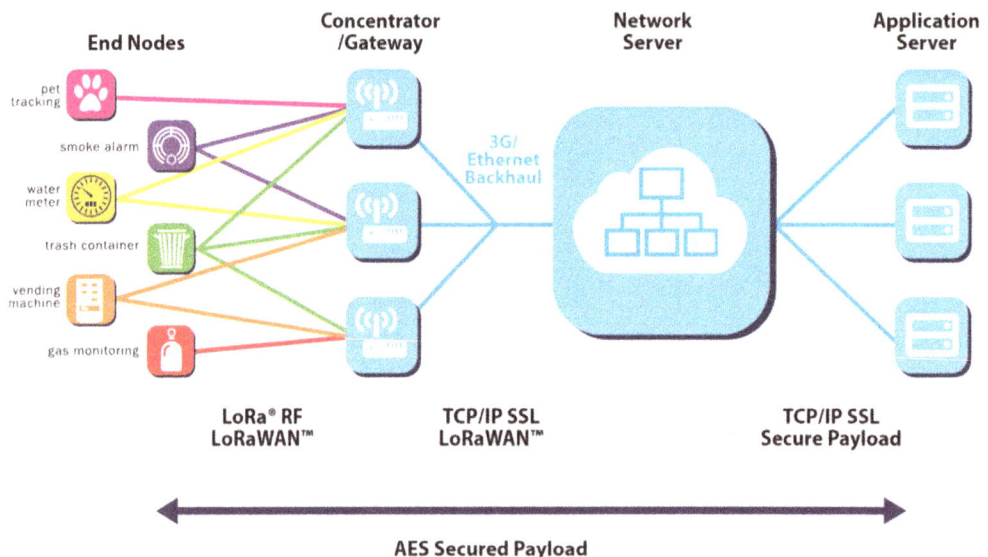

Figure 7.15: LoRaWAN Architecture (Source: LoRa Alliance)

In summary, LoRa offers long-range connectivity, a potential 10-20-year battery lifetime, and minimal infrastructure costs. These features make it very appealing when low-rate communications are required.

Because LoRaWAN gateways operate in unlicensed bands, you can establish a base station in an area and many devices, in a radius of around 5-7 kilometers or more, can connect to the base station in many cases. With the low-rate communications from sensors and other IoT devices, a simple 10 Mbps Internet connection can serve several dozen end-devices easily.

Some areas have public network coverage through The Things Network. The network consisted of 8,146 gateways at the time of writing.

ZigBee

The ZigBee Alliance was created to "enable reliable, cost-effective, low-power, wirelessly networked, monitoring and control products based on an open global standard." The ZigBee specification (currently, ZigBee Pro 3.0 in common production) defines the network, security, and application layers that reside above the PHY and MAC layers of the 802.15.4 standard for monitoring and control devices. This specification is used as an embedded technology in many consumer, health care, commercial, and industrial devices.

According to the ZigBee Pro 3.0 specification:

> *The ZigBee network layer (NWK) supports star, tree, and mesh topologies. In a star topology, the network is controlled by one single device called the ZigBee coordinator. The ZigBee coordinator is responsible for initiating and maintaining the devices on the network. All other devices, known as end devices, directly communicate with the ZigBee coordinator. In mesh and tree topologies, the ZigBee coordinator is responsible for starting the network and for choosing certain key network parameters, but the network may be extended through the use of ZigBee routers. In tree networks, routers move data and control messages through the network using a hierarchical routing strategy. Tree networks may employ beacon-oriented communication as described in the IEEE 802.15.4 specification. Mesh networks allow full peer-to-peer communication.*

Solution	Description
Network Protocol	Zigbee PRO 2015 (or newer)
Network Topology	Self-Forming, Self-Healing MESH
Network Device Types	Coordinator (routing capable), Router, End Device, Zigbee Green Power Device
Network Size (# of nodes)	Up to 65,000
Radio Technology	IEEE 802.15.4-2011
Frequency Band / Channels	2.4 GHz (ISM band) 16-channels (2 MHz wide)
Data Rate	250 Kbits/sec
Security Models	Centralized (with Install Codes support) Distributed
Encryption Support	AES-128 at Network Layer AES-128 available at Application Layer
Communication Range (Average)	Up to 300+ meters (line of sight) Up to 75-100 meters indoor
Low Power Support	Sleeping End Devices Zigbee Green Power Devices (energy harvesting)
Legacy profile support	Zigbee 3 devices can join legacy Zigbee profile networks. Legacy devices may join Zigbee 3 networks (based on network's security policy)
Logical device support	Each physical device may support up to 240 end-points (logical devices)

While ZigBee operates in multiple frequency bands, including 868 MHz, 915 MHz, and 2.4 GHz, it is most commonly used in the 2.4 GHz band as this range is available worldwide. Manufacturers can more easily support their devices operating in a single band. ZigBee uses a 2 MHz wide channel, and a total of 16 channels are available in 2.4 GHz. The channels are separated by 5 MHz. The maximum data rate for communications is 250 kbps.

ZigBee fits into the short-range, low-power, low-rate categories. Indoors, the range is between 75-100 meters with outdoor line-of-site ranges up to roughly 300 meters.

Many other specifications are useful in relation to ZigBee devices, and most of these are provided on the next page. Note the ZigBee Green Power Devices. These are devices that use energy harvesting for power. Rather than leaving energy harvesting as an afterthought, which most IoT-type protocols do, the ZigBee Alliance chose to certify devices for energy harvesting to help promote its use. Use of energy harvesting both eases deployment and results in far less energy consumption from the power grids of the world.

6LoWPAN

IPv6 over Low Power Wireless Personal Area Networks (6LoWPAN) allows for the use of IPv6 over 802.15.4 networks. The Thread Group (www.ThreadGroup.org) manages a specification for the use of 6LoWPAN on 802.15.4 devices. 6LoWPAN is based on RFC 4944, 6282, and 6775, which are both based on 802.15.4. Thread is a layer above 6LoWPAN that provides for IP routing, UDP communications, security, and commissioning.

According to the Thread Group:

Thread is designed to address the unique interoperability, security, power, and architecture challenges of the IoT.

- *Thread is a low-power wireless mesh networking protocol, based on the universally-supported Internet Protocol (IP), and built using open and proven standards.*
- *Thread enables device-to-device and device-to-cloud communications and reliably connects hundreds (or thousands) of products and includes mandatory security features.*

- *Thread networks have no single point of failure, can self-heal and reconfigure when a device is added or removed, and are simple to setup and use.*
- *Thread is based on the broadly supported IEEE 802.15.4 radio standard, which is designed from the ground up for extremely low power consumption and low latency.*

A key enabler for the IoT is interoperability. Thread addresses this challenge by providing a certification program that validates a device's conformance to the specification as well as its interoperability against a blended network comprised of multiple certified stacks.

Time will tell how this technology is adopted. As a recent entrant into the IoT space and with much existing competition, we will see. However, it does offer the ability to perform IPv6 communications on sensor networks and other low-power device networks. So, this advantage may play in its favor.

NB-IoT and LTE-M

LTE-M was the first cellular network protocol designed for IoT-type solutions. NB-IoT came after LTE-M. Both are available at this time in various regions. LTE-M (LTE Machine Type Communication) is an LPWAN technology effective for use with IoT and providing extended coverage. LTE-M provides for battery lifetimes of up to 10 years and more in some scenarios. When compared to NB-IoT, LTE-M has a wider bandwidth and supports higher data rates.

NT-IoT, while supporting lower data rates, it offers a battery life of more than 10 years because of the narrower bandwidth and lower data usage. In the end, it is a tradeoff between higher data rates (LTE-M) and extended battery life (NT-IoT).

Many industry professionals predict that NB-IoT will be the cellular IoT network of choice in the future rather than LTE-M. In many cases, a choice will not be required. If a service provider offers both LTE-M and NB-IoT, an organization can contract for both services and use the preferred service for each device.

In April of 2019, it was announced that AT&T had enabled its NB-IoT network. It launched an LTE-M network in May of 2017. So, this is an example of a service provider with both networks available.

Chapter Summary

In this chapter, you learned about the important factors related to range, rates, and power. You then explored several different wireless solutions that can fit into one or more of the low-rate, short-range, and low-power categories. In the next chapter, you will explore the details of wireless sensor networks (WSNs), which are primary networks that may utilize these and other technologies.

Review Questions

1. What is the time slot in a modulation scheme?

    a. The window of time to watch for symbols

    b. The place where the time is stored

    c. The place where the date and time are stored

    d. The number of bits in a symbol

2. What is an example of a coding rate?

    a. 3/4

    b. OFDM

    c. DSSS

    d. BPSK

3. What is not a factor impacting the range of wireless links?

    a. Output power

    b. Antenna gain

    c. Encryption

    d. Frequency band

4. What is a common output power level for low-power wireless solutions?

    a. 1 W

    b. 10 W

    c. 500 mW

    d. 10 mW

5. Which one of the following 802.11 PHYs does not operate in 2.4 GHz?

    a. HT

    b. VHT

    c. HT

    d. ERP

6. What new capability was added in HT (802.11n)?
    a. 40 MHz channels
    b. 80 MHz channels
    c. 54 Mbps data rates
    d. 11 Mbps data rates

7. Which one of the following is not an 802.15.4-based protocol?
    a. MiWi
    b. 6LoWAN
    c. NB-IoT
    d. Zigbee

8. What frequency band is most popular for use in ZigBee deployments?
    a. 915 MHz
    b. 868 MHz
    c. 5 GHz
    d. 2.4 GHz

9. What locationing solutions were added to Bluetooth to provide accuracy levels of 10 centimeters?
    a. Trilateration
    b. AoA and AoD
    c. Beacons
    d. Mesh

10. What is the protocol for single-hop communications in a LoRaWAN?
    a. LoRaWAN
    b. LoRa
    c. Bluetooth
    d. Wi-Fi

Review Answers

1. The correct answer is **A**. The time slot is the window of time or duration of time to watch for symbols.

2. The correct answer is A. Coding rates are ratios of useful bits to coded bits. 3/4, 3/5, and 5/6 are all valid examples.

3. The correct answer is **C**. Of those listed, only encryption is irrelevant to the range of a wireless link.

4. The correct answer is **D**. Low-power wireless generally rests in the range from 1 to 100 mW.

5. The correct answer is **B**. Of the listed 802.11 PHYs, only VHT does not operate in 2.4 GHz.

6. The correct answer is **A**. 802.11n (HT) added support for 40 MHz channels. 80 MHz channels were not added until 802.11ac (VHT).

7. The correct answer is **C**. NB-IoT is a cellular technology and does not run on 802.15.4. All other listed protocols do run on 802.15.4.

8. The correct answer is **D**. While ZigBee can operate on other frequency bands, due to its universal availability, 2.4 GHz is the most popular.

9. The correct answer is **B**. Bluetooth added Angle-of-Arrival (AoA) and Angle-of-Departure (AoD) for more precise locationing.

10. The correct answer is **B**. LoRa is the PHY/MAC for LoRaWAN single-hop end-device communications.

Chapter 8: Wireless Sensor Networks

Objectives Covered:

1.3   Define wireless network types

3.6   Administer the wireless solution while considering the implications of various vertical markets

Wireless Sensor Networks (WSNs) are among the most exciting and interesting areas of wireless networking. These networks allow for the gathering of information from the real world and analysis of that information in the digital world resulting in new understanding, enhanced business operations, improved health care, reduced costs, and so much more. If this is your first introduction to WSNs, welcome, we're sure you'll enjoy the learning process.

This chapter provides an overview of WSNs followed by an explanation of their architectures, components, and design processes. The best news is that WSNs are just wireless networks that happen to have clients that include various sensor types. Therefore, the knowledge you've gained so far in this book also applies to them. We will begin with a more thorough definition of a WSN.

What Is A Wireless Sensor Network (WSN)?

You have learned about the five senses: seeing, hearing, smelling, tasting, and touching. Using these senses, we can experience the world around us. You can see the beauty of a sunny day. You can hear the melodic sounds of music. You can smell that delicious dinner just before you taste it and you can touch a newborn baby's cheek or the soft fur of a kitten. These experiences make up our lives and add the ability to interact with our world.

In much the same way now, computer systems can experience the world through the senses of sight, hearing, feeling, and other senses that are not even available naturally to us as humans. They accomplish this action using sensors. Sensors have been around for many years, with some of the earliest designed nearly a century ago. However, the significant introduction of networked computing devices with sensing abilities has changed everything. This change is delivered today in WSNs.

A *sensor* is a device that can perceive varying states related to the physical world. They are implemented through the natural responses of particular elements, metals, and other components, to changes in state. These responses can be measured and reported for analysis or action.

Connected sensors and actuators are sometimes called *cyber-physical devices*. Collections of these devices and the entire solution, including monitoring and control, are sometimes called *cyber-physical systems*. A cyber-physical system (CPS) is an orchestration of computers, machines, and people working together to achieve goals using computation, communications, and control (CCC) technologies. Although the term CPS was coined only in 2006 by Helen Gill of the National Science Foundation (NSF), the CCC core technologies of CPS have had a rich and long history. Significant milestones for CPS include control theory in 1868, wireless telegraphy in 1903, cybernetics feedback in 1948, embedded systems in 1961, software engineering in 1968, and ubiquitous computing in 1988 (Hausi A. Müller, *The Rise of Intelligent Cyber-Physical Systems*, ComputingEdge, October 2018).

A WSN is a collection of wirelessly networked sensor devices, effectively forming a core component of a CPS. These devices may connect to the network through direct connections to access points, through ad-hoc networks, through mesh networks, or even through LTE or 5G networks.

WSNs are often considered a subset of the Internet of Things (IoT). Indeed, not all IoT devices are sensors, but sensors, particularly those reporting to a central cloud, may be IoT devices. It is very common to discuss Industrial IoT (IIoT) as a concept without ever using the phrase wireless sensor network even though many IIoT implementations are indeed WSNs connected to the cloud. At the same time, it is essential to realize that the WSN is the local network of sensor devices that may or may not participate in a complete IIoT solution.

A WSN may be under the umbrella of IoT in the minds of many engineers; however, it is important to remember that IoT does not equal a sensor network. Sensor networks may be categorized as IoT, but not all IoT devices are sensor devices. For example, a smartwatch may be an IoT device, but it may not have any health sensors or other sensor components in it. Therefore, it would be an IoT device, but it would not be a sensor device. Similarly, a smart coffee maker may be an IoT device, as it can receive instructions to brew coffee at certain times or on-demand through an app, but it may not sense and report anything back. Therefore, it is an IoT device, but it is not a sensor device.

More importantly, a sensor device is not necessarily an IoT device. IoT is about getting "things" onto the Internet or network. In many cases, there is no "thing" for the sensor to sense until it is actually connected to or positioned relative to the "thing" it is intended to monitor. Therefore, the "thing" that is brought onto the network is the sensed component, and it is brought onto the network by the sensor, which is already a network device.

To make this clearer, consider an accelerometer sensor. Without it, a security vehicle moving through a campus is not network connected. With it, the security vehicle can be monitored for movement, sudden impact, and possibly even location. The security vehicle has been converted to a connected vehicle through the use of the accelerometer and location-sensing components in the wireless sensor device.

The concept of Industry 4.0 is also related to wireless sensors and WSNs. Part of the fourth industrial revolution, Industry 4.0 is focused on the use of systems (machines, sensors, robotics, and more) that can monitor and control industrial processes and make decisions without human interaction through decision trees, artificial intelligence (AI), machine learning, and deep learning.

> The term *industry* refers to manufacturing and factory operations. *Industrial operations* are the operations that allow for manufacturing processes to be implemented, monitored, controlled, and evolved. With Industry 4.0 (originally spelled Industrie 4.0 when translated from German), the focus is on smart manufacturing, smart factory management, or IIoT.

As you can see, multiple concepts (CPS, IoT, IIoT, Industry 4.0, and more) integrate with or depend in wireless sensors and WSNs.

Wireless Sensor and Actuator Networks (WSANs)

As WSNs grew, the next evolution was the addition of interaction with the physical world. Sensors experience the world around them. Actuators interact with the world around them. A WSAN is a wirelessly networked collection of sensors with the ability to take action or direct that actions be taken in the real world.

For example, with your sense of touch, you may determine that a surface is hot enough that it will cause you harm. That is sensing and sensing alone. However, you also have the ability to actuate a change in response to the sensed stimulus. You can quickly move your hand away from the hot surface. Your sense of touch has actuated your movement away from the heat.

In a WSAN, an actuator may cause an item to move from an area that may cause it damage, change the thermostat settings to reduce or increase the temperature in an area, or stop a conveyor belt when human danger is detected. The key is to understand that actuators can take actions in the real world.

So, the evolution of these systems has taken place as organizations realized that more and more manual processes could be automated. When the first non-wireless or non-connected sensors were used, they were simple *measurement instruments*. Such devices used some form of sensing to monitor pressure, temperature, and other environmental elements. However, they were connected to gauges and eventually digital readouts that a human had to read physically. To do this, the human had to go to the location of the sensor.

The next step was to add alarms based on electrical circuits. If the sensor passed beyond a particular level (high or low), a circuit could be closed triggering alarms, flashing lights, or other warnings so that engineers would know the urgency of the system.

Next, sensors used special wired connections to send varying electrical signals back to distributed or central monitoring stations. Engineers could go to a few stations to see the health of the entire factory, oil refinery, electric plant, or other facilities.

Finally, sensors were implemented with wireless communications allowing signals to be transmitted over longer distances and without expensive cable runs. Today, data can be passed through the organization's networks to any location – even the cloud. Full centralized monitoring and control are available. The engineers who used to have to walk or drive around to several locations for system monitoring can now use a central dashboard to do the same work and spend more of their time improving efficiencies, enhancing quality, and performing other related tasks.

WSN/WSAN vs. Other Wireless Networks

WSN/WSAN implementations shave some unique features that do not exist in all other wireless networks, and it is valuable to explore these features. We will consider the following unique or non-universal characteristics:

- The use of sensors
- Self-forming and self-healing networks
- Low-rate data harvesting vs. high-rate data harvesting
- Energy management requirements
- Local storage requirements
- Local processing requirements
- Correlation of sensor data

The first, and most obvious, distinction of WSNs is the use of sensors in the network devices. Some other network devices have sensors as well, such as GPS radios for location tracking and gyroscopes in cell phones, but the devices containing these components are not primarily designed to sense, the sense functions are extra features or capabilities. WSN devices are there to sense as a priority.

Many, if not most, WSNs consist of self-forming and self-healing networks. *Self-forming* networks build the links and routes through the network without a central controller dictating the configuration. *Self-healing* networks make automatic adjustments to the configuration when required based on nodes going offline or moving to a new location.

Many sensor devices are low-rate data harvesting devices, which means that they do not require several Mbps every second or minute to transmit the data they gather. Instead, most of them operate with the need for some number of Kbps and many sit silent much of the time. However, some sensor devices are high-rate data harvesters and will require stable and fast connections. Very few would require more than 10 Mbps today. One exception, of course, would be video cameras with sensing capabilities (such as thermal cameras) that are also sending the video stream to a central monitoring system in real-time.

The devices in a WSN often have extreme energy management requirements. Many of these devices are battery operated, and the desire is that the batteries last months or preferably years and that the batteries be as small as possible. Therefore, strategic power management solutions must be implemented. Instead of batteries alone, some sensors will use solar power or other sources with energy harvesting. Solar power is seen frequently with stationary sensors in oil and gas, agriculture, traffic management, and other applications. In general, *energy harvesting* simply means that the device can use energy provided from interactions with the environment rather than through a power line. These sources may include solar, wind, temperature variations and electromagnetic fields.

> An *energy harvesting wireless sensor network (EH-WSN)* is a WSN that uses interactions with the surrounding environment to harvest energy. Solar and wind energy harvesting are common examples. In most cases, such components are coupled with batteries for reserved energy storage for use when the environmental source is not available, such as the loss of sunlight or wind.

Local storage requirements often exist for WSNs. If the sensor stores data for a period of time before transmission, it is best to have a long-term storage method in the device for recoverability in power failure events. If such functionality is desired, the device must provide for permanent storage. Most wireless sensors use only flash memory, which is cleared on the loss of power. Consideration must be made related to local long-term storage when it is required.

Local processing requirements are also an essential consideration in a WSN. In typical WSN deployments, the sensors report back through the WSN to a sink, which reports to local or cloud-based systems. If data is transmitted from sensors in real-time or near-real-time and communications have low latency, remote processing may suffice. However, in the case of a sensor/actuator integrated device, it may be necessary for specific actuated actions to occur immediately and this will typically require local processing.

To understand the possible need for local processing (beyond the processing which is available for wireless communications), consider a conveyor belt in a bottling factory. Assume that the bottled beverages reach an end location in the conveyor system and are lifted with a robotic arm to another location (a pick-and-place machine). Usually, the conveyor can continue moving at a defined pace that will ensure the robotic arm is ready for the next bottle. If a sensor/actuator detects a problem with the robotic arm, it must be able to stop the conveyor system (and any systems behind it in the bottling line) immediately. There is insufficient time to wait for communications back to a cloud server or even a server on the local network. By the time the response to stop comes back from the server, several bottles may have been damaged. This example is just one simple example that could be given on the frequent need for local processing.

Local processing may be performed using some form of low-capability CPU or specially designed processors (Application-Specific Integrated Circuits (ASICs)). An ASIC is, effectively, a processor designed for a specific use case rather than for general use like a CPU. ASICs are seen in network switches and other networking devices and they may be designed for sensors as well. Such limited use processors can accomplish the needed outcome with lower power consumption than a general-purpose processor.

Many sensor networks perform distributed correlation of the sensor data. They will share the sensed data with each other and use this shared information to correlate and analyze the data, to some extent, and report the findings of the correlated data rather than individual data points. This is something somewhat unique to WSNs. Such data correlation will require time synchronization among the sensors so that the analysis is accurate. If the sensors are mobile rather than stationary, location detection may also be required. In more complex scenarios, time synchronization is used among all sensors, and the data is passed (through a device called the sink) to a central processing system on the local network or in the cloud. This same central processing system may receive location data in real-time from the sensors or as archived location data at a point-in-time for proper analysis against a floorplan or location map.

> The phrase *sensor fusion* is used to reference the correlation of data from multiple sensor types so that knowledge may be acquired that would not be available from a single sensor type. For example, knowing that the temperature is rising in an area and the sun is shining on that area allows for the determination of cause.

WSN Applications

Before we explore the components that make up a WSN and the architectures available for implementation, we will consider the typical applications or use cases of these solutions.

Home and Office

Home and office solutions are often quite similar with the exception of component quality. Home, or consumer-grade, devices may be implemented using less-expensive and lower quality components to keep the price down. Office, or enterprise-grade, devices use improved quality components, much like the more expensive and lower noise figure LNAs discussed in Chapter 5. However, small business and small-office-home-office (SOHO) implementations are known for using consumer-grade devices quite frequently.

Regardless of the price, quality, or support associated with the product, similar sensor network solutions may be used in the home and the office as they are both locations where humans spend significant amounts of time. These solutions include:

- **Environmental Monitoring:** These systems include smart thermostats, wirelessly controlled lighting, humidity monitoring, CO_2 monitoring, smoke detection, motion detection, facial recognition, and more.
- **Device Control:** These systems allow humans to control the sensors and actuate changes in their settings using cloud-based web sites, mobile phone apps, and other remote-control solutions.
- **Home or Facility Automation:** When environmental monitoring and device control are brought together, you have automation potential. Thresholds can be configured, and automatic adjustments can be made based on user-defined actions.

As an example, consider a video monitoring system that uses thermal sensors in the cameras to detect heat signatures and machine vision to match these signatures to human shapes, animal shapes, and more. The system can respond differently if a dog walks onto your front porch than if a human walks onto your front porch. Dogs may not know better, but you may want to know when a human, assumed to be culturally aware as to the norms of society, has walked onto your front porch. More so, you may wish to know how long that human is in front of the door and be notified if it is beyond the usual times of a delivery driver or a solicitor getting no answer from their knock or ringing of the doorbell. As you can see, such a system can be invaluable, and they can be built with various sensors interconnected and reporting to a central system for intelligent analysis.

Industrial

For industrial operations, WSNs can provide for cost reduction, processing efficiency improvements, reduced error rates, and more. Typical benefits of industrial WSNs include:

- Improved asset management through monitoring and tracking of critical devices or equipment.
- Improved worker safety through monitoring and control of machinery, workflow, and environmental factors.
- Improved efficiency through the identification of under-performing equipment.
- Reduced energy costs through automated control of lighting, heating, cooling, machinery, and systems.
- Improved efficiency through automated monitoring of sensors instead of requiring workers to travel to the location to read the gauges and displays.

Health Care

In health care, WSNs are deployed for doctor and nurse location tracking, medical device location tracking, environmental monitoring, drug administration monitoring, vital patient monitoring, and more. The sensors may be fixed throughout the facility, attached to devices, or wearable for patient monitoring and faculty tracking. Today, ingestible sensors are even being used such that a patient may swallow a sensor that then wirelessly reports back its findings. The CWSA need not understand the medical facts related to these components, but it is important to understand how WSNs are used in health care and some of the unique applications.

Non-wearable health IoT or sensor devices may include:

- Weighing scales
- Drug administration machines
- Patient vital monitors
- Environmental sensors

Wearable health IoT or sensor devices may include:

- Patient vital monitors
- Drug administration devices
- Patient and faculty location tracking devices

Agriculture

In Smart Agro, WSNs are playing a progressively more active role. These roles include animal monitoring and environmental monitoring. Animal monitoring may be implemented through motion detection and video monitoring with or without thermal imaging. It may also be implemented using RFID or implanted sensors to track location. In less common scenarios, health monitors similar to those used in human health care may be used with animals as well. This is more common in research projects but has been used in animal farming as well.

Virtual fencing is another common solution. Such systems are commonly seen in the consumer market, known as invisible fencing. The animal will receive an acoustic warning, in many cases, when nearing the boundary, which is followed by a light electrical shock if the animal proceeds outside the boundary. These systems are also used in livestock farming in various parts of the world. The sensor detects the electrical signal in a buried or exposed wire and begins the warning process. More advanced sensors can include location tracking capabilities so that the farmer/owner can track the location of the animal.

Environmental monitoring may be related to animal farming, but is also related to produce farming. Sensors may be used to track air temperature, wind (anemometer), rain (pluviometers), soil moisture and mineral levels, insect presence, and more. In plant and produce applications, the administrator must consider the impact of foliage. If the WSN is deployed before plant growth, and the introduction of foliage and plant stems was not considered, the network may fail to operate afterward. Therefore, the administrator must implement the appropriate distribution of sensors to allow for changes in SNR, sometimes in excess of 10-12 dB or more, after plant growth. Thankfully, many such sensors require less SNR than, say, an 802.11 wireless laptop used to watch streaming videos from the Internet.

Transportation

When most people think of transportation and sensors, they immediately think of speed detection. This occurrence is likely because of the time (or times) when their speed was detected by a law enforcement officer and they received a fine. It certainly makes it a memorable moment. However, many other areas provide an opportunity for the application of sensors and WSNs in transportation, including:

- Structural health monitoring

- Traffic flow monitoring and management
- Autonomous vehicles and assisted driving
- Video monitoring of intersections with automatic collision detection

Before discussing these four examples, and many more could be given, it is important to realize that any video camera that is wirelessly connected can be categorized as a wireless sensor assuming it can be provided with intelligence internally or through backend processing. Machine vision can be used to automatically detect objects, people, and events. It can be integrated with alert and action systems for automatic response. Granted, it is not perfect today, but it is seeing continual improvements in accuracy and, soon, we will be able to automatically detect an accident on a highway and send the proper emergency response teams instantly with better than 99.9% accuracy. This will be because sensors in the vehicles will communicate with the WSN along the highway and other sensors participating in the WSN will provide input as well. This sensor fusion will result in very accurate summations of accidents and other events on the highways and roadways. Now, onto the discussion of the preceding three applications in transportation.

If you haven't realized it yet, WSNs and WSANs will be a huge market in the future. Call them IoT, IIoT, digitalization, or whatever you like. Those who know sensors and wireless will be well-positioned for the future.

Structural health monitoring can be applied to much more than transportation, such as commercial buildings and homes, but it is often a reference to bridges and railways. In a tragic event on August 1, 2007, a bridge spanning the Mississippi River collapsed killing many people and injuring even more. When the bridge was reconstructed in was implemented with sensors to monitor the structural integrity and provide early warning of problems.

Wired structural health monitoring was originally used in such implementations, but wireless sensors introduce new benefits. They do not require running wires along or through the bridge structure, and they can operate for years on battery power or be provisioned with local energy harvesting. For structural health monitoring, a special kind of sensor known as a strain gauge is likely to be used. Vibration gauges are also utilized in these environments. To conserve energy, such sensors will use sensor-triggered and radio-triggered wake-up events. The majority of the device can go into a deep sleep mode and be triggered to wake when the sensor detects vibration or strain beyond a threshold or when the radio detects a request from another radio.

Traffic flow monitoring and management are also implemented in transportation. Sensors can detect traffic by motion-sensing today and, in the future, they will be able to detect traffic based on beacons or other signals from the vehicles. With this information, automatic traffic light adjustments can be made as well as future planning for road construction to enhance traffic flows.

Autonomous vehicles and assisted driving are in their infancy; however, this is an area where we will continue to see growth. To function effectively, the vehicles must have intelligent sensors in them, which will communicate with each other through wired and wireless networks within the vehicle. Also, they will need to communicate with external roadside networks so that they can interact with the environment more fully. While this concept is young, it will be an area to watch as is experiences continued evolution.

Environmental

Environmental monitoring solutions may monitor indoor or outdoor environments or both. For outdoor monitoring, solutions include:

- Flood detection
- Volcano monitoring
- Forest fire detection
- Wind detection
- Rain detection
- Temperature monitoring
- Humidity monitoring
- Hazardous particle monitoring

For indoor solutions, Indoor Air Quality (IAQ) is an important factor in facilities management. A WSN will allow for the monitoring of IAQ and provide for automatic responses, such as increasing external ventilation, raising or lowering temperatures, increasing or decreasing humidity, and more.

In this section, we have explored just a few of the application areas for sensors and WSNs. We will see continued expansion into new areas in the coming years and the market for both the hardware and individuals who understand how to network them effectively will grow exponentially.

Sensors and Actuators

Let's begin our discussion of wireless sensors by first discussing the general concept of a sensor. In industrial monitoring, automation, and control, sensors have been used for decades. They were not equipped with wired or wireless communication capabilities in the beginning. These sensors were and are part of industrial instrumentation and control (IIC) solutions.

A *measurement instrument* is a component that is able to detect variations in a process or measurable value. For example, it can measure pressure in a liquid flow process, air pressure in tires, the temperature in a container or the environment, and more. Traditional measurement instruments indeed used sensors, but they were not connected to any remote monitoring solution. Instead they were connected to gauges and eventually electrical displays (such as LEDs or digital displays). Next, they were connected through wires to remote display consoles and now they can connect through wireless links for remote monitoring and control. Figure 8.1 illustrates the concept of a measurement instrument.

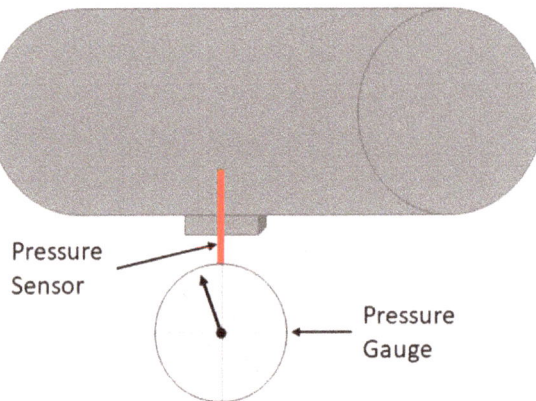

Figure 8.1: Traditional Measurement Instrument with Local Gauge for Monitoring

Traditional measurement instruments, whether attached to machinery or other components or portable hand-held devices, required the individual to go to the physical location to view the *indication* provided. The indication is the momentary reading from the sensor provided through the gauge or LED/LCD display.

For such localized measurement instruments, four primary components provided their functionality:

- **Sensor:** The element that detects states related to some physical property.
- **Amplifier:** The element that increases the values detected, as changes in state can be very small so that they are useful to humans.
- **Conditioner:** The element that ensures proper structure in the signal to all for display.
- **Display:** The pressure or electrical gauge, LED, or LCD display of the metric at the moment.

Figure 8.2 illustrates these components and how they work together. The sensor detects states and passes this to the amplifier as a signal. The amplifier increases the state value as required. The conditioner ensures proper structure or reformats to that structure for display. The display shows the value as it should be understood by the human viewer.

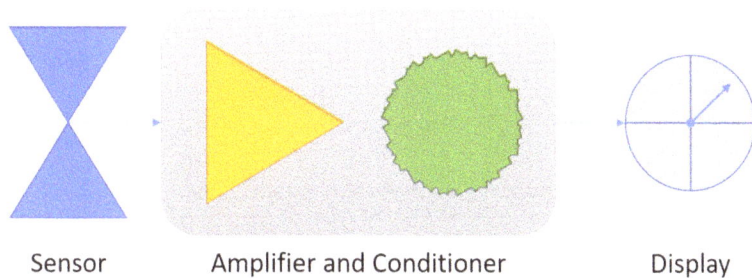

| Sensor | Amplifier and Conditioner | Display |

Figure 8.2: Parts of a Measurement Instrument (Traditional Sensor)

The next step with measurement instruments was to add a recorder to the solution. This allowed the user to view historical values over some time and to possibly even connect a device to the instrument and transfer the values. Figure 8.3 illustrates this added element to the measurement instrument.

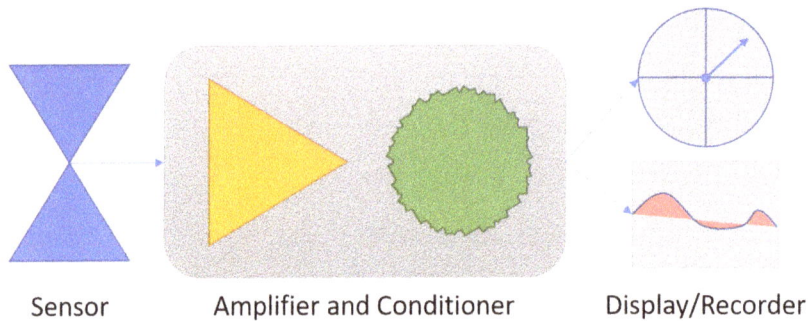

Sensor | Amplifier and Conditioner | Display/Recorder

Figure 8.3: A Measurement Instrument with a Recorder

Finally, these traditional sensors were integrated with communication relays allowing the detected states to be transmitted to a control center or control room where they could be centrally monitored. These communications were originally transmitted across proprietary wired links and eventually came to use Ethernet links and now have evolved to use wireless connections, which results in the modern WSN. Figure 8.4 illustrates this concept.

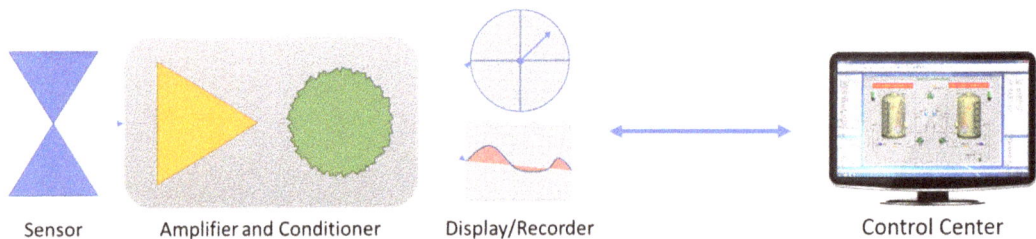

Sensor | Amplifier and Conditioner | Display/Recorder | Control Center

Figure 8.4: Sensor Connected to a Control Center

Wireless Sensors

With an understanding of traditional sensors, it is a simple leap to wireless sensors. Wireless sensors are traditional sensors, with new types being developed all the time, with wireless communications capabilities. However, because the sensors can communicate wirelessly, new capabilities are added that were more challenging with wired sensors:

- **Mobile sensors:** Special sensors that can be placed on mobile units for tracking and monitoring. Because they are wireless, as long as they move within the range of the overall WSN, they can continue to transmit metrics. If they leave the range of the WSN, they can transmit recorded metrics upon connection.

- **Remote area sensors:** Many areas are in difficult locations from a wired cabling perspective. These can be challenging areas in buildings or outdoors. Wireless sensors with batteries and/or energy harvesting can be placed in these areas.
- **In-ground sensors:** As long as they are not too deep wireless sensors can be placed beneath the Earth's surface for monitoring and reporting. These are sometimes called Wireless Underground Sensor Networks (WUSNs). Of course, depending on the characteristics of the soil (percent sand, silt, and clay) and the moisture levels, propagation becomes a challenge.

At 2.4 GHz, a distance of less than .5 meters is usually required for effective communications. Lower frequencies provide greater range, just as they do in free space. At 400-500 MHz, a range of 1 to 1.5 meters is often acceptable. WUSNs require a significantly saturated network of sensors to cover a large area. Depending on the soil, a sub-1 GHz signal will attenuate by 60-120 dB per meter.

- **In-structure sensors:** These are sensors embedded in structures. They may be embedded with cable runs for energy harvesting, or they may be embedded in accessible cavities so that batteries can be replaced every 3 or more years as needed. Those that are battery-only have a significant advantage over wired sensors; however, they have the obvious disadvantage of requiring charged batteries for operation.

Wireless Actuators

Actuators can aim antennas, reposition cameras, or even refocus or move the sensors themselves. In addition, they can interact with other physical objects. For example, an actuator coupled with a sensor may detect an object has been placed on a conveyor belt. Next, the actuator element is triggered to push a button that starts the conveyor belt. Then the actuator waits a period and pushes the button again to stop the conveyor belt. Of course, the actuator could use electrical signals for the same process, but the ability to use electromechanical elements, such as servo motors, to push a button means that legacy equipment can receive input actions from the modern sensor/actuator.

Figure 8.5 shows the common basic model for an IoT-based sensor/actuator. In some implementations, the sensor will exist as one device and the actuator as another. In other implementations, the sensor and actuator will be integrated into a single device. New, smart machinery is being developed with sensors and actuators integrated, but many legacy machines can be retrofitted with add-on sensors and actuators as well.

Whether implementing a WSN or a WSAN, it is useful to understand the various types of sensors that are available. In the next section, you will explore many sensor types and gain a basic understanding of how they function. This understanding will help you in selecting the appropriate sensors or, if the sensors are selected by a facility engineer or another person, it will help you assist them in selecting appropriate locations and ensuring proper signal coverage for connectivity and communications on the network. If a sensor is implemented without the ability to transmit and receive data effectively, it has become no more than a traditional measurement instrument.

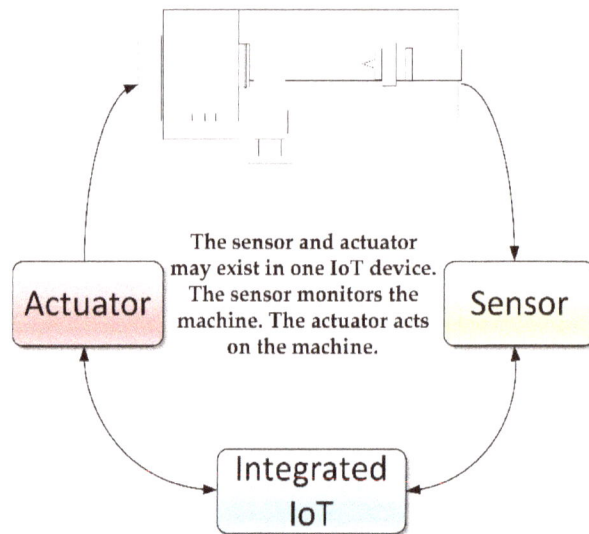

Figure 8.5: Common Basic Model for IoT Sensors/Actuators

Sensor Types

Hundreds of sensor types now exist with several variations among many of the sensor types resulting in thousands of options. This statement is a reference to the sensors alone. When you add in the networking functionality of the sensors, the variations can easily pass into the tens of thousands. However, in most cases, selecting the appropriate sensor comes down to the following three factors:

1. Does it measure what I need to measure and report it in the way I require?
2. Does it function in the environment in which I need it to function?
3. Does it communicate on the network that I support (or plan to support)?

If the answer to these three questions is yes for several sensors, then the decision comes down to support, cost and enhanced features. For the next few pages, several sensor types will be described to help you understand the capabilities commonly provided.

Temperature Sensors

Temperature sensors are typically implemented with thermistors, thermocouples, resistance temperature detectors (RTDs), or infrared. Thermocouples measure temperature changes over a wide range, when that is required, but are not as accurate as thermistors and RTDs. RTDs have a moderate range in temperature changes and are more stable than thermocouples, but are also more expensive. Thermistors are the most accurate, but have a narrower temperature range than even RTDs and are subject to self-heating (as are RTDs). Infrared sensors have the advantage of measuring temperature without surface contact but are not as accurate; however, with the use of fiber optic cables, they can measure temperatures outside of the line of sight. Ultimately, the chosen sensor type will depend on temperature ranges, required accuracy, and the ruggedized design of the solution. Table 8.1 provides a comparison of the characteristics of these sensor types.

Sensor Type	Common Ranges	Common Accuracies
Thermocouple	-250 to 1250 Celsius	1 to 2.2 Celsius
RTD	-200 to 850 Celsius	.15 to 1 Celsius
Thermistor	0 to 200 Celsius	.1 to .5 Celsius
Infrared	-50 to 600 Celsius	2 Celsius or more

Table 8.1: Temperature Sensor Types Compared

Optical Sensors

Optical sensors either detect changes in ambient light in the surrounding area or they detect optical beam crossing. Most residential garage door openers use optical beam crossing sensors. When the beam is interrupted, it sends a signal that stops the lowering of the garage door to prevent injury to the person crossing under it.

Ambient light sensors detect the current level of light in the area and can be used to signal changes in temperature control, operational machinery, and more based on the light levels. Figure 8.6 shows the ncd.io long-range wireless light sensor. This particular sensor senses light in the range from 0 to 65k lux at a resolution of 1 lux. It offers a range of up to 28 miles or 45 kilometers for wireless communications and, depending on configuration, can last up to 10 years on 2 AA batteries. It is based on the proprietary DigiMesh (XBee) protocol, which is loosely based on 802.15.4 and can work with the various components manufactured and sold at Digi.com. DigiMesh is similar to ZigBee, except that it is a proprietary protocol and the only devices that can participate in a DigiMesh network are those with embedded modules from Digi. Some modules are available from Digi that can be flashed to function as either ZigBee or XBee (DigiMesh) modules. *Flashing* is the process of loading a different firmware (software for the hardware) onto the module.

Figure 8.6: ncd.io Ambient Light Sensor

Lux is a measurement of luminous flux (the perceived power of light by the human eye) per unit area. One lux is equal to one lumen per square meter. Without going too deep, here is the critical thing to know: a moonless night provides ambient light of just about 0.0001 lux and direct sunlight on a clear day provides somewhere between 32k and 100k lux. Therefore, an optical sensor that can detect between 0 (0.0001 is greater than 0) and 65k lux can tell the difference between night and day. Some optical sensors have an upper range closer to 100k lux.

Proximity Sensors

Sensors that can detect range or the loss of contact between two items are called *proximity sensors*. They include sensors that detect open and closed windows and doors as well as ultrasonic range detectors. Door and windows sensors typically used magnets to detect the state of the enclosure. A wireless door/window sensor would trigger an alert for an open or close event depending on the configuration of the system. This alert would be communicated with the WSN. Dry contact sensors also exist, which simply detect whether a contact exists (shorted) or does not exist (opened) between two wires.

The ultrasonic range detectors work by transmitting sound waves above the level of human hearing (high frequency) and measuring the time it takes for them to be reflected back. These devices can detect the presence of new items (something has moved in front of the sensor changing the time of reflection) and are also used to detect levels in containers (such as large silos or storage tankers). They can be used for many scenarios, including:

- Factory bin level detection
- Waste management garbage levels
- Snow level monitoring
- Tank level monitoring
- Smart parking
- Rangefinder measurement
- Environmental change detection (new objects)

Figure 8.7 shows the RADIO Bridge Armored Sensor designed for outdoor or industrial use cases. It provides a 5- to 10-year battery life and can function on Sigfox, LoRa/LoRaWAN, and NBIoT networks.

Sigfox is a subscriber-based network existing in 45 countries around the world. It is a low-bandwidth network used by many (when coverage exists in their area) to implement IoT solutions and is currently most popular in Europe though they have several coverage areas in South America at the time of writing.

LoRa/LoRaWAN are covered in chapters 7 and 9.

Movement Sensors

Movement or motion sensors detect movement in the target area or movement by the monitored machinery or component. These sensors can be acceleration-based, tilt-based, or several methods of motion detection. Movement by the monitored machinery is typically acceleration-based of tilt-based. Motion detection may use Passive Infrared (PIR), MicroWave (MW), ultrasonic, or area reflective.

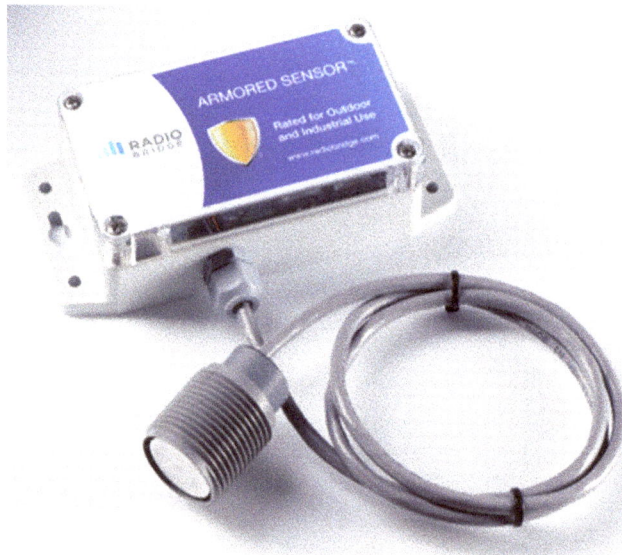

Figure 8.7: RADIO Bridge Armored Sensor Ultrasonic Range Sensor

Acceleration-based sensors use an internal accelerometer to detect movement. Tilt-based sensors use accelerometers or gyroscopes to detect movement from vertical to horizontal and vice versa. These methods can be used to determine if an attached device or item is stationary or moving. Alerts can be sent through the WSN for monitoring of movement or notification to the appropriate personnel if the item is not supposed to be moving. Coupled with a tracking solution, such sensors can aid in the location of a machine and track the use of that machine (based on movement) over time.

For motion detection, PIR is the most widely used in home or consumer devices and it is very popular in enterprise and business settings as well. It detects body heat or infrared energy. The MW sensors work by transmitting microwaves and measuring reflection back from objects. Ultrasonic sensors, as we've seen, use high-frequency sound waves and detect movement through variations in the reflected round-trip-times. Area reflective sensors also use reflection response times based on infrared emissions from LEDs. As you can see, much of motion sensing is about reflections of various waves including sound and electromagnetic waves.

Some motion sensors will combine multiple types, such as infrared and MW, in order to limit false positives. When both sensors are tripped, it is far more likely to be a true positive for the type of motion being detected.

Liquid Sensors

When you need to detect water, fuel, and other liquids, various liquid sensors may be used. One such sensor is the water rope sensor. Such sensors can be placed with the rope extending into a container or area. If any part of the rope senses a liquid an alert is triggered. They work well for many scenarios, including:

- Basement or area flooding
- Sump pump failures
- Water leak detection
- Toilet leak detection
- Frozen water pipe detection
- Server room protection
- Tank fill detection

Water rope sensors work by using two sensing wires wound around the "rope" material. When the material gets wet, it closes the circuit and water, or another fluid has been detected.

An additional liquid sensor type is a spot leak detector. These devices have probes that extend down from them and if water rises to the level of the probes it closes the circuit and results in fluid detection.

In the category of liquid sensors, you might include flow sensors. These sensors are often used to indicate leaks in pipes or hoses due to excess flow. They may also have freeze sensors in them to detect temperatures below the freezing point for a given fluid.

Air Sensors

Air sensors monitor the air for air quality purposes, such as CO_2 detection, humidity levels, and the detection of other contaminants. The Edimax AI-2002W sensor shown in Figure 8.8, is a 7-in-1 air quality sensor. It detects the following:

- PM2.5
- PM10
- CO_2
- TVOC
- HCHO
- Temperature
- Humidity

Figure 8.8: Edimax AI-2002W Air Quality Sensor

The PM2.5 and PM10 sensors detect airborne particles. The CO_2 sensor detects Carbon Dioxide levels. Total Volatile Organic Compound (TVOC) sensors detect organic chemicals from paints, cleaning supplies, and other possible harmful sources. HCHO is a formaldehyde sensor, which can be diffused into the air from remnants left in the manufacturing of furniture and other items. Clearly, this sensor can detect many possible detriments to air quality, while also tracking temperature and humidity levels. The AI-2002W connects to your Wi-Fi network and can be monitored and controlled through the Edimax cloud service.

Strain Gauge Sensors

Often uses in structural health monitoring, these sensors detect strain at very low levels. Figure 8.9 shows the RESENSYS SenSpot strain gauge sensor, which is simply fasted to the structure with adhesive and runs for up to ten years on the batteries.

Figure 8.9: RESENSYS SenSpot Strain Gauge Sensor

Strain is the amount of deformation a material experiences under an applied force. It is measured by ratios comparing the current state to the original state. Specifically, it is the change in length of a material compared to the original state. Axial strain measures the stretching of a material when compressed (like pressing against a wall). Bending strain measures the stretching of the material in a direction based on downward pressure (like the bending of a stick when you press down on it). These are the two most common measurements.

When metallic structures are placed under mechanical strain, they exhibit subtle changes in their resistance. Strain gauges measure this change in resistance in microstrains, and this is the basic building block of a strain gauge sensor.

Lord Kelvin was the first to report that metallic conductors change their resistance when they are deformed under strain. He discovered this in 1856, but it was not put into practical use until the 1930s. (Source: www.omega.co.uk, Strain Gauge, Transactions)

Sensor Connector Nodes

Sensor connector nodes are used to connect various sensors to the WSN. They provide standard analog interfaces for common sensor types. The LORD SG-Link-200 is shown in Figure 8.10. Various sensor types can be connected with a device like the SG-Link-200 and participate in the WSN. This particular connector node also includes an on-board temperature sensor so that you can harvest temperature information as well as the information provided by the connected input sensor. The device uses an AMPSEAL 16 14-pin connector to connect external sensors and operates using 802.15.4 for wireless communications in 2.4 GHz.

Figure 8.10: LORD SG-Link-200 Sensor Connector Node

WSN nodes are often called motes in honor of the first WNS nodes created at the University of Berkeley by Rene and Mica Motes. They are more often simply called modules, nodes, sensors, or wireless sensors, but you will encounter the term mote from time-to-time.

BEYOND THE EXAM: I Found a Mouse in My Salad!

In Australia, customers were buying loose leaf spinach for salads and other delicious recipes. The only problem was that mice like these spinach leaves too. Several customers reported finding mice in the sealed plastic bags of spinach leaves. Of course, this was a big problem for the packager, and so they sought a solution.

The solution was to implement a sensor to detect mice as the spinach leaves passed through the conveyor system. The question was, how do you accomplish such a task.

The answer was found in the use of a video camera with a 90-degree viewing angle monitoring the conveyor as the spinach leaves passed by. The selected camera was the FLIR A65, which is a thermal imaging camera. Using machine vision, the system was able to detect thermal patterns in the spinach leaves that were above the normal temperature of spinach. When such items were detected, the system triggered an alert so that a handler could locate the apparently living entity (like a mouse) and remove it from the conveyor.

This is just one example of a creative solution using sensors. Of course, this could be taken further such that the monitoring system used a wireless thermal imaging camera to transmit the video feedback to a remote central monitoring system, but whether localized, wired or wireless, you can see that sensors can be implemented in creative ways to solve real-world problems.

This section has provided an overview of some of the most common sensor types. With this understanding, you can better comprehend the value introduced by placing these sensors onto a WSN. Aggregating disparate data from multiple sensors to a central dashboard can provide valuable insights into activities, incidents, processes, and events within a facility or outdoor environment.

WSN Architectures

At the highest layer of abstraction, we can consider WSNs from five basic architectural perspectives (Senouci, Mellouk, *Deploying Wireless Sensor Networks*, Elsevier, 2016):

- **small-, medium-, large- and very large-scale WSNs:** the size of the WSN varies depending on several factors such as the sensors' characteristics, the Return on Investment (RoI), and the user's requirements. In practice, the number of sensor nodes in a WSN may be in the order of tens, hundreds, thousands, or even tens of thousands.

- **homogeneous versus heterogeneous WSNs:** a WSN may be homogeneous or heterogeneous. A WSN is homogeneous if all sensors of the network have the same capabilities (sensing, processing, communication, etc). A heterogeneous WSN consists of sensors endowed with different capacities, which may serve for different applications. Typically, some sensors will have more resources available, such as processing and energy, than the rest of the sensors.

- **stationary, mobile, and hybrid WSNs:** a WSN may be stationary, mobile, or hybrid. A stationary WSN is a network consisting of stationary sensor nodes that cannot move once deployed. With the advances in mobile devices, some of the sensors are able to move on their own; this is generally achieved by embedding the sensors on mobile platforms. A mobile WSN comprises only mobile sensors, while a hybrid WSN consists of both stationary and mobile sensors.

- **flat versus hierarchical WSNs:** in flat WSNs, all the sensor nodes are assumed to be homogeneous and play the same role. However, in hierarchical WSNs, a sensor node can be dedicated to a particular special function. For instance, a sensor could be designated as a cluster-head, in charge of communicating with adjacent clusters.

- **single-hop versus multi-hop WSNs:** in a single-hop WSN, sensor nodes transmit their data directly to the sink. In a multi-hop WSN, multiple relaying sensor nodes exist between sensors and sinks. A multi-hop WSN can be flat or hierarchical.

The remainder of this section will focus on conceptual and actual architectures used in WSNs and WSANs.

IIC IIoT Architecture Framework

The Industrial Internet Consortium has developed the IIoT Architecture Framework as a model for successful industrial IoT implementations. If explains the components of the network and how they interact and important considerations in their implementation. Figure 8.11 illustrates this architecture.

This framework document is designed *"to aid in the development, documentation and communication of the IIRA. The reference architecture uses a common vocabulary and a standard-based framework to describe business, usage, functional and implementation viewpoints that it has defined."* Further it states that, *"A reference architecture provides guidance for the development of system, solution and application architectures. It provides common and consistent definitions for the system of interest, its decompositions and design patterns, and a common vocabulary with which to discuss the specification of implementations and compare options."* (Industrial Internet of Things Volume G1: Reference Architecture v1.9, 2019)

The point of a reference architecture for IIoT is similar to that of the OSI Reference Model. It provides a conceptual way of thinking about the technologies and a shared language that professionals can use to communicate. This is much needed in the high variety IoT space, and this reference model can be beneficial beyond the scope of industrial IoT into health, retail, and agricultural IoT as well, though some systems and concepts may not universally apply.

The introduction of IoT into the industrial world brings an intersection between Information Technology (IT) and Operational Technology (OT). IT has been the realm of programmers, systems administrators, and network engineers. OT has been the realm of mechanical engineers, chemical engineers, and other control engineers. With IIoT, the two must work together to present the analog nature of OT in the digital world of IT and to convert the digital world of IT into the analog world of OT. For this reason, it is essential to have teams, including OT engineers and IT engineers, when deploying IIoT or WSNs.

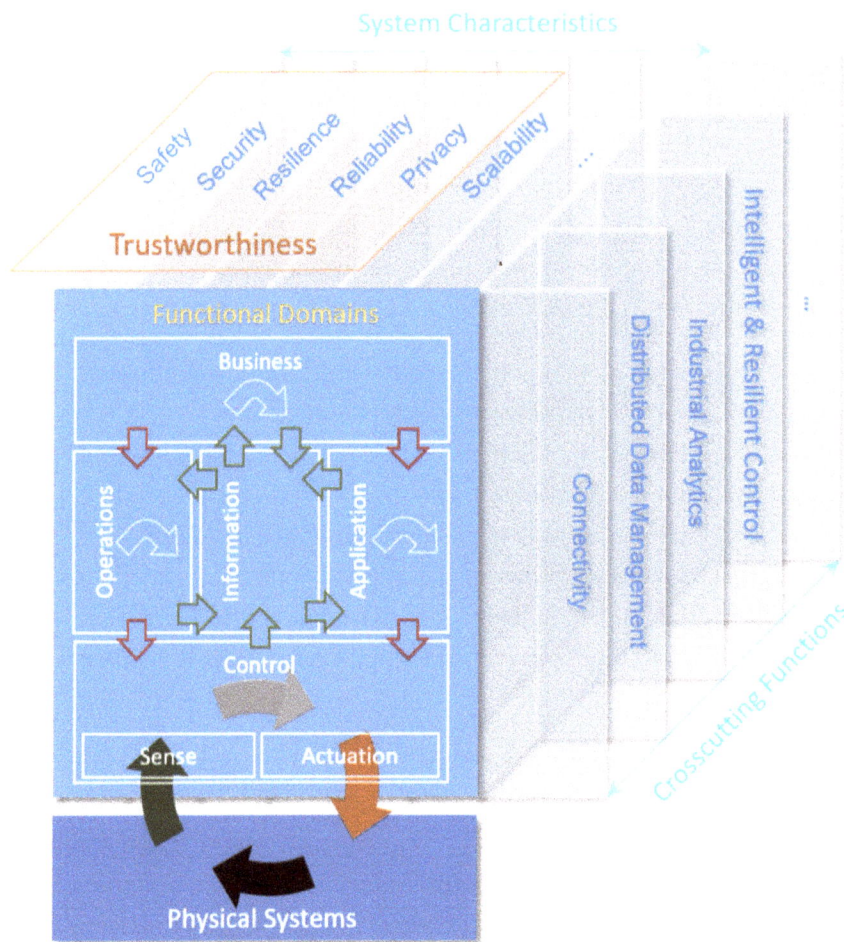

Figure 8.11: Industrial Internet Consortium IIoT Architecture Framework (Source: Industrial Internet of Things Volume G1: Reference Architecture v1.9)

At the foundation of the IIC IIoT framework (called the Industrial Internet Architecture Framework (IIAF) in shorthand) is the physical systems or the actual sensors and actuators participating in the WSN and the larger network(s) supporting the WSN as seen in Figure 8.11. The Functional Domain includes sensing, actuating, and control. However, within the Functional Domain, data is passed upwards to become information that can be acted upon by operations or processed by applications based on business requirements.

Within the Functional Domain is the Control Domain (or sub-domain). It can be divided into several components, and this breakdown is shown in Figure 8.12.

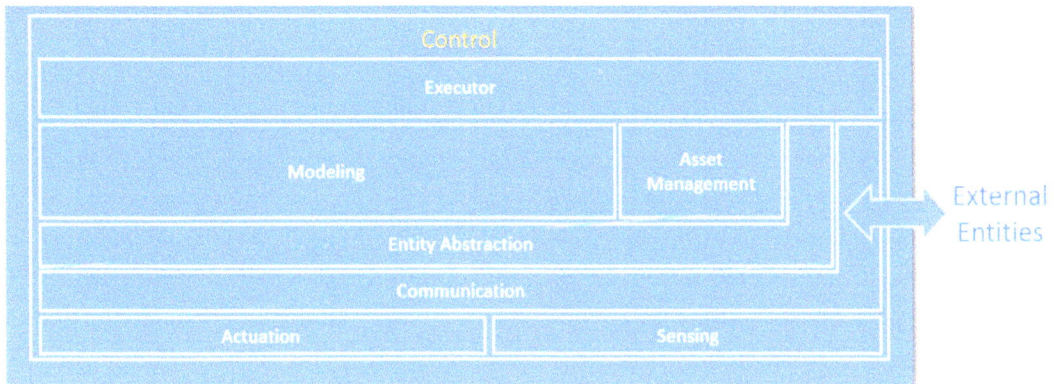

Figure 8.12: The Breakdown of the Control Domain

The components of the control domain include:

- **Sensing:** Reading sensor data from sensors.
- **Actuation:** Writes data and control signals to an actuator to enable actuation.
- **Communication:** The system that connects the sensors, actuators, controllers, coordinators, gateways, and other edge systems. May include a proprietary bus or a networked architecture (such as in a WSN) like mesh, point-to-point, hierarchical, or hubs and spokes. The communication function may use common APIs to abstract differences among systems.
- **Entity Abstraction:** Uses a virtual entity representation to allow sensor data from many types to be understood and processed as information.
- **Modeling:** Understanding the states, conditions, and behaviors of the systems under control. May include simple models and very complex models based on AI.
- **Asset Management:** Allows for operations management such as system onboarding, configuration, policy-based management, software and firmware updates, and other lifecycle management operations. Hierarchically under the executor so that higher-level policies can control lower-level actions.

303

- **Executor:** Executes control logic that may be straightforward or sophisticated. Straightforward control logic would be given directives that are static in nature based on predetermined thresholds. The sophisticated control logic may be based on cognitive learning capabilities with some level of autonomy within maximum and minimum thresholds.

These components allow for communications, ultimately, with external entities and a level of abstraction so that disparate systems can interact.

Above the Control Domain sits the Operations, Information, and Application Domains within the Functional Domain (see Figure 8.11). You can think of the Control Domain as management of operations within a single plant or facility and the Operations Domain as management of multiple facilities or an entire organization at all locations. It provides top-level policies and controls to be enforced downwards. Additionally, consider outsourced operations for small companies. The Operations Domain may exist in the service provider's network, and the Control Domain exists in the small business network so that multiple customers with similar management needs may be serviced more easily.

The Information Domain is about converting data to information. It is a primary intersection point between OT and IT. According to the IIAF, the Information Domain is *"a functional domain for managing and processing data. It represents the collection of functions for gathering data from various domains, most significantly from the control domain, and **transforming**, **persisting**, and **modeling or analyzing** those data to acquire **high-level intelligence** about the overall system. The data collection and analysis functions in this domain are complementary to those implemented in the control domain. In the control domain, these functions participate directly in the immediate control of the physical systems whereas in the information domain they are for **aiding decision-making, optimization of system-wide operations and improving the system models** over the long term."*

The Application Domain is the functional domain for the implementation of application logic. The logic implemented here should not be fine-grained, such as that at the Control Domain, but focused on realizing business goals or functions.

Finally, within the functional domains is the Business Domain. Here is where you would find Enterprise Resource Planning (ERP), Customer Relationship Management (CRM), Product Lifecycle Management (PLM), and Human Resource Management

(HRM). For example, ERP may be required to ensure that hardware is ordered for a new machine that will be implemented on a factory floor. Hence the need for integration of business units with the IT/OT operations.

Given that the functional domains comprise the foundation of IIAF for the wireless solutions administrator, the remaining components of the architecture framework will not be evaluated in detail here. The document can be freely downloaded at the IIC website at www.iiconsortuum.org in the Technical and Whitepapers section of the site. However, it is useful to summarize the remaining elements:

- **Trustworthiness:** This element is about ensuring that the WNS/IIoT implementation is as robust and capable as traditional OT systems. It includes the assurance of safety, security, resilience, reliability, and privacy. The various architectural models actually provided by vendor solutions (whether ZigBee, 802.15.4, DigiMesh, NB-IoT or any other) should be implemented with trustworthiness in mind. As we go forward into the fourth industrial revolution, we do not want to lose all of the gains we have achieved in operational efficiency and effectiveness over the past fifty years.
- **Scalability:** This element is provided by nature with many WSN solutions. As they are based on mesh and peer-to-peer architectures, the result is that scalability is available in the WSN. However, the wireless solutions administrator should be careful to ensure that scalability is also provided by the higher layers, such as in the operations, information, application and business domains.
- **Crosscutting Functions:** These components are those that may be required across functional domains while providing trustworthiness and scalability. Connectivity is a given requirement, but the CWSA should understand how this connectivity occurs in the selected solution. Distributed Data Management allows for the collected data to be made available where it is needed and when it is needed. Industrial Analytics allows for centralized or distributed reporting and analysis of operations. Finally, Intelligent & Resilient Control ensures that control functions are available to each functional domain as required.

Three-Tier Architecture and the IIAF

A common architectural model, also specified within IIAF, that is a practical implementation model is the three-tier architecture shown in Figure 8.13.

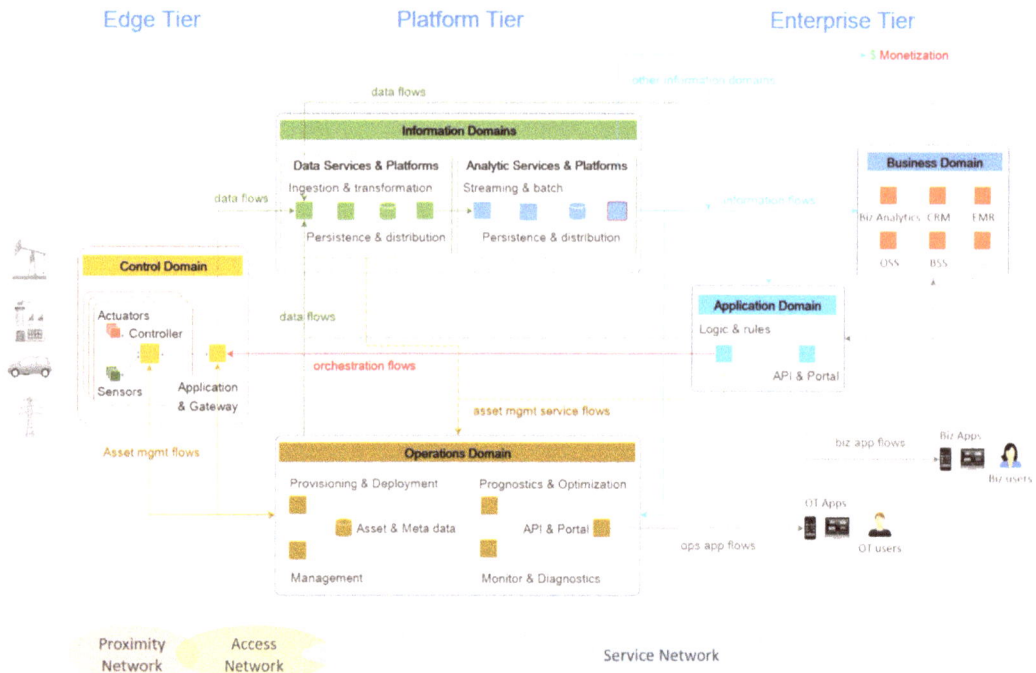

Figure 8.13: Three-Tier Architecture Model (Source: IIAF)

This model includes a proximity network, an access network, and the service network. In Figure 8.13, the service network spans the platform tier and the enterprise tier. The access network spans the edge tier and the platform tier. The edge tier and, precisely, the proximity network can be thought of as the actual WSN in this model. The access and service networks (the platform and enterprise tiers) provide the supporting operations, monitoring, analysis, and control. Notice that the sensors and actuators exist in the edge tier as do the gateways or coordinators that provide access to the supporting services.

It is also worth noting that the enterprise tier, shown in Figure 8.13, is often provided entirely within cloud services today. It may be that an organization runs their own enterprise tier in a cloud service such as Amazon Web Services (AWS) or Microsoft Azure or it may be an implementation depending on the WSN vendor's cloud service. Many vendors now provide such services so that the implementation of servers and primary application code is already provided through their cloud. We have seen much movement towards the cloud in the enterprise wireless LAN market, and we now see this as well in the WSN market.

It is important to consider the impact of cloud-based WSN management and this is illustrated in Figure 8.14. In the image, the cloud would be synonymous with the enterprise tier, the on-premises component would be synonymous with the platform tier, and the on-sensor component would be synonymous with the edge tier.

Cloud-Controlled WSNs

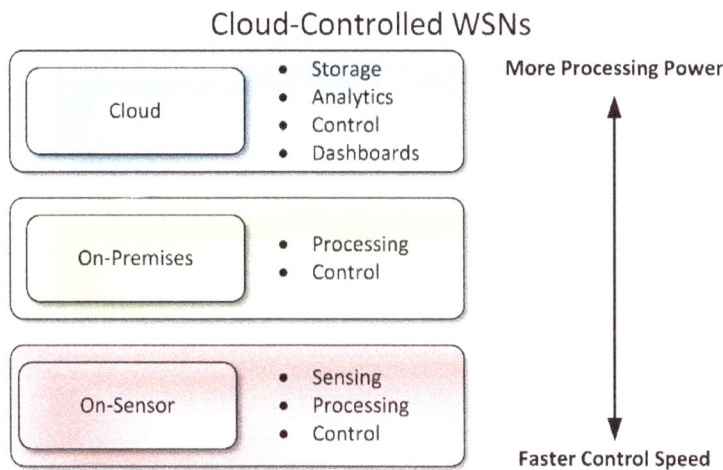

Figure 8.14 Considerations for Cloud-Controlled WSNs

An important revelation from Figure 8.14 is that the cloud can provide much more processing power, but it also results in slower control speed due to latency in communications with the cloud. As a wireless solutions administrator, you may be called upon to evaluate a cloud-based solution and you should determine what is provided in the cloud, on-premises, and on-sensor in the decision-making process. Processing at the sensors is far less capable, but can give near-instant responses and is particularly beneficial when actuators are involved. However, the cloud services can provide much more processing power for machine learning and AI so that long-term decision making is more efficient and accurate. Generally speaking, absolute decisions are made closer to the sensors, and business logic decisions or variable decisions can be made on-premises or in the cloud.

The remaining content in this section will describe two general architectures. These include the general architectures of a hierarchical architecture and a mesh architecture. For specifics on ZigBee and LoRaWAN architectures, see Chapter 7.

Hierarchical Architecture

In a hierarchical architecture, the wireless sensors connect to a specific node that is either the gateway onto the network at large or is a router providing connectivity to the gateway. A simple hierarchical architecture connects all sensors to a single gateway and is sufficient for WSNs in smaller facilities or small-scale outdoor deployments. A complex hierarchical architecture may use clusters with all sensors in a cluster connecting to a cluster head, which is then connected to the gateway. Other terms for gateway include base station, coordinator, hub, and controller.

The simple hierarchical architecture is shown in Figure 8.14. The complex hierarchical architecture is shown in Figure 8.15. Variations on these architectures exist, and you should check your vendor literature to see the options available for building a WSN with their solution. Many vendors support mesh or hierarchical implementations depending on your requirements.

The terminology used by different vendors will vary, but the concepts remain the same. A hierarchical architecture generally depends on all sensors connecting to a specific kind of device, which is, in turn, either connected to the rest of the business network or connected to a gateway that is connected to the network. Of course, the gateway may connect directly to a WAN solution, such as a low-power WAN (LPWAN) for connectivity back to the enterprise network or a cloud service provider.

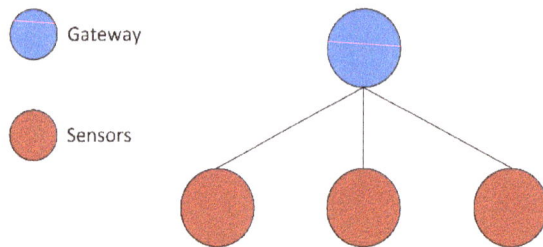

Figure 8.14: Simple Hierarchical Architecture Illustrated

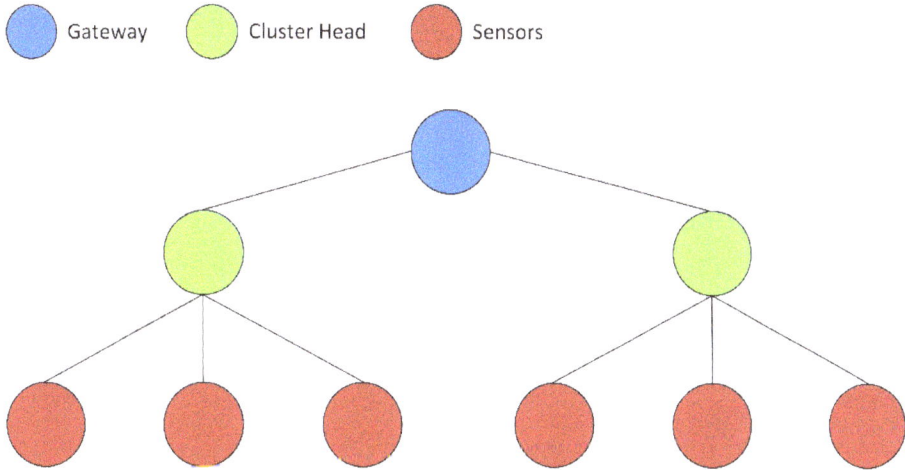

Figure 8.15: Complex Hierarchical Architecture Illustrated

Mesh Architecture

Unlike the hierarchical architecture, in a mesh architecture, multiple nodes can use multiple routes to reach the gateway. Mesh architectures support full mesh and partial mesh implementations. Some vendors support both full and partial models, and other vendors only support one model. Figure 8.16 illustrates a full mesh and a partial mesh architecture. The full mesh is on the left, and the partial mesh is on the right.

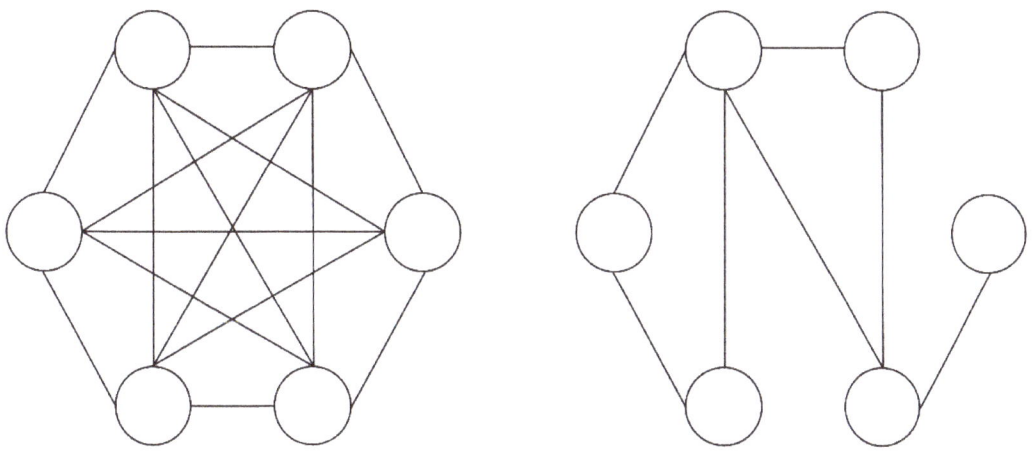

Figure 8.16: Full and Partial Mesh Architectures

In a full mesh architecture, every node connects to every other node. In a partial mesh architecture, some nodes have only one connection to the mesh, others have multiple, but every node is not connected with every node.

Planning a WSN

Planning a WSN is a similar process to any other wireless solution and includes defining requirements and constraints, selecting appropriate solutions, and planning for ongoing support (automation, integration, monitoring, and management). This final section of the chapter will outline the considerations for WSNs in these areas. Chapters 1 through 3 provided a more in-depth overview of these considerations as they apply to all wireless solutions.

Identify and Document WSN Requirements

To determine the basic requirements for your WSN, consider the following questions:

- What kind of sensors are required?
- Does a single vendor provide all of these sensors?
- If multiple vendors must be used, are sensors available that interoperate directly using the same technology (ZigBee, LoRaWAN, 802.15.4, etc.)?
- Does the solution require a pre-built cloud service for monitoring and control, or will local servers be utilized?
- Is it a single-site deployment or a multi-site deployment?
- What is the size of the area requiring sensor connectivity?
- Will power be available at the sensor locations or should they be battery powered?
- Will some sensors have to be custom-built to meet your needs? If so, do you have the personnel on staff to do this or must it be outsourced?

In specific environments, additional questions may need to be answered, but this list should be addressed in nearly all WSN planning projects.

Identify and Document WSN Constraints

To determine constraints related to your WSN, consider the following questions:

- What are wireless technologies already in use and in what frequency bands do they operate?
- Are specific security regulations, guidelines, or policies imposed upon you?

- Do existing sensors have to be integrated with the new WSN?
- Does the WSN have to integrate with existing ERP, CRM, HRM, or other systems?
- Does the WSN need to support integrations with SCADA or DCS solutions already in place?

> Supervisory Control and Data Acquisition (SCADA) and Distributed Control System (DCS) are existing OT solutions for monitoring, operating, and controlling many systems in industrial and other organizations. They are used in manufacturing, oil and gas, and even transportation systems. If they exist, they may need to be considered in your WSN deployment.

In specific environments, additional questions may need to be answered, but this list should be addressed in most WSN planning projects.

Select Appropriate WSN Solutions

The next step, once requirements and constraints are defined, which may involve use case development, is to select a WSN solution. This may involve assembling wireless sensors from varying vendors that support a shared communication protocol and management system, or you may be able to find a single vendor that can provide them all. Figure 8.17 shows the categories of sensors provided by RADIO Bridge (www.radiobridge.com).

WIRELESS PROXIMITY SENSORS

Wireless proximity sensors detect contact between two wires, proximity detection with a magnet, range with an ultrasonic signal, and more

WIRELESS INDUSTRIAL SENSORS

Wireless sensors for industrial standards such as 4-20mA current loops, DIO, and more.

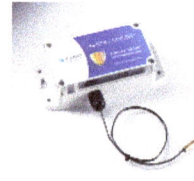

WIRELESS TEMPERATURE SENSORS

Wireless sensors to detect temperature thresholds or rates of rise

WIRELESS MOVEMENT SENSORS

Wireless movement sensors to detect acceleration, tilt, or other motion

WIRELESS LIQUID SENSORS

Wireless liquid sensors for detecting the presence of water, fuel, or other liquids

WIRELESS PUSH BUTTONS

Wireless push buttons that transmit on a button press event

WIRELESS AIR SENSORS

Wireless air sensors to detect ambient air temperature, humidity, or harmful gases

WIRELESS OPTICAL SENSORS

Wireless sensors that detect optical events such as ambient light detection or optical beam crossing

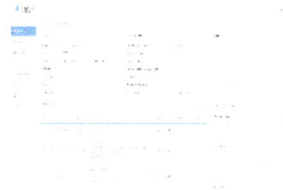

SOFTWARE APPLICATIONS

Software applications for provisioning, monitoring, and configuration of Radio Bridge sensors

Figure 8.17: RADIO Bridge – A Single Vendor's Offerings

As another example, Monnit (www.monnit.com) provides wireless sensors that operate at 433, 868, and 900 MHz as well as gateways for Ethernet, LTE, 3G, and 2G (though many cellular providers are discontinuing this in 2020). On the wireless link, Monnit gateways and sensors use the proprietary Alta protocol (FHSS modulation) for communications with their newest line of devices. They also provide what they call

Standard (Gen 1) devices that use the same frequency ranges, but implement DSSS modulation. However, they provide all of the following sensor types:

- Temperature Sensors: thermistor and thermocouple
- Humidity Sensors
- Fluid Sensors: water rope, water puck, and industrial spot leak detectors
- Proximity and Motion Sensors
- Carbon Monoxide Sensors
- Dry Contact Sensors
- Accelerometers
- Voltage Meters
- AC Current Meters
- Voltage Detection Sensors
- Pulse Counters (actuation counts)
- Light Sensors
- Resistance Sensors
- Pressure Sensors
- Activity Sensors
- Asset Sensors
- Vehicle Detection/Counter Sensors
- Air Pressure Sensors
- Air Quality Sensors
- Air Velocity Sensors
- Button Touch Sensors
- Seat Occupancy Sensors
- Strain Sensors
- Compass Sensors (returns orientation to magnetic north)

Remember, when considering a WSN solution provider, be sure to learn about the protocols they are using and answer the all-important question: Are you willing to be attached to a vendor that does not use standard (open) protocols? In some cases, it poses no problems. In others, it prevents doing business with that provider. Just be sure you know what you're purchasing.

An additional factor in selecting a WSN provider is the management methods. Some are managed only through the cloud. Others offer cloud management and local management. Still others offer only local management. Additionally, consider whether the vendor offers APIs (application programming interfaces) for integration with other systems. If they offer rich APIs, even a proprietary solution may be able to grow and expand beyond that vendor. For example, Monnit does provide the iMonnit API, which is a RESTful API. They also offer a Webhook API. Such APIs are discussed further in Chapter 12.

Plan for WSN Automation, Integration, Monitoring and Management

When it comes to automation, integration, and monitoring, it is important to evaluate the APIs that are available. A WSN solution that does not expose the data through APIs must be manually integrated through database access or conversion, which may require periodic exports of data and imports into other databases. With APIs, such tedious work is seldom required.

As a brief example, the following RESTful request of the iMonnit cloud may be used to retrieve readings from a temperature sensor:

```
https://www.imonnit.com/xml/SensorDataMessages/Z3Vlc3Q6cGFzc3dvcmQ=?
sensorID=101&fromDate=2011/01/01 6:22:14 PM&toDate=2011/01/02
6:22:14 PM
```

The following is an example partial result set, in XML structure, from the preceding request (notice the temperatures of 86- and 122-degrees Fahrenheit):

```
<APIDataMessage SensorID="101" MessageDate="1/1/2011 6:36:00 PM"
State="0" SignalStrength="-36" Voltage="3.1" Battery="100" Data="30"
DisplayData="86° F" PlotValue="86"
MetNotificationRequirements="False" GatewayID="1234"/>

<APIDataMessage  SensorID="101" MessageDate="1/1/2011 6:34:33 PM"
State="0" SignalStrength="-36" Voltage="3.1" Battery="100" Data="50"
DisplayData="122° F" PlotValue="122"
MetNotificationRequirements="False" GatewayID="1234"/>
```

Additionally, it is important to discover the frequency bands used for the solution so that you can acquire appropriate tools for monitoring and troubleshooting. For example, a 2.4 GHz spectrum analyzer will not assist in locating sources of interference related to a 900 MHz WSN. Generally speaking, the handheld spectrum analyzers that support from 9 kHz to 20 GHz (or more) cost several thousand dollars. Dedicated spectrum analyzers that operate in specific ranges, like the RF Explorer WSUB1G, which works from 240-960 MHz, can be acquired for under $200 US. However, the resolution (level of detail) will not be as high. Figure 8.18 shows the RF Explorer WSUB1G, which is sufficient for basic interference location.

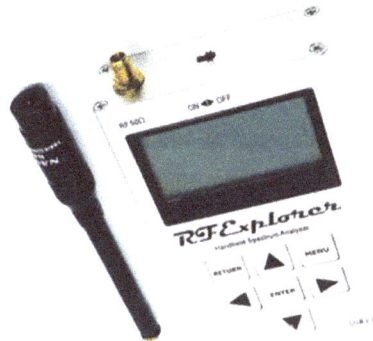

Figure 8.18: RF Explorer WSUB1G Spectrum Analyzer (240-960 MHz)

Another part of integration is ensuring proper frequency coordination. You must be sure that the WSN will not interfere with other wireless solutions and that the other wireless solutions in your coverage area will not interfere with it. If you have an existing 2.4 GHz WLAN, for example, the WSN may function better on 900 MHz or another frequency band. However, it is possible for 2.4 GHz WSNs, such as ZigBee to function alongside 2.4 GHz WLANs and other 2.4 GHz solutions. In the end, it will be a factor of airtime required by all of the solutions in place.

Finally, you should ensure that the management solution will work well for your environment. If it is cloud-based, it is likely that it will work well for any organization; however, if latency is an important factor in your WSN, you must ensure that proper local controls are available for actuators and alerts. If it is a local management solution, what are the requirements and how will you integrate it with your current IT and OT operations? This will be an important part of handover once the WSN is implemented.

Chapter Summary

This chapter may have seemed rather large based on the objectives covered. However, the concept of a WSN is a large part of IoT, particularly in industrial networks. It is a significant part of smart cities, smart agro, smart homes, smart offices, and Industry 4.0. As you can see, it actually spreads across many of the CWSA objectives and it provides an excellent case study in all of the considerations that must be made when planning a wireless network of any kind. Professional-level certifications will be offered by CWNP covering WSNs in even more detail. The Certified Wireless Connectivity Professional (CWCP) will address the complexities of IoT in its many forms, which is inclusive of IIoT and WSNs.

Review Questions

1. In a WSAN, what device can interact with other machinery to cause changes?
    a. Sensor
    b. Gateway
    c. Actuator
    d. Coordinator

2. What is the process of capturing energy from the environment called?
    a. Sensing
    b. Actuating
    c. Energy Harvesting
    d. Radiating

3. What might be used in a temperature sensor?
    a. Thermistor
    b. Ultrasonic transducer
    c. Water rope
    d. Strain gauge

4. Where is the control of sensors and actuators in the IIAF model?
    a. Information Domain
    b. Application Domain
    c. Operations Domain
    d. Control Domain

5. In the three-tier architecture, where are sensors placed?
    a. Edge tier
    b. Platform tier
    c. Enterprise tier
    d. None of these

6. When you need more processing power, where is the processing likely to occur?
    a. On-sensor
    b. On the radio waves
    c. In the gateway
    d. In the cloud

7. When clustering is used in WSN architectures, what device is added?
    a. Sensor
    b. Gateway
    c. Cluster head
    d. None of these

8. What can be used to connect a non-wireless sensor to a WSN?
    a. Coordinator
    b. Access point
    c. Connector node
    d. None of these

9. Which of the following may be used to detect motion?
    a. Thermocouple
    b. Accelerometer
    c. RTD
    d. Water rope

10. What is defined as a component that is able to detect variations in a process or measurable value?
    a. Gateway
    b. Coordinator
    c. Actuator
    d. Measurement instrument

Review Answers

1. The correct answer is **C**. An actuator can interact with other machinery or devices. Sensors only monitor them.
2. The correct answer is **C**. Energy harvesting consumes energy from the environment using solar panels, wind, and even vibration.
3. The correct answer is **A**. A thermistor is a temperature sensor, as are RTDs and thermocouples.
4. The correct answer is **D**. The Control Domain includes the control of sensors and actuators.
5. The correct answer is **A**. Sensors are located in the Edge tier on the proximity network.
6. The correct answer is **D**. Today, most complex processing takes place in the cloud.
7. The correct answer is **C**. When clusters are used, a cluster head acts as the controller of the cluster.
8. The correct answer is **C**. A connector node can connect an otherwise non-wireless sensor to the WSN.
9. The correct answer is **B**. An accelerometer is often used in motion detecting sensors.
10. The correct answer is **D**. A measurement instrument, the precursor to modern sensors, is so defined.

Chapter 9: The Internet of Things (IoT)

Objectives Covered:

1.3   Define wireless network types

3.8   Understand the basic features and capabilities of common wireless solutions and plan for their implementation

The Internet of Things (IoT) was added as a term to the Merriam-Webster dictionary back in 2017. This goes to show that IoT is a term that is expected to be used for a long time and that it has crossed the pre-requisite of it having been used for quite some time, gained a widespread presence, and has been written about in different forms from different sources. That's why it was added to the dictionary, and here within lies the argument towards defining what the Internet of Things means.

This chapter will define IoT and explain the many solutions available for its implementation. The solution range from theoretical models to actual protocols implements. While IoT has been addressed many times already throughout this book, this chapter will bring all of the various concepts together.

Internet of Things (IoT) Defined

The IoT has gained popularity ever since it was first coined (more about that in the following section) to an extent that it became one of the biggest buzzwords without necessarily being referred to with the same meaning. In addition, many sources discuss IoT from different perspectives including its hardware, software, security, communication, use-cases, architecture, business outcomes, forecast, and others that also show IoT in a different perspective.

The Merriam-Webster itself approaches IoT to define it practically as "something that's promising to make all kinds of tedious tasks go down more smoothly with information being sent to and received from household objects and devices—say, your bedroom fan or your toaster oven—using the Internet." [1]

Dissecting the term, we can see that it has to do with connectivity – the Internet – and connected things, or connected objects (CO). But how does that make it different than the Internet itself? Is a personal computer connected to the Internet considered an IoT CO? What about mobiles? IoT has certainly been around for a while and has grown fast enough to create this blur, which might make it difficult to define.

As put by B.Russell and D.Van Duren in the "Practical Internet of Things Security,"[2], the Internet of Things is far more than just mobile or computer connectivity, and as a technical book, we can look into how the Institute of Electrical and Electronics Engineers (IEEE) as well as the International Telecommunications Union (ITU), who set technical communication standards, define IoT:

"The ITU's member-approved definition defines the IoT as "A global infrastructure for the information society, enabling advanced services by interconnecting (physical and virtual) things based on existing and evolving, interoperable information and communication technologies."

The IEEE's small environment description of the IoT is "An IoT is a network that connects uniquely identifiable "things" to the Internet. The "things" have sensing/actuation and potential programmability capabilities. Through the exploitation of the unique identification and sensing, information about the "thing" can be collected and the state of the "thing" can be changed from anywhere, anytime, by anything."

The IEEE's large environment scenario describes the IoT as "Internet of Things envisions a self-configuring, adaptive, complex network that interconnects things to the Internet through the use of standard communication protocols. The interconnected things have physical or virtual representation in the digital world, sensing/actuation capability, a programmability feature, and are uniquely identifiable. The representation contains information including the thing's identity, status, location, or any other business, social or privately relevant information. The things offer services, with or without human intervention, through the exploitation of unique identification, data capture and communication, and actuation capability. The service is exploited through the use of intelligent interfaces and is made available anywhere, anytime, and for anything taking security into consideration."

The ITU refines the definitions for IoT further in their ITU Internet Reports 2005:

The Internet of Things, as being a new dimension added to the world of ICT which evolves "anytime, any place connectivity for anyone," to "connectivity for anything. Connections will multiply and create an entirely new dynamic network of networks – an Internet of Things." [3]

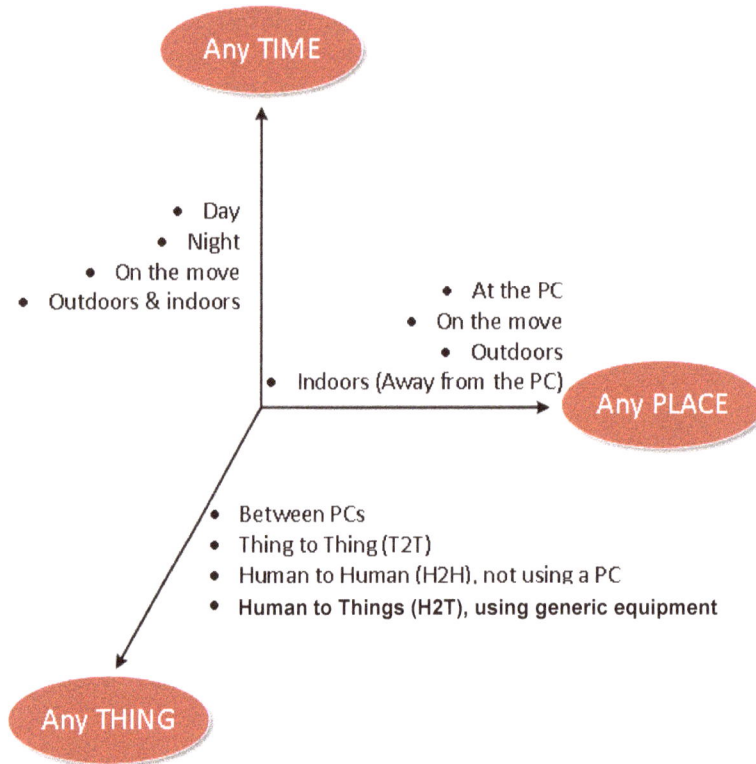

Figure 9.1: Connection New Dimensions Re-drawing - based on the ITU's adaptation from the Nomura Research Institute

These definitions overlap and complement each other in including anything physical or logical that can be connected to other things over a diversely connected world.

Being discussed and cited in multiple references, IoT is also referred to as the "The Internet of Everything," "The Internet of Relevant Things," and "The Internet of Useful Things."

History of IoT and Its Definition Revisited

As you can see, several definitions of the Internet of Things render defining the term IoT a bit vague. However, this is not the only indefinite aspect of the term! Even the numbers surrounding statistics about the growth of IoT or the total number of connected devices are varied between different IoT reference books, research institutes, IoT vendors or evangelists. To summarize all market research findings and references, the total number of connected things is expected to be anywhere between 20 and 50 billion devices by 2020. The rapid evolution of IoT might have contributed to this variation over time of the reported growth.

To help give more precision to put some tabs on IoT, at least to the definition itself, we have to go back in history as a frame of reference to see how IoT was shaped and how it, in turn, is helping shape the industry.

"The term IoT can most likely be attributed to Kevin Ashton in 1997 with his work at Proctor and Gamble using RFID tags to manage supply chains. The work brought him to MIT in 1999 where he and a group of like-minded individuals started the Auto-ID center research consortium (for more information, visit http://www.smithsonianmag.com/innovation/kevin-ashton-describes-the-internet-of-things-180953749/)." [4]

IoT has evolved a lot since being a catchy definition for an RFID system to a large ecosystem and industry that will have a potential impact sized between $4 and $11 trillion on the markets. [6]

Industry 4.0 and the History of the Internet of Things

Technology is acting as the catalyst to change the way we do things around to the extent that it's affecting the economy, society, businesses, and even individuals. We are experiencing shifts across all business models, including those in enterprise, municipal, industrial and consumer markets alike.

Revolutions have occurred throughout the system that has triggered deep changes in economic systems and social structures. In the historical frame-of-reference, the industrial revolutions can be considered very recent, with the first industrial revolution spanned from around the mid of the 18th century and for about a century. Triggered by the construction of railroads and the invention steam engine, it ushered in mechanical production. The second industrial revolution started in the late 19th century and into the early 20th century made mass production possible fostered by the advent of electricity and the assembly line. The third industrial revolution began in the 1960s, usually called the computer or digital revolution because it was catalyzed by the development of semiconductors, mainframe computing (1960s), personal computing (1970s and 1980s), and the Internet (1990s). Mindful of the various definitions, today we are at the beginning of a 4th industrial revolution which began at the turn of the

21st century and builds on the digital revolution. Characterized by a much more ubiquitous and mobile Internet, smaller and more powerful sensors that are cheaper, and driven by AI and ML. Digital technology has fundamentally transitioned from being driven by its hardware, then software and networking cores to the more sophisticated and integrated market-driven platforms, and are, as a result, transforming societies and global economy with a break from the 3rd industrial revolution. Industry 4.0 is a term coined to lead discussions describing how this will revolutionize the organization of global value chains. [7]

At the core of it, the 4th industrial revolution, you will see the smart connected machines and systems. This is our IoT building block, whith the culmination of technology in all the previous industrial revolutions gearing us towards it. So IoT aims to help as a building block of digitization, or the evolution of the result of the third Industrial Revolution, the Internet, get into its next phase.

What milestones have marked the birth of the Internet of Things?

With the historical-frame-of-reference that we have started with, some key dates provided a contribution to defining what IoT is. Many discussions and forums on the Internet argue that IoT has been developing for some time following the 3rd industrial revolution. There are many milestones and dates that are significant pivot points in the history of IoT that have fast-tracked hardware changes, software introductions, connectivity, and the birth of whole new platforms that in turn have bolstered the advancement of IoT.

Consider the following dates [8]:

- 1999: The Internet of Things Coined: In an attempt to link the new idea of RFID in Proctor &Gamble's supply chain to the hot topic of the Internet back then, Kevin Ashton coined the phrase "Internet of Things" in his presentation "to get executive attention" (https://www.rfidjournal.com/articles/view?4986)
- 2005: International Telecommunications Union (ITU) published its first report on the Internet of Things that was "specially prepared for the second phase of the World Summit on the Information Society (WSIS)" during that year (ITU The Internet of Things)
- 2008: Marking the Birth of IoT: Different vendors, entities and groups label 2008 as the year IoT was born. It was an inflection point in history when the number of devices connected to the Internet exceeded the human population.

2008: IoT is made official: The European's Union recognition of the IoT is embodied with the first IoT conference held in Zurich, Switzerland.

One More Comment on the Definition of IoT

A more recent and general definition from [9] specifies the IoT as "a vision of a world in which billions of objects with embedded intelligence, communication means, and sensing and actuation capabilities will connect over IP networks." This is all happening and will continue to evolve due to the interconnection of seemingly disjoint intranets with strong horizontal software capabilities. Evolution is a very important keyword here. It's the transition and advancement of the 3rd industrial revolution that has made faster evolution of things: in size, capabilities, connectivity, and efficiency.

To come up with a simple definition for IoT, we can identify IoT as enabling the utilization and transmission of data from devices (sensors, controllers, actuators) that have been enabled with connectivity with constrained requirements (size, processing, memory) and that require efficient utilization of resources to communicate with other devices, humans and systems, in order to come up with informative decisions supporting business objectives. See, it's simple.

IoT Verticals

Many other verticals and use cases for IoT can exist. An analogy from a paper that tried to define the evolution of personal computing in 1991 describes how it will advance to integrate in our everyday life that we will stop thinking about it as a separate term.

"IoT applications are boundless. Volumes could be written today about what is already deployed and what is currently being planned. The following are just a few examples of how we are leveraging the IoT." [2]

The IoT is being utilized in different individuals and by different businesses. The goals of using IoT also vary between those different utilizers. Goals can range from optimizing operations, reducing risk, enhancing a service, developing a new product, and enhancing outputs to improving personal care. Those goals still apply to enterprises and individuals alike. For example, in optimizing operations, an individual might leverage his home assistant platform to plan his day efficiently. The same goal could be set for an industrial factory with the aim of reducing downtime and saving costs. These are the most general goal categories, but we will be discussing vertical-specific use-cases and their goals in the next sections.

We don't mean to add to the uncertainty of IoT given its broad definition. Many of the following verticals and their use-cases overlap in multiple ways. For example, a use-case for tracking school students in school buses could be classified under education and under transportation at the same time. In other cases, a specific vertical or industry category of use-cases can be discussed in a more specialized approach. For example, IoT in transportation can be split across fleet management, whose use-cases overlap with some of Industrial IoT, mass transit systems, and the connected car use-cases. The connected car can use a whole dedicated section of its own!

Industrial IoT

"Industrial IoT (IIoT) is one of the fastest and largest segments in the overall IoT space by the number of connected things and the value those services bring to manufacturing and factory automation."[4]

Having tackled the definition and history of IoT, we already had to show how IoT is one of the major building blocks for digital transformation and the fourth industrial revolution. Thus, it might be very natural to start off with the first business vertical that has involuntarily employed IoT and sparked the expansion and advancement of IoT onto other verticals.

Two important landscapes in industries: Information Technology (IT) and Operations Technology (OT). OT had mainly dominated industrial environments with technology mainly comprised of hardware and software tools to do physical monitoring and to output data around fields including but not limited to metrics, uptime, real-time data monitoring, systems response, and system/environment safety. IT, on the other hand, is focused on technical services, data delivery, systems security and logical segmentation of the industrial setup.

IoT has come a long way since it was a term coined to turn heads to what then was an RFID integration into the supply chain. Advances in processing, memory, data storage, connectivity and have pushed OT and IT forward, but IoT has evolved as well to bridge OT and IT together.

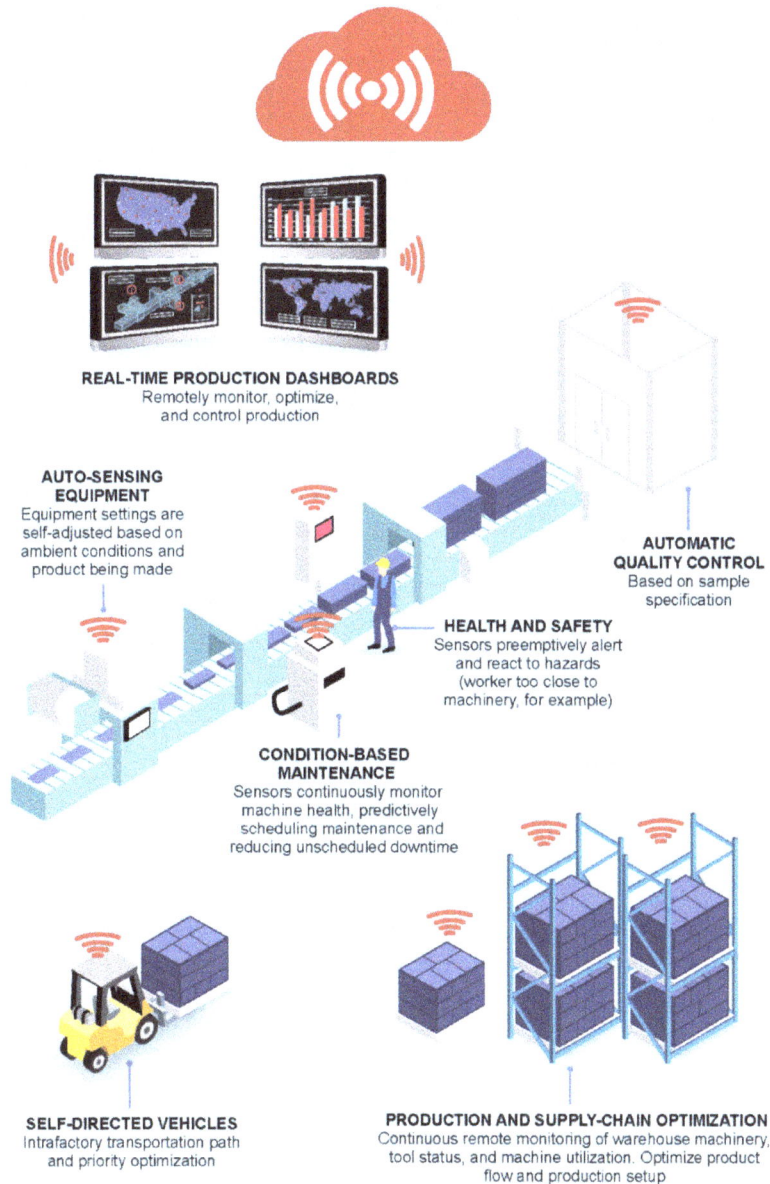

Figure 9.2: The Internet of Things: Mapping the value beyond the hype - McKinsey Global Institute Analysis

IIoT utilizes ruggedized devices and systems that are usually constructed for operating in harsh environments that are typically uncontrolled. Being in IIoT, specifically with manufacturing and supply chain means that ruggedizing IoT comes with the main goal of running systems with high mean time between failures (MTBF). Bearing in mind that they are running over or within the aforementioned OT and IT, IoT can often be found in the industrial verticals to communicate over a different range of wired and wireless physical sensor layers and network and communication protocols. Real-time, low-latency communications are essential for real-time decisions. Another characteristic of IIoT is that it's often deployed in siloed setups without connectivity to the Internet and with the core objective of specific operations such as running critical machine control feedback and monitoring.

The IIoT category covers many industries including automotive, bottling, food processing, oil, gas, manufacturing, mining, paper, petrochemical, pharmaceutical, power generation, power distribution, pulp, transportation, water treatment, and more.

This diversity means that there isn't a single go-to IoT solution but a variety of solutions that could utilize different sensors, communications protocols, services, and application. Each could require its own design approach, security standards, certification program, regulatory restrictions, health & safety standards, performance monitoring, and construction standards.

We can go deeper to classify IIoT in more specific categories, each with its own characteristics and requirements.

Manufacturing has driven a great amount of the industrial IoT use-cases. Most of us can envision robotic systems, assembly lines, manufacturing plants, and even manufacturing design and operation engineers getting to work on their enhanced production plans. Different types of connected sensors and actuators are driving these systems with the general objectives of controlling costs and improving efficiency.

According to an IoT study run by the McKinsey Global Institute in 2015, the total economic impact of IoT in industrial worksites and factories in 2025 will be $1.3T-$4.6T. The top areas they identified include operations optimization, predictive maintenance, inventory optimization, and health and safety.

 "After decades of squeezing costs out of production systems and the supply chain, manufacturers are recognizing that further cost containment may only impede customer service and open the door to competition." [5]

Economic changes have shifted industry models due to global competition, changing the primary focus toward innovation and improved business models. This change is primarily being led by digitization and IoT. With the production system no longer a competitive advantage, IoT data is being used to boost manufacturing performance. Connectivity, speed, accessibility, and anchoring are the 4 main areas IIoT is expected to enhance production systems, according to a more recent study by McKinsey. [10]

- Connectivity: Originally isolated, production systems are now, or eventually will be, connected over various communication platforms, intranets, and the Internet with the distributed functions, such as automation and control, being given the proper communication channels into management and monitoring applications, whether local or over the cloud.
- Speed: Production systems are looking at faster management of day-to-day operation, faster capability building with focused employee training, and faster improvements in the system itself by detecting and optimizing production gaps benchmarked against the company itself (over the intranet) or against global benchmarks (over the Internet).
- Accessibility: IoT connectivity will enable platforms that will help different employees from the workforce to tap into real-time data and analytics through role-based, secure and simple dashboards, eventually feeding back into the speed general use-case of IoT in manufacturing. Accessibility can also be extended beyond the network of the same manufacturer itself towards external suppliers, experts, and other companies which will in turn help optimize production and give feedback to the whole supply-chain cycle.
- Anchoring: A word that might carry the opposite meaning of what IoT is intended for in manufacturing is defined as "anchoring the production system in the organization's psyche" meaning to gear the company towards overcoming the most critical challenge they struggle with today: "sustaining change, so that the organization improves continually." [10] This is meant to leverage IoT to help adapt to a manufacturing company's vision as it constantly changing rate by adapting production, providing employee collaboration, and giving production visibility on where and how value is being created in the production cycle.

Oil & Gas

Oil & gas are considered some of the most critical resources in the world, not only to keep it running as a source of energy but also feeding back into the processing of basic manufacturing materials (e.g. plastics, paint among others).

The main focus of utilizing IoT in oil & gas companies is to control production (reduce cost, improve production efficiency/speed, better utilize facilities) while maintaining improved working conditions pertaining to health and safety in dangerous environments.

When we discuss oil and gas as an industry, it's not just limited to specific areas like oil fields or processing factories only. This usually covers all the locations that the value chain through which oil and gas are transformed from primary resources to final products. Oil & gas industrial locations extend from rigs for exploration and resource extraction, to offshore shipping, to factories for processing/refining, to pipelines for distribution, and to other distribution/selling networks. [5]

IIoT in oil & gas use-cases include:

- The connected oil field: Usually remote and isolated, a robust and self-contained communications infrastructure based on wireless technology is what supports use-cases for situational awareness for the operation of the drilling system, communications with head office locations, and health, safety, and environmental monitoring systems. In addition, use-cases for communications for entertainment are also required for employees stationed in remote oil fields for long periods. [2]
- The connected pipeline: Sharing some use-cases with the connected oil field, IoT use-cases here are usually utilized to overcome operational challenges related to the harsh environments, vast areas and long distances, network isolation, leaks, natural (earthquakes, landslides, storms) and human-caused disaster (vandalism, theft).
- The connected refinery: As big complexes with different types of work areas, buildings, and people working, the most important at refineries is to guarantee a 24x7 operation. Process control and monitoring, optimization, maintenance and repairs are constantly required to maintain an effective, efficient and safe environment. A communication system that enables these control, management and safety systems should be fast and reliable. [2]

For all of the above different characteristics and challenges of oil & gas locations, IoT is used to ensure:

- Personnel safety with real-time tracking and alerting systems
- Security, pipeline protection, health & safety through surveillance and video analytics
- Ruggedized Wireless for communication, data processing, and location services
- Video/voice-grade wireless communications for remote expert support
- Security and environment protection with Leakage, Intrusion and deformation detection
- Mobility with vehicle/fleet connectivity for mesh and long-haul communications
- Command and control center communications and monitoring
- Industrial control sensors
- Equipment location tracking
- Process optimization for fleet identification, management, and loading

Human IoT and Wearables

Some IoT systems are deployed for use-cases in the context of the human body as the main environment where the applications fall into two broad-categories that adjust the human habits. These are mainly focused on enhancing productivity and improving health. The contrast here with other IoT applications is that we are not turning off an Oil & Gas valve, but information and reminders are provided so insightful decisions could be made about exercising, the general state of health (sleep, heartbeats, movement), exercising and other productivity and health-related habits.

Wearables cover any devices that can come in touch with the human body (strapped or attached) that collects the individual's state, communicates information and alerts and triggers, or performs other functions on or around the individual. But IoT for humans doesn't stop at wearables.

Soon, IoT devices for human health can include implantables, ingestibles, and injectables, such as nanobots that can clear arteries or help detect early-stage cancer. Once these devices have cleared clinical trials and are properly certified/approved, we will see big adoption with significant impact on human health.

The largest source of value would be using IoT devices to monitor health and treat illness ($170 billion to $1.1 trillion per year). The value would arise from improving quality of life and extending healthy life spans for patients with chronic illnesses, and reducing cost of treatment. The second-largest source of value for humans would be improved wellness—using data generated by fitness bands or other wearables to track and modify diet and exercise routines. [6]

The use-cases for human IoT devices would include:

- Wearables: Traditionally mobile phones are the prior form of IoT that have extended to smartwatches and fitness tracker wearables. In brief, these could be any device that is designed to be worn or carried around. These are good examples of devices that are at the core networked sensors that can detect acceleration, heart rate, temperature, location, among many other data.
- Implantables, injectables, and ingestibles: highly overlapping with IoT in healthcare, these are devices that can be inserted, injected, or swallowed to enter the human body in order to enable real-time monitoring of body vitals (blood pressure, heart rate, glucose levels) and body organs. Examples of these could be cameras, pills, and other sensors. Other IoT that could be embedded in the human body to allow for physical or logical access and multi-factor authentication through RFID/Wireless/NFC communication.
- Non-wearable devices: These are IoT devices that collect and transmit data while detached periodically but are not attached continuously, such as Bluetooth enabled pulse oximeters or Wi-Fi-enabled scales.

Other wearable devices are now extending to different structures of sensors such as those embedded into glasses, clothes, wearable goggles for Virtual Reality/Augmented Reality, and advancement in the sensors' hardware capabilities are enabling those wearables to achieve attractive use-cases such as in education, entertainment, manufacturing, and also health. Apple has already rolled out its electrocardiogram (ECG or EKG) by the beginning of 2019, and it has already managed to effectively diagnose different users with positive symptoms so they can get preventive treatment.

Healthcare

According to McKinsey the total economic impact of IoT in healthcare in 2025 will be $170B-$1.6T, making it one of the top three verticals for IoT solutions.

Adoption of IoT in healthcare lags behind other industries, specifically in adopting technologies that engage end-user, the healthcare patients, in order to deliver a better user experience and retain customers

The use-case areas include monitoring and managing illness and improving wellness including preventive care. With added focus on optimizing spending and increasing profits, more attention is being given to increasing situational awareness surrounding the patient, hospital operations, asset management, predictive maintenance, and positive postoperative outcomes. In brief, these use-cases could be highlighted as [11]:

- Expanding patient care and monitoring through enabling IoT for skilled nursing
- Enabling into-the-home remote monitoring, improved wellness and preventive care with remote patient monitoring
- Monitoring and illness management through remote connectivity within secondary facilities (branch hospital, remote clinics)
- 24x7 data collection, analysis and action to decrease post-surgical readmissions, helping patients avoid complications while saving costs for hospitals and insurance providers
- Avoiding readmissions of post-surgical or chronically ill patients by predicting negative changes in their health and intervening in such a way that their conditions improve and they stay out of the hospital, especially intensive care units.
- Asset tracking for medical equipment to help save effort and time for medical practitioners, accurate inventory tracking, and budgeting/capacity planning for acquiring or leasing new equipment.
- Mobile engagement for patients and practitioners to help enhance the experience around healthcare premises, optimize operations for doctors and nurses, and maintain regulatory compliance

Retail
Digitization has shifted the traditional brick-and-mortar scene of retail and shopping into different frontiers. On-line retailers are managing to deliver more enhanced "digital experiences" for shoppers today, and the traditional retailers are trying to catch-up with that. Since shoppers are seeking out different experiences, not just buying the products, retailers are seeking new ways to engage customers while

transforming operations and seeking out the advantages of digitization in the whole sales cycle, not just for the shoppers themselves.

McKinsey estimated back in 2015 that the total economic impact of IoT in retail will be $410B-$1.2T by 2025, making it one of the top spaces for IoT solutions. [12]

IoT in retail is helping in different use-cases including:

Mobile engagement for bringing the online shopping experience into the brick-and-mortar setup with features including:

- Automated checkout
- Real-time advertising and promotion
- Wayfinding for product shopping and planned shopping routes
- Shopping tips, special offers, and smart marketing via shopper personalization and profiling
- Product placement and layout optimization through shopper analytics
- Real estate optimization via presence analytics and location context
- Transforming store operations by making them more efficient through the proper resource (goods as well as HR) allocation and operation

Perhaps the biggest assortment of IoT and its use-cases setup in a single place could be found nowhere else than the Amazon Go grocery shops. It automates the tracking of shopping item tracking, both on the shelves and in the virtual shopping cart, even in the inventory store. IoT sensors and cameras use different algorithms including computer vision, deep learning, geofencing, and others in order to enhance the customer experience by also automating purchase, checkout and payment [13].

Geofencing is implemented by providing a floorplan or location map to a system that includes the mapping of the physical location of sensors. Then, the administrator defines the areas within which a tracked device can exist without triggering alerts.

Education

The rate of technological development in education is significantly slower compared to the wearables or consumer product industry because funding keeps fluctuating. The low-end of this fluctuation is evident in schools that try to utilize their infrastructure beyond its capitalization and even beyond end-of-support periods.

A primary objective of IoT in Education is to cover the technological challenges involved in learning and collaboration solutions. Long IT infrastructure lifecycles are the result of different solutions utilizing different protocols and software that, in turn, challenge the IT supporting any school or university. IoT must break the challenges posed for data access, device management, networking and security via interoperability and simplification instead of adding complexity.

IoT in Education and for learning and collaboration objectives can:

- Connect buildings and campuses, both centralized and remote with the proper security layer and monitoring for feedback on performance and capacity planning
- Provide wayfinding around campuses with the engagement of students, visitors, staff and parents
- Help power learning management systems (LMS) and integrating the different blocks by enabling bring your own device (BYOD), mobile device management (MDM), rolling out and distribution of 1-to-1 learning software and electronic books material
- Optimizing the operation of smart classroom equipment and collaboration spaces with analytics, and proper resource allocation and reporting

Another major area for IoT is the educational facility itself, ranging from a single building to a large campus, to distributed campus and even extending to remote campuses.

IoT can help here via:

- Exercising industrial-like controls over lab spaces, central steam plants, power generating facilities, and overall campus buildings
- Sensor telemetry for safety monitoring and IP surveillance solutions
- Student tracking and safety
- Facility management

Transportation

With the advent of the 3rd and 4th industrial revolutions, we have strived to enhance the way we communicate and interact. A more efficient way to travel around with better means of transportation has also revolutionized those. The exponential growth and explosion of the means of transportation have brought along with them many challenges around safety, traffic management, efficiency, as well as environmental challenges. IoT is already helping out overcome a few of these challenges through traffic optimization, trip conditions, vehicle state, driver experience, and infrastructure management. Whether it's on a smaller scale of a connected vehicle/car or a massive one like inter-city or inter-country railroads, IoT brings the advantages of data collection, sharing it across distances of any transportation solution scale, and provides the tools to process that data and as a result optimizes the underlying transportation system.

The car reflects the advancement of industrial revolutions on a human invention. It has shifted from mechanical to electrical and over the past few decades, electronically infused that shows how much technology has advanced this industry. A car by itself is an island of data collected from different systems and sensors within the car and shown to the user on its dashboard and digital screen(s). Wireless communications can enable the car to integrate within the IoT transport system to enable it to share valuable information with the driver, other vehicles, service providers (car dealership and other 3rd party insurance, safety and security systems), and the transportation infrastructure.

Some of the use-cases for connected vehicles with IoT are:

- Autonomous or self-driving cars
- Dealership notification for due maintenance calls, service, software upgrade, or call-backs
- Tracking of stolen cars
- Relaying safety or security of car and driving conditions to car vendor, other cars, the transportation infrastructure system and other 3rd parties
- Relaying vehicle utilization to optimize cost for insurance companies as well as new models of carsharing and car subscriptions/leasing

- Enhancing the driver experience with entertainment and productivity applications as well as integration with other IoT services such as a smart garage or smart home

If the use-cases mentioned before can be applied on a single car or vehicle, imagine what could be done to a bigger group of connected vehicles. But in contrast to a regular consumer car, fleet vehicles could be part of a bigger industrial, commercial or city-wide network of vehicles focused on specific tasks. Examples of these would be a fleet of mining trucks, actually doing the drilling themselves or in charge of transporting the ore. Another would be a fleet of garbage collection trucks or municipal utility repair trucks.

A third example can be a fleet of shipping, mail, goods or any logistics services. Investment in big fleets, usually the investments being long term ones, have a lot of challenges. One of the most basic challenges is having the vehicles retrofitted with the required technology to include them in the transportation IoT because they usually serve for a long-term and didn't come initially with support for the specific sensor or required communication. In general, enabling IoT with fleets serves the goal of creating a more efficient service model with enhanced machine and human productivity at lower costs.

Use-cases for a connected fleet can cover:

- Condition-based maintenance that is relayed from onboard sensors, both local and remote
- Improved safety and security with preventive and predictive analytics
- Fleet tracking and route optimization
- Optimization of fleet distribution based on required services

Connected vehicles and fleets are the things in the IoT since they have the sensors and connectivity to relay the proper data to the user/driver, operator, other vehicles, and the external transportation infrastructure. What if close the loop by adding sensors and connectivity to the transportation system itself? A full mesh communication system exchanging actionable data can now be sent not only from vehicles to the infrastructure but also out from the infrastructure itself.

IoT in the broader reference of transportation can enable the following use-cases in real-time:

- Readings on the utilization of capacity onboard public transit
- Updates for all vehicles and users about the state of transportation including the weather, traffic congestion, and rerouting, journey duration, arrival/departure information of public transit system vehicles
- Dispatching emergency services in case of accidents
- Proactively monitor the infrastructure utilization by measuring and providing an understanding of its state and need for maintenance

Smart Buildings

IoT enabled offices to become smart workplaces. Underlying IT infrastructure can be integrated and used to connect IoT devices for different work functions including physical access, security, printing, scanning, space management among others. The objective from making the workplace smart is to enhance the experience of the end-users (employees and visitors) as well as the line-of-business owners and decision-makers. Enhancing the experience is not just about providing more excellent interactions; it is also about increasing productivity while decreasing cost of resources. Use-cases for IoT around the workplace can be:

- Utilizing IoT for energy and environment management to optimize the operation of lighting and cooling/heating
- Understanding the usage of the workplace spaces (desks, offices, public areas, and meeting rooms) and optimizing the overall area utilization and tenancy costs
- Enhancing the productivity of the workforce by enabling mobility and smart work solutions such as meeting room booking and management systems, wireless printing/projection, knowledge exchange tools, mobile apps, collaboration software, and others
- Streamlining the guest management system and experience to admit guest securely while also introducing a subset of the workforce enabled IoT use-cases

The connected home is a home that is enabled by IoT. What's changing? It's the way people spend time around their home and how they interact with their connected "things."

Digital assistants have enabled speakers and screens to interact with humans via application platforms, voice, and video to help organize an individual's workday and carry out chores around the house. Connected appliances not only plug into the home network to reflect data about their status but collect analytics and trigger alerts based on the owner's habits and living preferences.

The role of IoT in the connected home can spur off several use-cases including:

- Automation of domestic chores with appliance programming and extending to the devices automatically adapting to behavioral and optimal operating conditions to operate such as waiting for the home to be empty or for the off-peak electricity time
- Optimization of energy management around homes through sensors and local/remote control of lighting, power outlets, and thermostat settings
- Securing the home with data that are analyzed from cameras, entry access, and sending out event alarms and triggers.

It's worth mentioning that this IoT space is one of the few where the integration of the IoT devices is the easiest. Beside human wearable technology, smart homes are the largest consumer-based market for IoT. It is very easy to develop your own platform based on basic computer kits and single-board computers (SBC) that you can buy from the market or choose one of the prevalent ecosystems such as those built by Amazon, Google, and Apple. Devices like smart bulbs, light switches, power outlets, speakers, screen, televisions, cameras, doorbell, thermostat that are shipped today have the capability to be tied with each other on at least one or two platforms above. Even wearable IoT devices can integrate with these for the sake of providing control or exchanging data and notifications.

Agriculture (Smart Agro)

With the demand for a big population for cultivated products, including livestock animals, technology is also a major player in the enhancement, monitoring, and sustainability of many agricultural practices. Out of all the natural resources and effort that go into agriculture, water is the most valuable one, and allocating water for growing livestock and for irrigating crops properly guarantees that water usage is optimized.

IoT can help in this area as well as in other conditions including:

- Improving crop yields with smart irrigation and lighting in addition to controlling growing conditions to maximize production (eg: greenhouses)

- Utilizing programed vehicles and robots in farming to increase efficiency with monitoring
- Monitoring livestock for activity and health patterns to prevent disease and promote its well-being and yield
- Tracking livestock and farm-to-market supply for more efficient allocation of resources
- Monitoring plants and crops for the proper growing conditions such as temperature, soil moisture, as well as the climate and its changing variables
- Saving resources and driving a more efficient cycle by reducing water usage and decreasing product losses

Agricultural practices can also extend to urban and other settings such as monitoring public park status with irrigation control/reduction as well as programming irrigation for golf courses, sports, and playing fields.

Smart Cities

If we combine all the above use-cases from different verticals, we could create a larger city-wide ecosystem. Many governing authorities have had the vision of integrating IoT technologies into districts, towns, municipalities, cities and even countries with different goals towards achieving more efficient metropolitan management, economic development, sustainability, innovation, and citizen engagement.

Smart cities can combine use-cases from different verticals to achieve city-wide managed, monitored, and facilitated IoT use-cases such as:

- Addressing city issues related to public transportation, roads, tariffs, parking, maintenance and so on to adjust them for better conditions for the cities and their people both economically, socially and experience-wise
- Providing more efficient delivery of security and safety services to help protect citizens, law enforcement personnel, and assets to make the city a safer place to live in which would also help boost attract more people to visit or move to those cities
- Improving the quality of the environment and living conditions by monitoring environment-related conditions such as pollution levels, water quality and safety levels, air quality monitoring, and forest fire detection.
- Improving the efficiency of the power-grid utilization through smarter municipal street lighting while adding monitoring for maintenance

- Optimize public parking space utilization and improve parking operations, organize traffic, automate payment processes, and help traffic in the city flow more freely

These are just a few of the use-cases that might borrow from different verticals to extend to other areas, all with the objective of enhancing the quality of living for the people while making cities more sustainable.

IoT Structure/Architecture Basics:

The core of IoT is about enabling devices, sensors, and general things with connectivity to communicate relevant data from and to the connected devices, systems, and platforms.

IoT Models

Connectivity, mainly wireless, including Wi-Fi and other technologies, is a main building block of the IoT stack. But there are other components too without which IoT would not be achieved. Different sources from literature, research, and vendors consortiums in IoT provide different depictions of the IoT architecture reference model. For example, the IoT World Forum (IoTWF), defined the reference model shown in Figure 9.3 in 2014.

We can look at different representation from the IoT-A that defines the starting key building blocks which create the foundation of IoT. Based on an experimental paradigm, it combines top-down reasoning about architectural and design principles and guidelines with simulation and prototyping that result in specific technical consequences based on architectural design choices. It results in a reference model composed of sub-models that set the scope for the IoT design space and that address the technical consequence-based guidelines and perspectives. Figure 9.4 shows what this model looks like.

Different reference models could be considered from the Purdue Model for Control Hierarchy, the Industrial Internet Reference Architecture (IIRA) by the Industrial Internet Consortium (IIC) as well as others. While different models vary with their depiction from a high-level approach to a more detailed and inter-linked one, as in the above IoT-A model, the main purpose of setting a reference model is to tackle how to establish common grounds with focused development in an exponentially growing space. Technically, this would mean setting models to tackle scalability, interoperability & integration, communication, governance and standards compatibility and more.

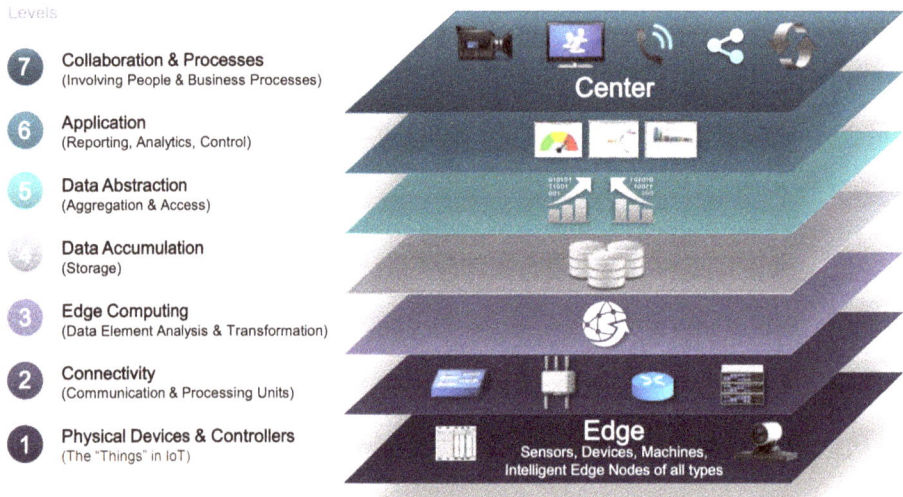

Figure 9.3: IoT Reference Model Published by the IoTWF

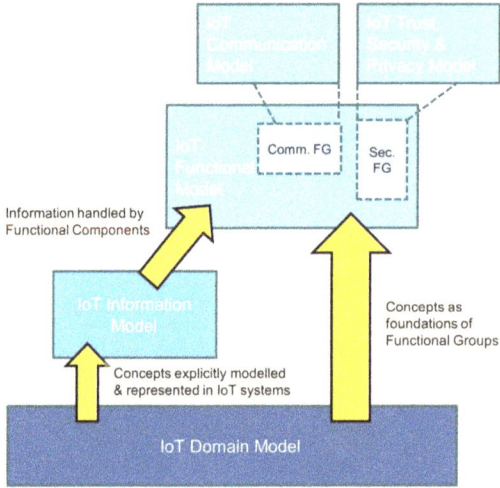

Figure 9.4: Interaction of all sub-models of the IoT-A Reference Model [14]

While different entities or sources can give us different reference model, they all include the basic components, requirements, and answers to the challenges that the IoT needs to overcome in order to operate. To simplify the representation, a 3-layer model will be used to represent the core IoT functional stack [15]. These components, along with wireless communication, can be listed as:

- The sensing or "Perception" layer at the edge of the IoT where the IoT hardware or sensors is attached to the communication, usually wireless links, where data is gathered from the source relevant to a specific IoT use-case
- The communication or "Network" layer which handles data transportation. This could be data transmitted between the sensors, gateways, networks, servers and end-user devices
- The interface or "Application" layer, with the various applications and interfaces that handle the insights and decisions made in accordance with the business use-case at the top of the stack

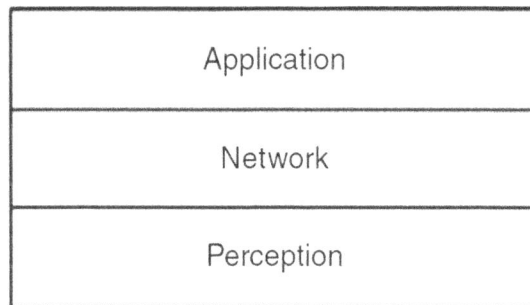

Application
Network
Perception

Figure 5 3-layer Model for Wireless IoT [15]

As a Certified Wireless Solutions Administrator has to be able to implement and administer and troubleshoot technologies that heavily rely on wireless, the Network layer of IoT is our most significant focus where we can talk about the underlying wireless communication technologies that a CWSA must know. This will be discussed next, as we dive deeper into the IoT architecture requirements, including hardware, connectivity, security, and applications.

IoT Hardware

The most familiar component, responsible for the sensing/perception layer of IoT, is the IoT hardware itself. Professionals and consumers alike are fairly knowledgeable about IoT hardware, especially in non-industrial use-cases. With the expected explosive growth of IoT devices of up to 50 billion devices by 2025 cited by different sources, we should note the hardware characteristics and requirements in IoT.

On a high-level, the components of a wireless IoT device must include [9]: a communication module, a microcontroller, sensors, and a power source to power up the components.

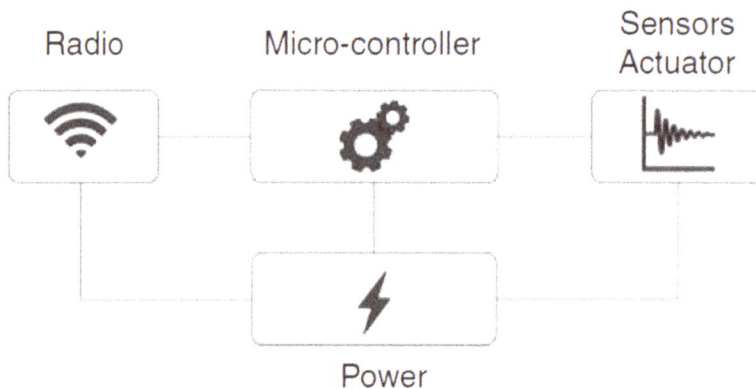

Figure 9.6 Wireless Interface IoT Hardware [9]

Communication module: enables a smart device to communicate over a wired, or more commonly, wireless connection

Microcontroller: defines the function and behavior of the smart device. This usually contains two different types of memory, the Read-only Memory (ROM) and Random-Access Memory (RAM). The ROM stores the software code that defines the functions that the device carries out while the RAM is utilized for temporary variables and data that is processed by the software. IoT devices can be categorized across different classes to help differentiate between them, and this is usually done based on memory and processing capabilities.

- Sensors/actuators: give the smart object an interface to collect data and interact with the system(s)
- Power source: provide power for the electric/electronic operation of the smart object. Different IoT devices require different forms of input power. The most common form of power source is batteries, usually Lithium cell, but there are other sources such AC, DC (eg: AC/DC adapter, USB), solar cells, RF/radio-transmitted energy, Power over Ethernet (PoE), mechanical movement (eg: based on vibration or body motion), and other forms of power. Most consumer IoT utilize rechargeable batteries or plug into a limitless source of power. A smart-bulb plugs into AC power fittings, whereas a fitness tracker employs button cell Lithium batteries and is usually recharged every few days.

Some of the most common IoT hardware platforms available on the market today serve multiple purposes. The platforms can be IoT gateways, SoCs, and development kits. These platforms are readily designed so IoT can be built for different purposes like research and education, industrial testing and prototyping, DIY projects and other general purposes. Some of the companies who provide these platforms are Advanticsys1, Ardiuno2, Banana Pi3, BeagleBone4, Espressif5, Intel6, OpenMote7, Particle8, Raspberry Pi9, and Zolertia10.

Figure 9.7 Raspberry Pi 4 Model B SBC (2019)

When selecting a platform, multiple factors can help decide which to use. From a technical perspective, those could be form factor, analog and digital I/O interfaces and pins, chipset capabilities, RAM/ROM/flash memory, power requirements, sensors, and RF capabilities and certification. From a business perspective, those could be price, use-cases, customizability, re-usability, future-proofing, open vs. closed platforms, and OS and applications flexibility.

It is also important to consider how you will update firmware or software in the IoT devices. Less-capable devices may require that you connect them to another system, such as a laptop, to update the firmware/software. More-capable systems may support FoTA updates.

> *Firmware over the Air (FoTA)* allows IoT devices and wireless sensors to be updated either manually or automatically through the radio interface. When selecting IoT devices, be sure to consider support for this feature.

Security Concerns

The nature of IoT requires minimal hardware and power consumption with optimized communications. This poses some trade-offs for the hardware and software, and at the same time raises the risk of deploying a less secure network of smart objects.

The small form factor of hardware and the limited capabilities it enforces, there could be a limitation on computational power and memory. In addition, to save consumption power would also mean to rely on lightweight operations and minimized communication between IoT nodes. What are the security implications of using IoT devices, and how do we balance between security requirement and hardware, software, and communication restrictions?

Security at all layers must be guaranteed, from the hardware layer to the application layer. From physical access to the IoT hardware to insecure web or app-interfaces is a big range of a very easy attack surface for an attacker.

Authentication and Authorization, in addition to Confidentiality, Integrity, and Availability (CIA) of the data and the services going over IoT is a must, or we risk data spoofing, un-allowed access, privacy invasion, loss or tampering of information, or the loss of service. Other examples of security concerns that IoT design needs to be wary of could be [9]:

- cloning of smart objects by unauthorized manufacturers
- malicious substitution of smart things during installation
- firmware replacement attacks
- extraction of security parameters
- eavesdropping attacks if communication channels are not adequately protected
- man-in-the-middle attacks
- routing attacks
- denial-of-service attacks

Threats could be related to the physical nature of smart IoT devices and their location of deployment having to be physically accessible, having small form factors, their provisioning methods, connectivity setup, and their wireless communications.

Wireless for IoT Connectivity

The convergence of terminals and servers that are connected onto a single global network is what we know as the "Internet." In the same manner, the convergence of the devices that have become enabled with data and connectivity to integrate within the web of all devices and platforms is how IoT is coming together. Since the Internet has paved the way for the most common protocols based on IP communications, convergence of IoT is also happening on an IP-based architecture. The readers of this book are assumed to be most familiar with the OSI model as a guiding principle for communications. To dissect the IoT communications into its different building blocks, or layers, and focus more on the wireless aspect in it, we can map it to the OSI model.

From an IoT perspective, we are most interested in how the technologies of wireless communications for IoT operate. TCP/IP is the base of Internet communication. Referring to an IP-based reference model of the Internet would help us simplify representation of the OSI model further as this is a more popular stack model. The central interest on the "Network" layer represented in the IoT model above would have to extend from the "Perception" layer, starting with hardware, sensors, and physical layer, and connecting it up to the "Application" layer. The IEEE breaks down different wireless technologies on the lower layers to clearly describe interoperability. In relation, we can look at how the IoT "Network" layer relates to the lower layers of the IEEE models of wireless technologies while its other layer components link them to the Physical and upper Network/Internet and Application layers.

Many wireless technologies are utilized in IoT. Each of these protocols has a stack with common features. At the lower layer, the PHY and MAC are standardized by neutral standards bodies (eg: IEEE and ITU) whereas the upper layers are maintained by an industry group. Being based on neutral standards means that these are readily available for developers to use, while the specific industry group protocols must be licensed or paid for to be accessed and used to certify operations and interoperability by developers [15].

In many cases, it may be essential to implement Virtual LANs (VLANs) on the Ethernet side of the network to segregate IoT or wireless sensor traffic from the rest of the network. In many cases, IoT and sensor devices need only to communicate directly with a cloud service. If this is the case, use VLANs to allow the IoT or sensor traffic to communicate with the Internet and prevent them from being used as a point of ingress to your network.

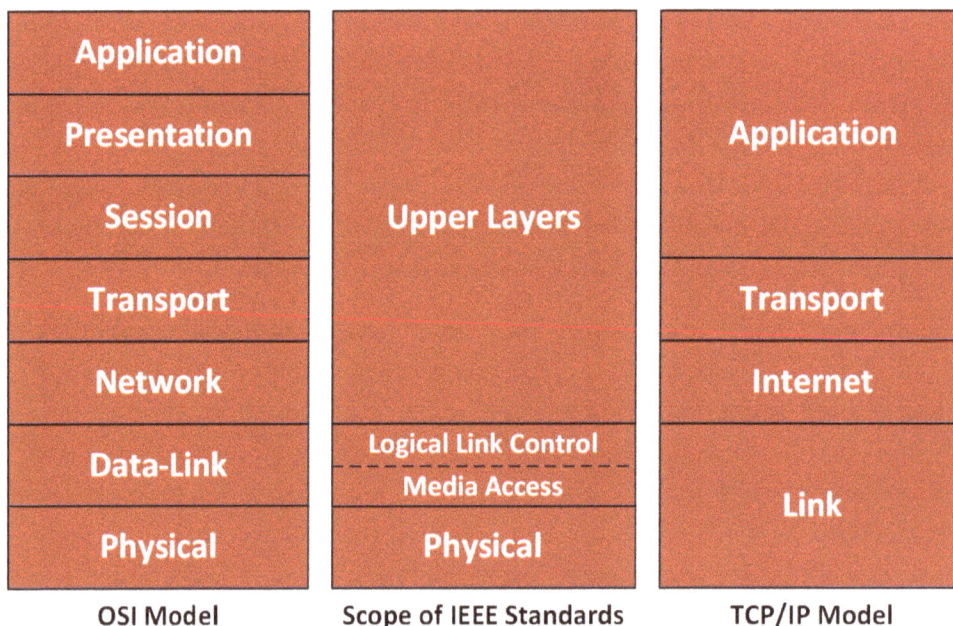

OSI Model	Scope of IEEE Standards	TCP/IP Model
Application	Upper Layers	Application
Presentation		
Session		
Transport		Transport
Network		Internet
Data-Link	Logical Link Control / Media Access	Link
Physical	Physical	

Figure 9.8 Mapping between OSI, IEEE, and TCP/IP Models

Wireless technologies discussed in this chapter are not necessarily IP-based. We are simply adopting the TCP/IP and OSI models of representing different communication layers to make the task of mapping different components of the several wireless technologies to a familiar reference. Most wireless personal area networks (WPANs) utilize protocols, like Bluetooth, ZigBee, and Z-Wave that don't inherently communicate over TCP/IP but are similar to a true TCP/IP protocol. However, adaptations of a few of these protocols that are based on IP do exist (eg: ZigBee-IP, 6LowWPAN on IP over Bluetooth).

Wireless Communications Characteristics

Why do we have so many standards and protocols that different IoT devices and networks could use for communication? The reason is that there are different criteria that can lead to a selection of a specific communication standard. We have already discussed how IoT hardware can be constrained to specific form factor, power, chipset, and other defining capabilities. Wireless communication is one of these capabilities, and these factors also come into play when a specific communication technology is utilized which in turn affects the hardware itself. The most important factors to consider for wireless communications for IoT are:

Frequency: the choice between licensed and unlicensed bands greatly affects the technology, complexity, and service guarantee of the IoT communications.

The unlicensed industrial, scientific, and medical (ISM) frequency bands are free to utilize, often regulated by different national and regional authorities and bodies to set device compliance when it comes to transmit power/gain, channel allocation, channel selection and hopping mechanism, duty cycles and dwell times. This makes it easy to deploy on the ISM bands but usually restricts the technology to a short or medium range wireless communication. At the same time, this does not guarantee a quality of service for the frequency space since it will be busy with interference from different standards and employed by many other deployments. Wi-Fi, Bluetooth, and 802.15.4 based protocols are some examples of wireless technologies that can utilize the unlicensed bands.

The licensed bands are usually regulated for exclusivity, but to use them, the license should be provisioned and paid for in order to enable operations. This adds to the overhead of any solution from a logistics and commercial perspective. Wi-Fi, WiMax, cellular and narrow-band IoT (NB-IoT) are some examples of the technologies that utilize these licensed bands.

Range: the distance that the communication within the IoT components greatly affects the choice of the underlying wireless technology. This can scale on a short-range from a few meters, with technologies like Bluetooth, to medium-range technologies extending to 10s and 100s of meters such as 802.11 and 802.15.4, to an even longer range with 802.11 and cellular networks extending in kilometers of connectivity.

The range can be subject to variation and can never be considered independently from throughput and other coverage requirements. How far a signal can reach highly depends on the signal strength, receiver sensitivity, noise, line-of-site, antenna design and gain, regulatory constraints, and whether the proper error/control functions and other enhancements of a specific protocol are being utilized.

Power Consumption: regardless of the power source, there is a balance between mobility and a reliable power source. A continuous or perpetual power source means there is a wired powered source, such as AC, DC, or PoE power. While this provides a more significant benefit than a temporary power source such as batteries that need to be recharged or serviced, it ties down the mobility aspect of any IoT node wherever this flexibility is needed. When we look at battery-powered devices, this ties it back to consumption and connectivity constraints, since lower consumption technologies must be used.

Topology: the communication topology also affects the choice of technology as well as the power source and consumption.

In a star topology, the hub could be a wired power source device with high processing capacity to sense or communicate to all other connected objects. The connected objects, in this case, could be mobile, battery power, and less processing-intensive or communication-oriented than the hub itself.

In a peer-to-peer connection, establishing the link is a primary factor of the communication and usually requires peers or devices of equal capabilities utilizing most of the same functions.

The mix of a star and peer-to-peer topologies lead to a mesh topology, with different hubs connected to other hubs or end-nodes of communication. Mesh communications usually require different or more advanced protocols than single-hop/link communications. Mesh devices could either be connected at the edge, as mesh points, or they could be connected to other nodes that extend communication hops throughout the mesh layout. Similar to a star topology, a mesh node that connects at the edge, which doesn't interconnect nodes, can have fewer functions or limited capabilities (Reduced Function Devices) compared to mesh nodes, or portals, that interconnect one or more mesh nodes (Full Function Devices). This leads to different protocol approaches and processing, not only for establishing communication but for adapting the mesh for time-sensitive applications and changing links.

Wireless IoT Connectivity Protocols

The above factors are considered for lower layer IoT wireless access protocols. The protocols that determine the majority of wireless communication technologies in IoT are:

- Bluetooth (previously 802.15.1): The IEEE standardized the lower layers of Bluetooth based wireless personal area networks (WPANs) on the PHY and MAC layers as the 802.15.1 standard. This is maintained now by the Bluetooth Special Interest Group (SIG).
- ITU-T G.9959: provides a standard for the PHY and MAC layers for short-range narrow-band digital radio communication transceivers. The Z-Wave protocol utilizes the ITU-T G.9959 while the upper layers are maintained by the Z-Wave Alliance.
- IEEE 802.15.4: provides PHY/MAC standards for low-rate WPAN. Wireless IoT protocols, like ZigBee, use a subset of the PHY and MAC layers it defines.
- IEEE 802.11: provides PHY/MAC standards for wireless local area networks (WLAN). The certification of 802.11 as Wi-Fi to guarantee compliance and interoperability is maintained by the Wi-Fi Alliance (WFA).

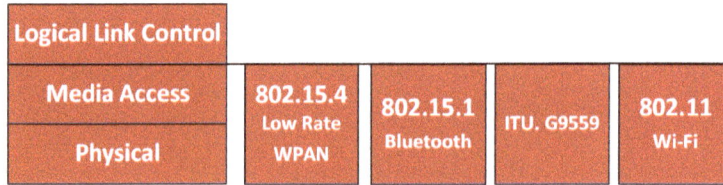

Figure 9.9 The Lower Layer Standards Supporting Wireless IoT Technologies

We will next highlight technical details about the lower protocols and the industry standards that they support. Some of the factors for choosing a specific protocol or technologies outlined before will also be highlighted for each of these different standards.

Bluetooth

Developed by Ericsson along with IBM, Intel, Nokia and Toshiba formed the Bluetooth SIG to develop the Bluetooth standard which was approved by the IEEE 802.15 committee in 2002 as a WPAN based on Bluetooth protocol dubbed 802.15.1-2002. In 2005, the IEEE amended the standard with additional improvements as 802.15.1-2005. After 2005, the Bluetooth SIG maintains the standard as well as the upper layers to serve different applications.

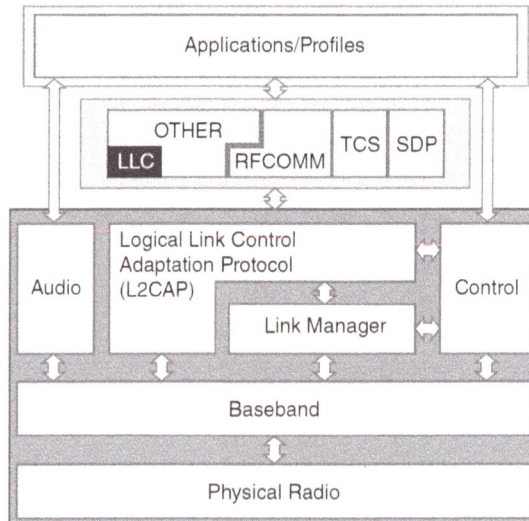

Figure 9.10 IEEE 802.15.1 Bluetooth WPAN [15]

The lower layers of Bluetooth corresponding to the PHY and MAC layers are the radio, baseband, and link manager layers where the:

- Radio specifies the transceiver and modulation requirements
- Baseband handles error control and checking methods including the forward error correction (FEC), cyclic redundancy checks (CRC), and automatic repeat requests (ARQ)
- Link manager handles the wireless link management and control including power control, connection establishment/management, authentication, and functions.

Bluetooth has evolved over multiple version with different enhancements out of which BLE, or Bluetooth Low Energy (LE), that came out with Bluetooth 4.0, was important for IoT since it used lower energy and could be used in a range of wireless sensors and IoT devices. The most recent version of Bluetooth as of the writing of this chapter is Bluetooth 5.1.

Operating Frequency:

Bluetooth operates in the 2.4 GHz (2.402 – 2.480 GHz) ISM band only. Bluetooth BR/EDR uses 79 frequency channels with 1MHz apart while Bluetooth LE uses 40 channels with 2 MHz spacing (3 advertising channels/37 data channels). Bluetooth uses Frequency-Hopping Spread Spectrum (FHSS) across these defined frequency channels as the channel hop plan.

Transceiver:

Bluetooth identifies different "classes" of Bluetooth transmitters. These classes dictate the limits on the transmit power. The transmit power for BLE ranges from 0.01mW to 100mW, as shown in the table below.

Power Class	Minimum Tx Power	Maximum Tx Power
Class 1	1 mW (0 dBm)	100 mW (+20 dBm)
Class 1.5	0.01 mW (-100 dBm)	10 mW (+10 dbm)
Class 2	0.01 mW (-100 dBm)	2.5 mW (+4 dBm)
Class 3	0.01 mW (-100 dBm)	1 mW (0 dBm)

The Bluetooth standard specifies –70 dBm as the minimum reference receiver sensitivity. The receiver sensitivity is defined as the signal strength at the receiver. At this minimum, the receiver should be able to achieve a Bit Error Rate (BER) of 0.1%, or one error in 1000 bits.

Range:

Specified as 1 to 100 meters, depending on the Bluetooth device class. This, however, could be subject to variation depending on the signal strength, receiver sensitivity, noise, and whether the proper error/control functions and other enhancements or versions in the protocol are being used. For example, the LE Coded enhancement introduced in Bluetooth 5 adds forward error correction (FEC) that decreases the rate but makes communication less susceptible to errors. This can extend the range up to 4 times with tests showing reception of Bluetooth notifications over a mobile phone at a distance of 350 meters! [16]

ITU G.9959
Originally created based on the lower layers of Z-wave, ITU G.9959 is not a wireless IoT technology by itself, but a standard to guide the conformation of technologies. The International Telecommunication Union created the standard to allow inoperability with different hardware of "short-range narrow-band digital radio communication transceivers." [17]

Operating Frequency:

ITU G.9959 operates on sub-1GHz bands. Different frequency ranges are allocation in different regulatory domains. This disallows using devices across different regions freely and mandates that specific ITU G.9959 be created specifically for the different regions. ITU G.955 can operate in the 900MHz ISM band in the Americans and Australia, while it operates on the 800MHz Short Range Devices (SRD) band in Europe. The advantage of operating in these lower frequency bands is avoiding the congested 2.4GHz ISM bands and better propagation features.

Transceiver:

In the calculations of a link budget for proper ITU G.9959 operation, the standard allows for the below data rates at the nominal power level with the minimum receiving sensitivity as long as it is within regulatory bounds.

Data Rate	Minimum Receiver Sensitivity
19.2 kbps	-95 dBm
40 kbps	-92 dBm
100 kbps	-89 dBm

Range:

Depending on the link budget calculations with the give noise floor and target data rates, technologies that utilize ITU G.9959, like Z-Wave, can reach a range of 30 meters indoors and extends to 100 meters in outdoor deployments.

802.15.4

Similar to ITU G.9959, the IEEE 802.15.4 is a standard unto which many communication protocols used in IoT are based. Throughout its lifetime, the standard has been updated with the majority of additions adding to the channel planning, modulation schemes, new waveforms, and other features. The standard makes use of different frequency band allocations that make up its specified 27 channels, with a trade-off for each of these bands discussed below.

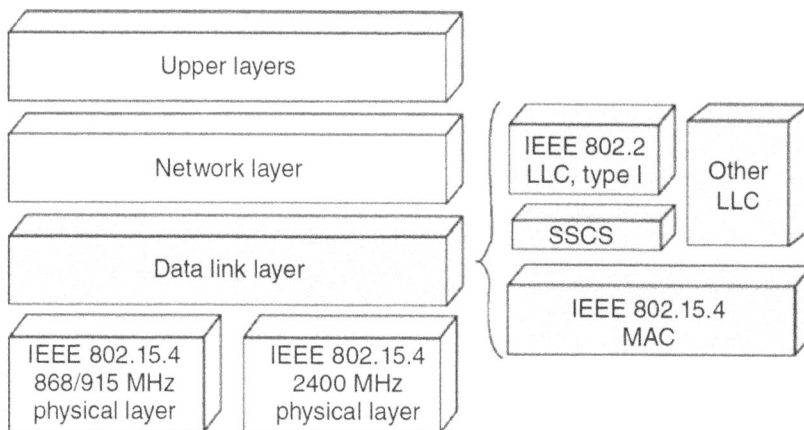

Figure 9.11: The IEEE 802.15.4 Protocol Stack [18]

Operating Frequency:

IEEE 802.15.4 can operate in the 2.4 GHz ISM band as well as the sub-1 GHz unlicensed bands. Operation in the 2.4GHz band makes it possible for vendors to manufacture IoT devices with transceivers that can be globally used. This, however, has the drawback of operating in a highly congested band, but the protocol allows to choose between up to the 16 channels highlighted below. The sub-1 GHz operation gives the same advantages as the ITU G.9959, less congested channels, but restricted by the frequency bands specified in different geographies.

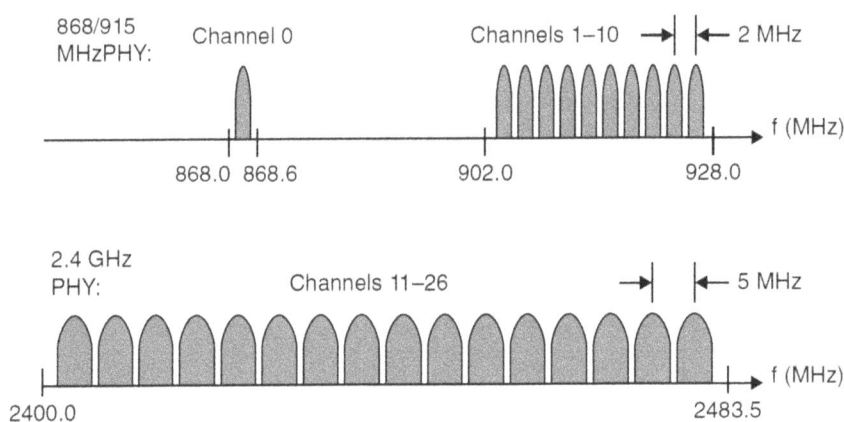

Figure 9.12: 802.15.4 Channel Frequencies Allocation [18]

Transceiver:

Receiver sensitivity depends on the modulation scheme and operation frequency band and are listed in the following table:

Band	Modulation	Receiver Sensitivity	Data rate
2.4GHz	Offset QPSK (OQPSK)	-85 dBm	250 kbps
Sub-1GHz	BPSK	-92 dBm	20 kbps
Sub-1GHz	ASK	-85 dBm	250 kbps
Sub-1GHz	OQPSK	-85 dBm	100 kbps

Range:

Depending on the link budget calculations with the give noise floor and target data rates, technologies that utilize 802.15.4, like ZigBee, can propagate up to 10-100 meters.

802.11

The role of the IEEE802.11 standards is to replace the wired cables 802.3 uses to create a local area network (LAN). It was first established in 1997 and has since undergone many revisions and enhancements. 802.11 is referred to as Wi-Fi, which is the certification maintained by the Wi-Fi Alliance (WFA). Although used interchangeably, Wi-Fi does not utilize all of the IEEE 802.11 specifications defined by its committees.

For example, on the PHY layer, 802.11 can operate over different frequencies including 900MHz, 2.4GHz, 3.6GHz, 4.9GHz, 5GHz, 5.9GHz, and 60GHz bands. The majority of Wi-Fi that is certified by the WFA operates on the 2.4, 5GHz, and 60GHz bands. As outlined earlier, 802.11 operates on the lower PHY and MAC layers of the communication stack. New updates that every new 802.11 standard/revision introduces brings changes to how the PHY and MAC layers are adopted with new frequency spreading functions, modulation and coding schemes, rates, range, power consumption, topology capabilities and other factors that decide the best 802.11 standard for each different application.

More 802.11 and Wi-Fi details are covered in the CWNP Certified Wireless Network Administrator Official Study Guide.

ZigBee

With hundreds of companies backing up the standard forming the ZigBee Alliance, this technology has been driven for interoperability between different vendors as a major solution to interconnect IoT devices. Although many people refer to ZigBee and 802.15.4 as the same, one should always remember that ZigBee is the standard that makes use of a subset of the 802.15.4 specifications. This is analogous to the relation between the WFA certifying Wi-Fi based on the 802.11 protocols. The certification guarantees interoperability between devices manufactured by different vendors.

ZigBee is tailored for sensors and devices with low bandwidth and power needs. You can usually find these devices in home automation and smart energy devices and systems.

While the lower layers of ZigBee are built on top of 802.15.4, while ZigBee specifies the network and security layer and application support layers which ties the lower PHY and MAC to networking to the upper application profile layers. These application profiles are pre-defined based on the specific industry use-cases, but vendors can create their customized profiles as well.

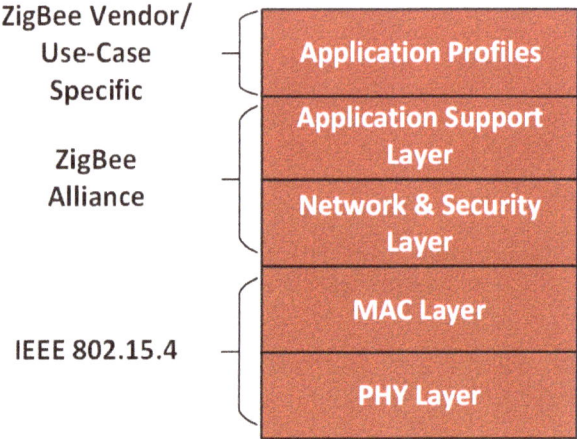

Figure 9.13: High-level View of the ZigBee Protocol Stack

While the ZigBee Alliance has guaranteed interoperability between different vendor ZigBee devices, this hasn't provided the same with other IoT solutions. This gave birth to the ZigBee IP to support open standards based on the

ZigBee IP [5]

ZigBee IP was built on the same lower layers as 802.15.4 on the PHY and MAC layers, but also incorporates IP, TCP/UDP and other IETF standards on the upper network and transport layers. The layers specific ZigBee are only at the top of the protocol stack for applications.

To provide integration and interoperation with any other 802.15.4-based IoT network based on open and current standards coming from the IETF (Internet Engineering Task Force), ZigBee IP supports 6LoWPAN (IPv6 over Low -Power Wireless Personal Area Networks) as an adaptation layer for fragmentation and header compression schemes. "At the network layer, all ZigBee IP nodes support IPv6, ICMPv6, and 6LoWPAN Neighbor Discovery (ND), and utilize RPL (IPv6 Routing Protocol for Low-Power and Lossy Networks) for the routing of packets across the mesh network. Both TCP and UDP are also supported, to provide both connection-oriented and connectionless service." All of these are aimed at providing lower bandwidth and power and more cost-effective communications in IoT.

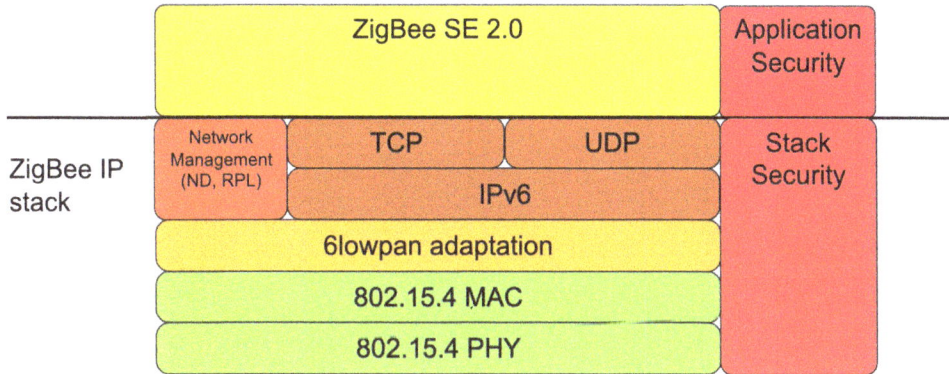

Figure 9.14: ZigBee IP Stack (ZigBee Alliance)

ZigBee IP was initially explicitly designed by the ZigBee Alliance for the Smart Energy Profile 2.0 (SE 2.0) specification for smart metering and residential energy management systems, but it can be utilized for any other application that utilizes it as a standards-based IoT stack.

Thread

Nest Labs, a subsidiary of Alphabet, the mother company of Google, has led the development of Thread with the aim of creating secure, IP-based, low-power mesh networking protocol built on top of the 802.15.4 standard. The Thread Group was formed as an alliance between Nest and other industry partners to promote this protocol.

The Thread protocol stack is analogous to ZigBee IP as it is built on top of 802.15.4 and it follows the TCP/IP model. UDP fulfills the duties of the transport layer while the Internet layer is represented by IPv6 and 6LoWPAN.

For extensive information about Thread, you can download the specification document here: https://www.threadgroup.org/ThreadSpec

WirelessHART

WirelessHART originated from the Highway Addressable Remote Transducer (HART) communication protocol aimed at industrial process measurement and control. It is a self-organizing wireless mesh network communication protocol that uses the IEEE 802.15.4 radio standard.

WirelessHART has been designed by industry experts with security in mind. It is an open, multivendor, interoperable protocol that is secure out of the box with security always on and no user configuration needed. This makes WirelessHART a simple, easy to use protocol designed primarily for industrial and other applications. You will find WirelessHART deployed mostly in Industrial IoT settings, delivering the processes data through a wireless field device network (WFN) is also comprised of both OT and IT, and leveraging HART to even communicate both digital and analog data readings. Since it's a wireless extension to HART, this made the deployment of networks that utilize WirelessHART very easy and allowed deployers to extend the reach of the industrial use-cases to places that were either hard or too expensive to reach with wired equipment.

Z-Wave

ZenSys created Z-Wave to support home automation products like lighting controls, thermostats, and garage door openers. After marketing Z-Wave in 2003, other companies joined ZenSys to form the Z-Wave Alliance.

The Z-Wave PHY layer focused on low power consumption, battery-saving, and propagation in indoor environments. Its MAC layer controlled the framing and carrier sensing functions. These lower layers got standardized in 2012 by the ITU as the G.9959 standard.

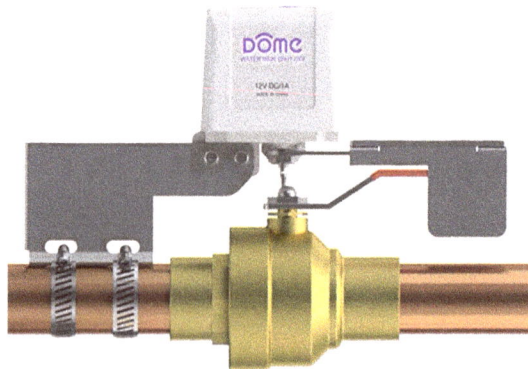

Figure 9.15: A Dome Water Main Shut-off with Z-Wave Support

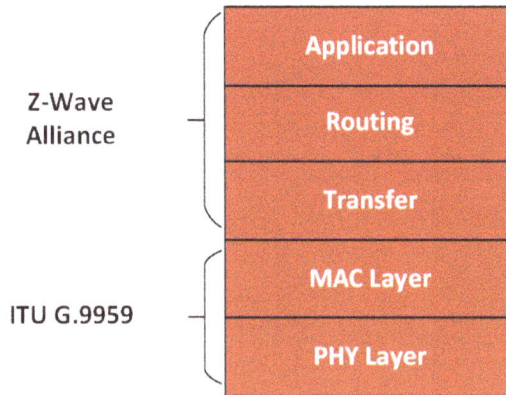

Figure 9.16: Z-Wave Protocol Stack

The specifications for Z-Wave upper layers of operation for Transfer, Routing, and Application are provided by the Z-Wave Alliance.

802.11ah

802.11/Wi-Fi protocols face some of the challenges for wireless communication in IoT because they might require more power, whether on being able to connect many nodes, or to achieve higher penetration, thus also being a challenge for battery operated nodes.

IEEE 802.11ah was published in 2017 as "industrial Wi-Fi" access technology to answer those challenges. It's based on the stack with the PHY layer adjusted to operate on the sub-1 GHz unlicensed frequency bands. Meanwhile, the MAC layer is also optimized to support the new PHY specifications while supporting a large number of endpoints (up to 8192/AP).

Region/Country	Frequency Band
Europe & Middle East	863-868 MHz
China	314–316 MHz, 430–434 MHz, 470–510 MHz, 779–787 MHz
Japan	916.5-927.5 MHz
North America & Other Asia-Pacific Regions	902-928 MHz

802.11ah is aimed at lower data rates, so while it does utilize the OFDM like other 802.11 protocols, it can easily reach further because of the lower rate requirements. The MAC is optimized with a shorter header. It also supports power saving through different control and contention mechanisms. Some of those are like sectorization, where areas are partitioned in order to limit collisions, null-data packets that makes management and control frames exchange more efficient, restricted access window (RAW) for reducing collisions, and target wake time to reduce power and also collisions.

As a result, 802.11ah offers a longer range than "traditional Wi-Fi" and provides excellent support for low-power devices at high numbers that need to send smaller bursts of data at lower speeds.

The Wi-Fi Alliance branded 802.11ah as Wi-Fi HaLow (hey-low) certification program to ensure interoperability between different vendors implementing this standard.

NB-IoT

With the objectives of meeting the IoT requirements, decreasing throughput and power consumption, while at the same time also decreasing the complexity and cost of equipment that relies on cellular technologies, the 3rd Generation Partnership Project (3GPP) set off to a new category of LTE devices under LTE-M 3GPP Work Item.

LTE-M went through different iterations as well as proposals from different vendors and companies, including Ericsson, Nokia, Huawei, Alcatel Lucent, Qualcomm, Sigfox as well as other standardization bodies, that were finally consolidated into a single Narrowband-IoT (NB-IoT) category of devices supporting Low Power Wired Area (LWPA) IoT.

"NB-IoT operates in half-duplex frequency-division duplexing (FDD) mode with a maximum data rate uplink of 60 kbps and downlink of 30 kbps. NB-IoT is defined with a link budget of 164 dB with the high link budget that should cater for better signal penetration in buildings while achieving battery life requirements." [5]

Compared to all the previous technologies and certifications, NB-IoT is considered as a long-range IoT communication technology.

LoRaWAN

"The LoRaWAN specification is a Low Power, Wide Area (LPWA) networking protocol designed to wirelessly connect battery operated 'things' to the Internet in regional, national or global networks, and targets key Internet of Things requirements such as bi-directional communication, end-to-end security, mobility and localization services." [19]

LoRaWAN is used as the term that defines the MAC and Data-Link OSI Layers standard by the LoRa Alliance who maintains it to distinguish it from the LoRa (Long Range) PHY modulation. LoRaWAN is at the same time based on the LoRa modulation, although Frequency-shift Keying (FSK) is also utilized as an alternative in some areas. LoRa utilizes the sub-GHz ISM bands to operate based on different frequency ranges in different regulatory domains. It can utilize different combinations of the frequency bands, spread factor, and channel bandwidth (125, 250, or 500kHz) to achieve data rates of up to 21,900bps.

Similar to NB-IoT, LoRaWAN is considered a long-range IoT technology.

IoT Connected Objects

All the devices connected to the network at the edge forming all these IoT setups that we have been discussing make use of the underlying wireless technologies.

In IoT communications, data can flow in many ways. A sensor or a node, or just simply a connected-object (CO) can be sending, receiving or relaying data to other components in an IoT setup. This could be a node telling another node or hub what command it has received from a user to adjust the temperature on the main building thermostat, seen on the factory production belt, or relaying an alert to that node that its application layer has detected such as dangerous fume levels that users can also see. Users can be interacting with those nodes by getting readings off them, inputting some data to send to another node or to the application layer through an interface, or receiving some form of alert that the application or IoT node has sent, either via application notification of sounded alarm. These are some examples of how communications and their interactions can happen in an IoT environment.

Breaking down those scenarios, we can look at the following simplified bi-directional flow of data:

- CO-to-CO: a primary connected device can be talking to a secondary connected device since the secondary device is offering a hop closer to the central system of the setup, or hub, or could be itself the hub of the setup. This is what most of the literature refers to as Machine-to-Machine communications (M2M).
- CO-to-service: a connected device can be talking to a service, usually in the application layer, in order to relay the data so that it becomes actionable. At the same time, a service can return data to the CO for an automated action or monitoring purposes.
- CO-to-user: this is where interface, the input and output, can happen, such as getting a reading from a device of entering specific information into the system through an application, wearable, hand-held, dashboard or terminal screen.
- User-to-service: a user can feed the service their specific preferences, KPIs, and program triggers/alerts through an application interface, application programming interface (API), or "If This Then That" (IFTTT) service among other ways.

Using the above data communication flows, we can come up with extended combinations to complete a system's interactions threw multiple flows like:

- User-to-service-to-CO
- CO-to-service-to-CO
- CO-to-service-to-user

At the level of wireless technologies in IoT, we can consider the role of a specific technology in the CO-to-CO communication since it is a lateral flow across the IoT stack as opposed to a vertical flow (up or down) across the stack for the other cases.

Many technologies discussed can form peer-to-peer, star, or mesh topologies. However, special attention should be given to how these communication links can be established when designing an IoT network/environment. For example, many ZigBee devices need an FFD node acting as a coordinator to establish the mesh between the nodes themselves. At the same time, a hub would be required if the communication involves integrating the ZigBee system with the Wi-Fi connectivity so that the traffic can flow throughout a full IoT environment like a smart home.

Chapter Summary

In this chapter, you learned about the details of IoT and brought together many concepts discussed earlier in this book. While the CWSA is exam is not focused on IoT alone, the reality is that more and more objects are being categorized as IoT today. Therefore, understanding these concepts is essential.

References

[1] Merriam-webster.com. (2019). Welcome to the New Words. [online] Available at: https://www.merriam-webster.com/words-at-play/new-words-in-the-dictionary-sep-2017 [Accessed 25 Jul. 2019]

[2] B.Russell and D.Van Duren, "Practical Internet of Things Security," Packt Publishing, 2016.

[3] "ITU Internet Reports 2005: The Internet of Things," International Telecommunication Union, 2005.

[4] P.Lea, "Internet of Things for Architects," Packt Publishing, 2018.

[5] D.Hanes, G.Salgueiro, P.Grossetete, R.Barton, J.Henry, "IoT Fundamentals: Networking Technologies, Protocols, and Use Cases for the Internet of Things," Cisco Press, 2017.

[6] J.Manyika, M.Chui, P.Bisson, J.Woetzel, R.Dobbs, and J.Bughin, D.Aharon, "The Internet of Things: Mapping the Value behind the Hype," McKinsey & Company, June 2015.

[7] 4th industrial revolution

[8] Postscapes. (2019). Internet of Things (IoT) History | Postscapes. [online] Available at: https://www.postscapes.com/internet-of-things-history/ [Accessed 01 Aug. 2019].

[9] S.Cirani, G.Ferrari, M.Picone, and L.Veltri, Internet of Things: Architectures, Protocols and Standards. JohnWiley & Sons Ltd, 2019.

[10] V.Gupta and R.Ulrick, "How the Internet of Things will reshape future production systems," McKinsey & Company, 2018.

[11] "2019 Trends in the Internet of Things," 451 Research

[12] "THE INTERNET OF THINGS: MAPPING THE VALUE BEYOND THE HYPE," McKinsey, 2015

[13] En.wikipedia.org. (2019). Amazon Go. [online] Available at: https://en.wikipedia.org/wiki/Amazon_Go [Accessed 25 Jul. 2019].

[14] M.Bauer, M.Boussard, N.Bui, F.Carrez, C.Jardak, J.De Loof, C. Magerkurth, S. Meissner, A.Nettsträter, A.Olivereau, M.Thoma, W.Joachim, J.Stefa, A.Salinas. Internet of Things – Architecture IoT-A Deliverable D1.5 – Final architectural reference model for the IoT v3.0. European Lighthouse Integrated Project, 2013

[15] D.Chew, "The Wireless Internet of Things – A Guide to the Lower Layers," JohnWiley & Sons, Inc., 2019.

[16] Bluetooth Technology Website. (2019). Exploring Bluetooth 5 – Going the Distance | Bluetooth Technology Website. [online] Available at: https://www.bluetooth.com/blog/exploring-bluetooth-5-going-the-distance/ [Accessed 05 Aug. 2019].

[17] Short range narrow-band digital radio communication transceivers—PHY and MAC layer specifications, Recommendation ITU-T G.9959, 2012.

[18] E. Callaway, P. Gorday, L. Hester, J. A. Gutierrez, M. Naeve, B. Heile, and V.Bahl, "Home networking with IEEE 802.15.4: A Developing standard for low-rate wireless personal area networks," IEEE Commun. Mag., vol. 40, no. 8,

pp. 70–77, 2002.

[19] Lora-alliance.org. (2019). About LoRaWAN® | LoRa Alliance™. [online] Available at: https://lora-alliance.org/about-lorawan [Accessed 05 Aug. 2019].

Review Questions

1. What year is often said to be the birth year of IoT?
    a. 1800
    b. 1990
    c. 2008
    d. 2018

2. What is the best definition of Smart Agro?
    a. Using sensors and digitization in farming and ranching
    b. Using sensors and digitization in transportation
    c. Using sensors and digitization in K-12
    d. Using sensors and digitization in higher education

3. How many levels are in the IoTWF Reference Model?
    a. 7
    b. 5
    c. 3
    d. 1

4. In the 3-layer IoT functional stack, what layer handles data transport?
    a. Application
    b. Network
    c. Perception
    d. None of these

5. What are the four primary components of a IoT sensor device?
    a. Radio, micro-controller, sensors/actuators, and power
    b. Radio, network cable, sensors/actuators, and power
    c. Network cable, micro-controller, sensors/actuators, and power
    d. Network cable, storage, sensors/actuators, and power

6. What standard is Z-Wave built upon?
    a. IEEE 802.11
    b. ITU-T G.9959
    c. IEEE 802.15.4
    d. Bluetooth

7. What version of Bluetooth introduced BLE?
    a. 1.0
    b. 1.3
    c. 4.0
    d. 5.1

8. What IoT network solution is provided directly by cellular service providers?
    a. Bluetooth
    b. ZigBee
    c. 802.11ah
    d. NB-IoT

9. What IoT connection scenario indicates that an IoT device communicates with an application on the Internet?
    a. CO-to-CO
    b. CO-to-service
    c. CO-to-user
    d. None of these

10. What is required in a ZigBee network to act as a coordinator?
    a. An FFD
    b. An RFD
    c. An 802.11 AP
    d. None of these

Review Answers

1. The correct answer is **A**. 2008 is often said to be the year of the birth of IoT because this is when the number of devices connected to the Internet exceeded the size of the human population.

2. The correct answer is **A**. Smart Agro is about farming, ranching, livestock management, environmental monitoring, and other areas in agriculture.

3. The correct answer is **A**. The IoTWF Rederence Model includes seven layers.

4. The correct answer is **B**. The Network layer handles data transport in the three-layer model.

5. The correct answer is **A**. The four primary components are the radio, micro-controller, sensors/actuators, and power.

6. The correct answer is **B**. Z-Wave builds on the ITU-T G.9959 standard.

7. The correct answer is **C**. BLE was introduced in Bluetooth 4.0.

8. The correct answer is **D**. Narrowband-IoT (NB-IoT) is a cellular service provider IoT network specification of the 3GPP.

9. The correct answer is **B**. The CO-to-service model indicates that the IoT device communicates with a service or application on the Internet directly.

10. The correct answer is **A**. A full-function device (FFD) must act as the coordinator. An RFD (reduced-function device) cannot.

Chapter 10: Securing Wireless Networks

Objectives Covered:

4.3   Understand and implement basic installation procedures

Wireless communication technologies are a key part of our lives. From the Wi-Fi networks we use at home or office to the more complex machine-to-machine communication in the robotics or manufacturing industry, we live in a world of wireless connectivity. It is almost impossible to spend a single day without using any wireless device. With all the blessings that these wireless technologies bring in terms of mobility and flexibility, many security concerns arise as well.

Implementing proper security controls for the selected wireless networks can no longer be an afterthought. Security must be incorporated and addressed from the initial planning and design phases throughout the lifecycle of the network. In this chapter, we will explain the fundamental security concepts that should be addressed to secure wireless networks. We will then explain the importance and need for authentication. Afterward, we will explain some vital cryptographic technologies that can be leveraged to achieve some of the security goals. A brief high-level description of the commonly used authentication methods is discussed next, followed by a brief explanation of authorization concepts. Finally, monitoring will be briefly discussed to ensure availability and integrity of the deployed network.

Confidentiality, Integrity, and Availability

Like any project, we need first to understand the requirements and objectives before we start designing the solution. Generally speaking, the attack vectors on wireless are aimed at adversely affecting one or more of these key security principles of confidentiality, integrity, and availability similar to a wired network. Therefore, it is essential to explain these principles which our security controls should address. Please note however that it is not mandatory to meet all of these requirements all the time.

- Confidentiality: You send a private message over a wireless link, and third-party systems are able to read and understand the content of the message without your consent or knowledge. Your documents and images on your favorite cloud storage website are made public without your consent or knowledge. All of these are examples of violation of confidentiality. Confidentiality ensures that only authorized people/systems should have access to information and this information shouldn't be shared with third party without your consent. The information must be protected from unauthorized disclosure.

- Integrity: You check your bank statement, and you realize that some transactions were added and you are charged for items you didn't order. You order an item online, but you get charged a different price on your card as compared to the invoice. You send a message over a wireless link, and it gets altered before reaching the intended recipient. You send an order to your smart lock to lock the door, but it gets as unlock instead. All of these are examples of integrity violation. Integrity is the guarantee of data non-alteration. Data and systems should be protected from intentional, unauthorized or accidental changes. If any alterations happen, the intended recipient should be able to identify that data was altered.

- Availability: You try to access your bank's online account, but it shows as down. You try to send an email, but it just doesn't get sent. You try to access the Wi-Fi network, but it is not available. You lost control of the IoT sensors controlling your smart home or your smart car. All of these are examples of violation of availability. Availability is the guarantee that data and systems are operating and accessible when required in a timely manner.

Confidentiality, Integrity, and Availability principles can be also complemented by other closely related security concepts. Privacy, non-repudiation, authenticity, and safety are four key concepts that strengthen the CIA concepts, and that might need to be considered as well in securing the networks.

Privacy, Non-Repudiation, Authenticity & Safety

Privacy: The terms "privacy" and "confidentiality" are often used interchangeably as they have a lot in common. However, "privacy" is more concerned about the right of the individual to keep his personal information to "himself" or "herself". It is the right of the individual not to be recorded or monitored. For example, do you want to share your browsing history? Your shopping history? Your location? Your social media information? Do you want your camera at home to start recording your conversations and uploading them to unknown party? Do you want everyone to have access to your bank account? Your health records? Do you want your car to share your location? Most likely, you consider this information to be private and you don't want to share it with everyone. The enormous use of sensors and connected devices in our lives, whether at home, in the car, or as wearable devices pose a serious privacy concern that should be addressed.

On the other hand, confidentiality is the guarantee of data privacy. The information must be protected from unauthorized disclosure. Confidentiality limits access to information to authorized entities so that it helps in achieving "privacy" for the consumers. This is very important concept nowadays where everything is connected, and the information collected from these systems can be used to identify or track a person.

Non-repudiation: You order an item online, and then you deny that you have ordered this item and refuse to settle the amount to the bank. While being angry, you send a harsh email to your colleague then you deny sending it. You launch a wireless attack from your device, and then you deny that you have done it. All of these are examples of violation of non-repudiation concept. Non-repudiation prevents a person or entity from denying having performed an action. Proper measures should be in place that prevents the subject from repudiating the claim against them. If an entity performs an action, proper unequivocal evidence should be available that confirms that this action was done by this entity.

Authenticity: You try to access your bank account, but you get redirected to an attacker website who crafted a website that exactly looks like your bank's website so you erroneously login there. You try to access the Wi-Fi network, but you connect to an "evil twin" network created by the attacker. These are example of violation of authenticity principle. Authenticity principle tries to confirm the identity of the parties that are involved in the transaction to make sure that each entity is whom it claims to be.

Safety: Safety is another very closely related concept that should be considered as well, especially in our world now where everything is connected. What if some attacker takes control of your connected car and disables the brakes? What is an attacker takes control over the connected door locks and unlocks them or keeps them locked? What if an attacker takes control of the heating systems? What if an attacker takes control of a power system and turns off the light in a city? Safety can be considered as part of availability and integrity, but we prefer to mention it here alone to highlight the importance of this aspect in our design. We are relying more and more on wireless networks, and the unavailability of a system can impact one's safety nowadays.

Now that we have explained the critical security principles that should be considered, we will now focus on the first step of authentication so that the users/devices are able to use the wireless network.

Importance of Authentication in Wireless Networks

Authentication is the first foundational step in the AAA model. Both authorization and accounting steps rely on having reliable and secure authentication. If the authentication is broken, both authorization and accounting steps are of no practical use. Authentication aims at verifying the identity of the person or object that is connecting to the resource. It is very critical to determine who or what will be authenticated? Will only one side, for example server-side, be authenticated or will both sides, for example client and server, be authenticated? When both sides (client and server) are authenticated, this is called mutual authentication. This helps prevent a third-party device from intercepting the authentication and pretending to be the other party.

Unlike wired networks which are directly linked to a particular location, wireless networks span a larger location. For example, in a wired network, you might be able to trace a user/device to a particular wall outlet / switch port while in wireless networks you might be able to trace a user/device to a particular Access Point (AP) or Base Transceiver Station (BTS) so your location accuracy will be very different as compared to wired. It is true that triangulation systems exist and can help in locating a user/device, but due to mobility, it might harder to locate a device as it keeps moving.

Another important point to consider is the risk of eavesdropping and spoofing. An attacker might use a simple receiver operating at the same frequencies and listen to the communication between two users/devices. The attacker just needs to be in range of the wireless communication, and he can do this without even being detected. If the communication is not encrypted, it will be trivial for the attacker to steal the authentication credentials and use them to gain access. That's why it is very critical to have proper authentication systems to protect the wireless network and ensure only authenticated devices/users have access to the network.

There are many authentication methods that can be used for authentication, some of which can be considered basic authentication, while others are more advanced authentication methods. The exact method to be used depends on the technology and device capability, intended use-case, ease-of-use, cost and value of the asset being protected.

In terms of technology and device support, if you are deploying a solution to authenticate temperature or humidity sensors, you use authentication methods that are supported by these sensors and not biometric solutions, for example. The device capabilities play an important role in selecting which technology can be used. If you are using cellular networks, you can leverage the SIM cards for authentication, but you can't use the same for BLE. The technology that you are using dictates the supported authentication methods. Also, one critical point to consider is the support for the chosen authentication method by all parties involved. You don't want to end up deploying a solution that is supported by your server, but no clients support it! You also need to think about how this solution integrates with your other components / systems in your network.

In terms of the intended use-case, you need to understand what are you trying to authenticate? Are you trying to authenticate both ends of the communication or only one party? How will the device/user be authenticated? Based on something they know? Something they have? Something they own? A combination of the above? All of these will impact your logical choices of authentication methods.

In terms of ease-of-use, you can deploy a very secure authentication method, but it will be tough for the users to use it, or for the devices to be configured. For example, you might set a requirement of having a password of 30 characters that should be changed daily! Yes, this solution might offer stronger security, but it is not practically usable. You might design the solution based on certificates but installing certificates on the devices might be a very tedious task. You need to find the right balance between proper security and usability.

In terms of cost and value of the asset being protected, you need to consider the value of the asset being protected from both direct monetary and non-monetary values versus the cost of the solution being deployed. Let's consider this example to better understand the non-monetary value. If you buy a laptop for 1000$, its monetary value is 1000$, and this value will decrease with time due to depreciation. However, for the past six months, you have been working on a massive project, and all the design documents are saved on this laptop. Do you still value this laptop for 1000$ or is it way more expensive? What if the laptop was hacked and your documents were sent to your competitor? Is it more valuable now? What is the price of the data on this laptop? What is the price of time spent to come up with these results? What is the price for the "bad" reputation of having your computer/system hacked? Therefore, adequate authentication solutions should be deployed in place to protect the assets. Needless to say, this is an oversimplified example just to explain the concept.

Regardless of the authentication method used, the majority of authentication methods use some cryptographic technologies to achieve the required goals. In the next section, we will discuss the key cryptographic concepts of encryption, PKI, hashing, message authentication code, digital signatures, and nonce. As you read throughout this section, try to understand the goal of each technology, and understand how it can help achieve the security goals discussed in later sections. Afterward, we will discuss some common authentication methods that can be used.

Key Cryptographic Technologies & Concepts

Encryption for Confidentiality

Due to the nature of wireless communications, wireless signals are sent over the air. It is critical to protect these communications from an eavesdropper trying to listen or spy on the channel. The best tool to protect from a malicious eavesdropper is to use encryption.

Encryption is the method of converting a plaintext or any other form of data to another encoded format that can be only decoded by another party which knows the decryption key. Therefore, the main goal of encryption is to achieve confidentiality. Any third-party entity will only see the encrypted text or data and will not be able to understand the content of the message even if it knows the encryption algorithm used.

On the sender side, encryption happens, as shown in Figure 10.1. The message or content that the sender wants to send is encrypted and converted to a ciphered text. The ciphered text is sent over the unsecured wireless medium.

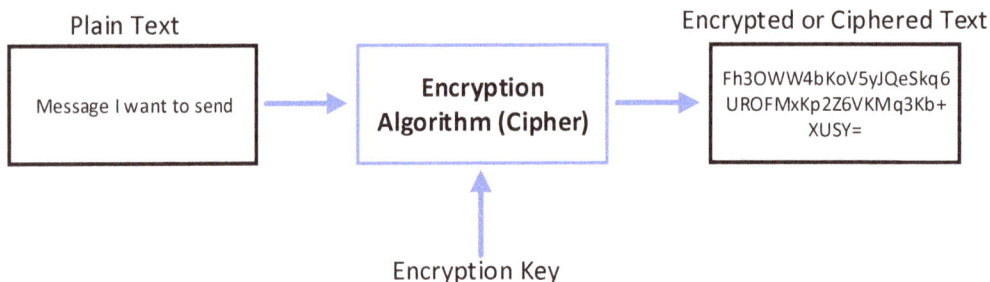

Figure 10.1: Encryption Algorithm Simplified

On the receiver side, decryption happens, as shown in Figure 10.2. The encrypted or ciphered text is converted back to the plain text.

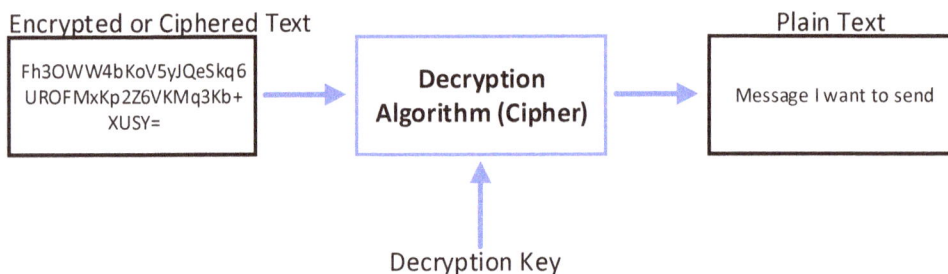

Figure 10.2: Decryption Algorithm Simplified

There are two main types of algorithms: symmetric key algorithms and asymmetric key algorithms as explained below.

Symmetric Key Algorithms

Symmetric key algorithms use the same secret key for both encryption and decryption. Let's say we need to send the message

I LOVE CWNP!

The sender can encrypt the plaintext message with the secret key he knows using the chosen encryption algorithm. The output will be an encrypted text, as shown in Figure 10.3. The sender will send the encrypted text

XA5iYo3iF5tEohMv5mgmMw==

instead of sending the actual plaintext "I LOVE CWNP!". As such, even if an eavesdropper is listening to the message, he/she will not be able to understand the content of the message. Even if the attacker records the message "XA5iYo3iF5tEohMv5mgmMw==" sent over the air, they will not be able to decipher it since the attacker doesn't know the secret key.

On the receiver end, the receiver will get the encrypted message

XA5iYo3iF5tEohMv5mgmMw==

and will use the decryption algorithm with the same secret key to decrypt the message and get the original message

I LOVE CWNP!

Note that for this example, we have used the website https://aesencryption.net/ with the secret key Lm70kCoPHl and 128 bits encryption. You can try it yourself to encrypt/decrypt some messages.

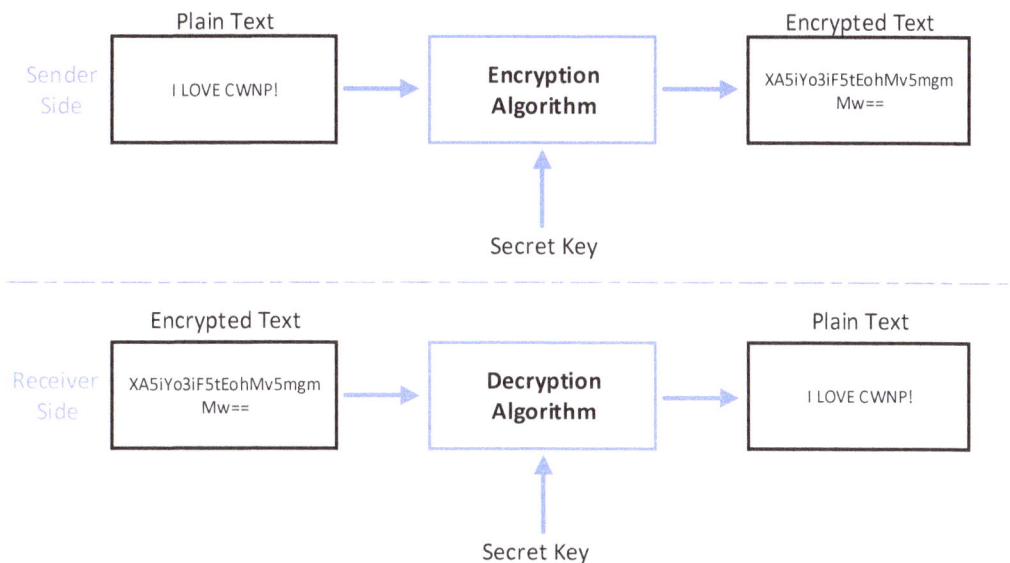

Figure 3: Encryption and Decryption Example (Symmetric Key Algorithm)

It is essential to keep this secret key secure as the entity that has this key can use it to encrypt or decrypt any message. This key is commonly referred to as single key, shared key, or session key.

Symmetric key algorithms are in general, faster than asymmetric algorithms. Symmetric key algorithms use either block ciphers or stream ciphers. Block ciphers take a predetermined number of bits, known as a block, and encrypt it. Stream ciphers encrypt data one bit, or one byte, at a time in a stream. There are many commonly used symmetric key algorithms as listed in Table 10.1. With advancement in quantum computing and computers processing power, some of these algorithms are considered weak nowadays like DES, Skipjack, Blowfish or even 3DES. It is very important to note that the strength of the algorithm doesn't depend on the algorithm being secret but rather on the mathematical strength of the encryption/decryption algorithms and the key length.

For example, the recommended symmetric algorithm to protect sensitive but unclassified information by National Institute of Standards and Technology NIST's latest report "Transitioning the Use of Cryptographic Algorithms and Key Lengths"[1] is AES. 3DES / TDES is now being deprecated.

Asymmetric Key Algorithms

Asymmetric algorithms, also known as Public Key Algorithms, use different keys for encryption and decryption, as shown in Figure 10.4. Every entity has a pair of keys: a public key and a private key. From its name, the public key is public and can be shared with any other entity. Similarly, from its name, the private key should be kept private. The public and private keys are mathematically linked by one-way functions such that if a message is encrypted with the public key, it can be only decrypted with the associated private key. Also, even if someone knows the public key, it is not possible for him/her to calculate the private key.

For the sender to securely send the message, the sender will encrypt it with the receiver's public key. As such, the sender is sure that no one decrypts it except the receiver since it is the only one that has the private key. The receiver will use its private key and decrypt the message.

Symmetric Key Algorithm	Block Size (Bits)	Key Size (Bits)
Data Encryption Standard (DES)	64	56
SkipJack	64	80
International Data Encryption Algorithm (IDEA)	64	128
Blowfish	64	32-448
Twofish	128	128, 192,256
Triple Data Encryption Standard (3DES or TDES)	64	112, 168
Advanced Encryption Standard (AES)	128	128, 192,256
Rivest Cipher 4 (RC2)	64	8-1024
Rivest Cipher 4 (RC4)	Stream Cipher	40-256 (Commonly 128)
Rivest Cipher 5 (RC5)	32, 64, 128 (Recommended 64)	0-2040
Rivest Cipher 6 (RC6)	128	0-2040

Table 10.1: Commonly Used Symmetric Key Algorithms

The same will happen when the receiver needs to send back a message to the sender. It will use the sender's Public Key (which is now the receiver) to encrypt the message. The sender (which is now the receiver) will decrypt it with its Private Key. Note that for the above example, we have used the website https://8gwifi.org/rsafunctions.jsp with the default Public and Private Keys. You can try it yourself to encrypt / decrypt some messages using asymmetric algorithms.

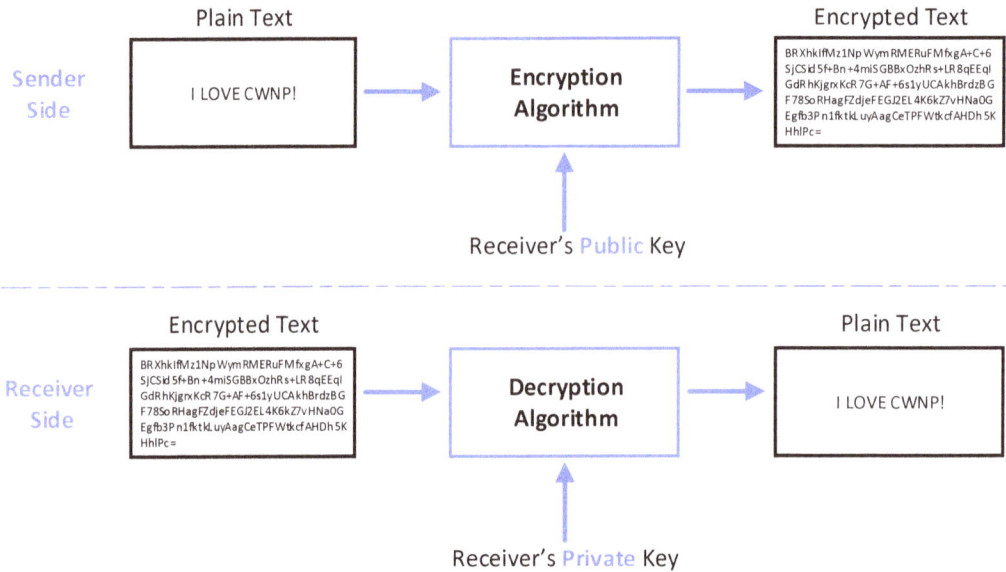

Figure 10.4: Encryption and Decryption Example (Asymmetric Key Algorithm)

Few of the commonly used asymmetric algorithms are listed in Table 10.2. It is important to note that not all asymmetric algorithms are used for encryption/decryption. Some of them are used for secure key-exchange and digital signatures, as explained in later sections.

Symmetric and Asymmetric Algorithms in Tandem

Symmetric key algorithms are in general faster than asymmetric algorithms. However, the main challenge with the symmetric key algorithms is the distribution of the common secret key between the sender and the receiver. Without having the common secret key, symmetric key algorithms can't work.

On the other side, asymmetric algorithms are slower however they don't require the initial distribution of a common secret key. To optimize the performance, it is very common to see hybrid systems utilizing both symmetric and asymmetric key algorithms.

Asymmetric Algorithms	Encryption & Decryption	Digital Signature	Key-Exchange	Description
Rivest–Shamir–Adleman (RSA)	Yes	Yes	Yes	Widely Implemented (SSL/TLS)
Elliptic Curve Cryptosystem (ECC)	Yes	Yes	Yes	Current US Government Standard, Requires Less Computing Resources (Fast)
El Gamal	Yes	Yes	Yes	Slower Compared to Others
Digital Signature Algorithm (DSA)	No	Yes	No	Used for Digital Signatures
Diffie-Hellman	No	No	Yes	Widely used with IPSEC, SSH, PGP…etc.

Table 10.2: Common Asymmetric Algorithms

- Phase 1: Asymmetric Key Algorithms are used to exchange the secret or session key.
- Phase 2: Symmetric key Algorithms are then used to encrypt the data using the secret key generated in phase 1

Figures 10.5 and 10.6 explain this concept in a simplified manner. Please note however that the generation of the session key is not implemented as such and more secure key-exchange protocols will be used like Diffie-Helman Key Exchange. For simplicity purpose only, we assumed that the sender will generate the session key to be used in phase 2 and will share it with the receiver using asymmetric encryption.

The sender needs to send a series of messages to the receiver. Using an asymmetric algorithm to encrypt every message will be slow and inefficient. The first message that the sender wants to send is "I LOVE CWNP!" However, the sender and receiver don't have a shared key to use symmetric encryption. The sender and receiver can use asymmetric encryption to generate a common session. At the end of the first phase, both the sender and receiver have a common shared key, E1UPdRxDwb, that was securely exchanged using asymmetric encryption as shown in Figure 10.5. Therefore, to encrypt the message "I LOVE CWNP!" and subsequent messages, the sender can use this key, E1UPdRxDwb, with the symmetric based algorithms as shown in Figure 10.6.

Asymmetric encryption was just used to securely send the session key. Symmetric encryption was used to actually encrypt the data.

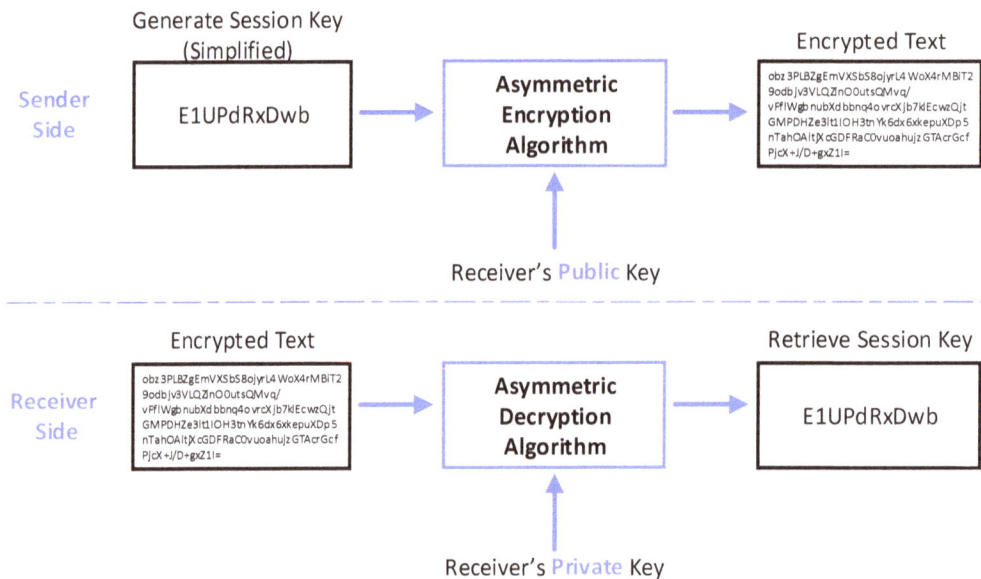

Figure 5: Exchanging Session Keys Using Asymmetric Algorithms (Phase 1 - Simplified)

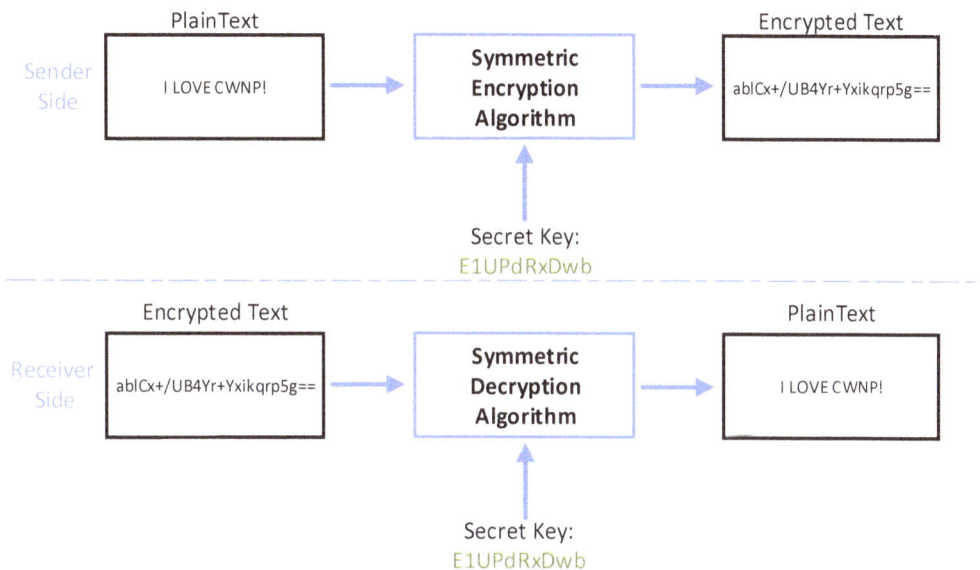

Figure 6: Data Encryption (Phase 2) using Symmetric Keys Generated in Phase 1

Lightweight Cryptography (Encryption) for Constrained Devices

It is important to note that it is challenging to apply conventional cryptographic encryption algorithms discussed above to small IoT devices like RFID tags, sensors, smart cards…etc. These cryptographic standards were optimized for desktops, servers, tablets and mobile phones that have better processing capabilities. Lightweight cryptographic techniques are proposed[2] for these systems to cater to the constraints related to "physical size, processing requirements, memory limitation, and energy drain"[3].

Below is a list of some the lightweight cryptographic encryption algorithms[3].

- Tiny Encryption Algorithm (TEA) [6]
- XTEA (Extended TEA) [6]
- Scalable Encryption Algorithm (SEA) [6]
- PRESENT
- CLEFIA
- SPONGENT
- SPECK
- SIMON
- Enocoro

- Trivium

The primary motivation of lightweight cryptography is to utilize fewer computing resources, less memory, and less power supply to provide security solution on these constrained. So, these methods are usually simpler and faster compared to conventional cryptography. However, the main disadvantage of lightweight cryptography is less security.

Public Key Infrastructure (PKI)

The strength of asymmetric algorithms discussed before relies on its ability to enable secure communication between parties that don't have a shared secret key. This is made possible by having a public key infrastructure (PKI). PKI is an infrastructure for the secure distribution of public keys that will be used in public-key cryptography, whether for secure-key exchange, asymmetric encryption or digital signatures. It consists of the collection of hardware, software, policies, processes, and procedures required to create, manage, distribute, use, store, and revoke digital certificates and public-keys.

A PKI implementation usually consists of the following elements:

- Certificate Policy
- Certificate Authority (CA)
- Registration Authority (RA)
- Digital Certificate
- Revocation Services

Certificate Policy

Certificate policy is the security specification that outlines the structure and hierarchy of the PKI ecosystem, along with the policies surrounding the management of keys, secure storage, and handling of keys, revocation, and certificate profiles/formats[14]. This is a very a key component as it outlines the role of every component in the PKI infrastructure and helps ensure the proper security posture of the whole PKI infrastructure. We can't rely on the PKI infrastructure if it is compromised. A sample certificate policy from Digicert CA can be checked at this link https://content.digicert.com/wp-content/uploads/2019/04/DigiCert_CP_v418.pdf

Certificate Authority (CA)

The CA issues digital certificates representing the ownership of a public key. Usually, a hierarchy of trust is created where there is one Root CA and several intermediate or subordinate CAs. The Root CA is usually kept offline after it signs the certificates for its intermediate CAs. This will help protect the PKI infrastructure from any attack against the root CA. The intermediate CAs issue certificates to other CAs typically known as Issuing CA or they can act themselves as Issuing CAs. Issuing CAs are the ones that maintain, issue, and distribute digital certificates. These certificates are stored in a certificate database. In case an issuing CA is compromised, its certificate and all the certificates that it issued will be revoked.

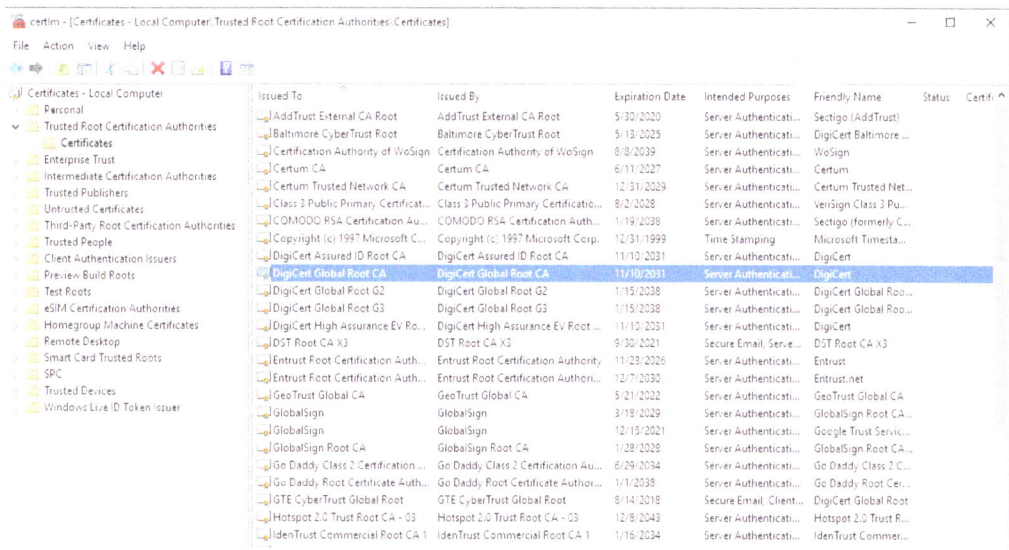

Figure 10.7: Sample of Trusted Root Certification Authorities

CAs can be internal to an organization or provided by a trusted third party. In case the CA is internal, the organization will have complete control over the lifecycle of the certificate including requesting certificate, verification, issuing, renewing, revoking certificates…etc. This will help the organization build its own internal PKI and define the certificate policies as per its needs and requirements. However, it will be hard to use this infrastructure to communicate with third-party systems since this root CA will not be trusted outside the organization. The root CA's public key needs to be shared with entities outside the organization to be able to trust this PKI infrastructure.

On the other hand, trusted third party Root CAs are well known, and they are already included as trusted root CAs in all modern operating systems. Below is a list of few well-known, trusted root CAs:

- IdenTrust
- Comodo
- DigiCert
- GoDaddy
- Global Sign
- Entrust
- Let's Encrypt

As such, an entity can request a certificate to be signed by one of these CAs. Once the entity receives the signed certificate from one of these Root CAs, it will be able to use it in the communication with other entities. The other entities will be able to communicate with this entity since they both have a common authority which they both trust.

Registration Authority (RA)

The Registration Authority (RA) validates the registration of a digital certificate with a public key. It is responsible to validate the identity of the requestor and approving or rejecting the certificate request. The RA handles issuance, revocation, or even renewal of certificates. Every time a request for verification of any digital certificate is made, it goes to the RA. RA can also issue certificates for specific use cases depending on the permissions granted to it by CA if it is acting as an issuing CA.

RAs usually support various certificate management protocols to enable faster certificate enrollment and device deployment like:

- Simple Certificate Enrollment Protocol (SCEP)
- Enrollment over Secure Transport (EST)
- Certificate Management over CMS (CMC)
- Enterprise API

X.509 Digital Certificate

A digital certificate offers the communicating party the assurance about the identity of the other party that they are communicating with. The certificate is used to associate a public key to a uniquely identified subject. The standard for digital certificates is X.509, which identifies the fields and values to be used in the certificate[11]. These fields include:

- Version of the certificate

- Unique serial number associated with the certificate
- Algorithm ID used to sign the certificate
- Issuer Name
- Validity Period
- Subject's Name
- Subject's Public Key
- ID of the issuing certificate authority
- Subject Unique

Optional extensions: These extensions, for example, include key usage, enhanced key usage, CRL distribution points, Certificate Policy, Subject Alternative Name, CRL Distribution points, and more.

Figures 10.8-10.11 show a sample digital certificate issued to the National Institute of Standards and Technology for the common name, nvd.nist.gov. The chain of trust for this certificate is shown in Figure 16; this certificate is issued by DigiCert SHA2 Secure Server CA whose certificate is issued by the DigiCert Root CA. The other details about this certificate are shown in Figures 17 and 18. It is important to note that the public key is part of this certificate.

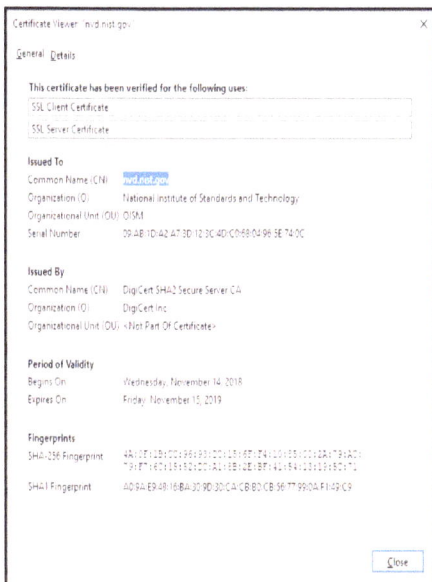

Figure 10.8: Certificate for nvd.nist.gov

Figure 10.9: Certificate for nvd.nist.gov - Certification Path

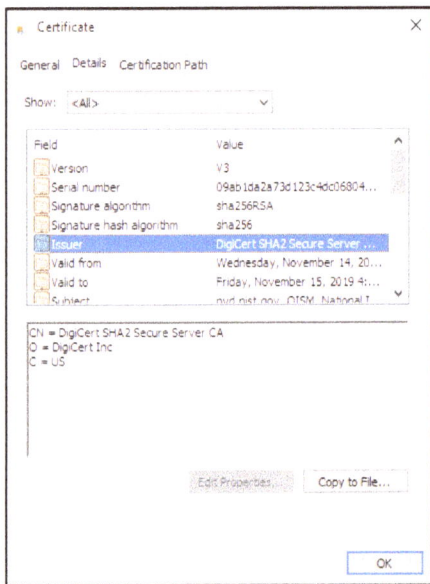

Figure 10.10: Certificate details

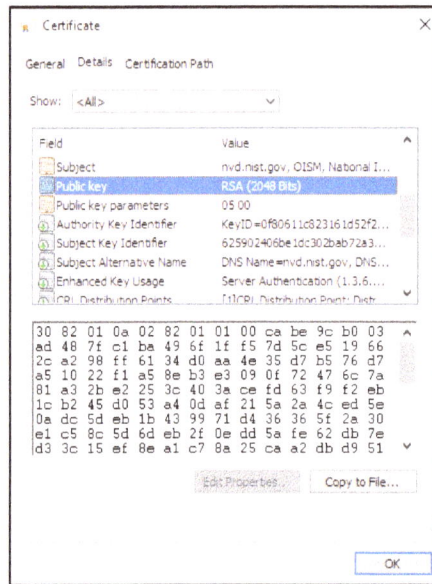

Figure 10.11: Certificate public key

Revocation Services

Revocation services offer the mechanism to terminate the trust relationship with an entity by revoking its certificate. This is a common task that the PKI should offer. There could be many reasons why a certificate should be revoked like the private key was exposed, a device was compromised or retired…etc. This is usually achieved through cryptographically signed certificate revocation lists (CRLs) which are periodically generated by the CAs. For example, in Figure 10.12, we can see the URL for the CRL Distribution point for Digicert. Accessing this URL will allow us to download the certificate revocation list file (.crl). This file contains a list of all revoked certificates, as shown in Figure 10.13. CRLs have many disadvantages including large overhead and delayed updates. The client device needs to search in the revocation list, which can grow quite large, to confirm that the serial number is not present. Also, CRLs are updated every 5 to 14 days which can leave a potential attacking surface until the next update.

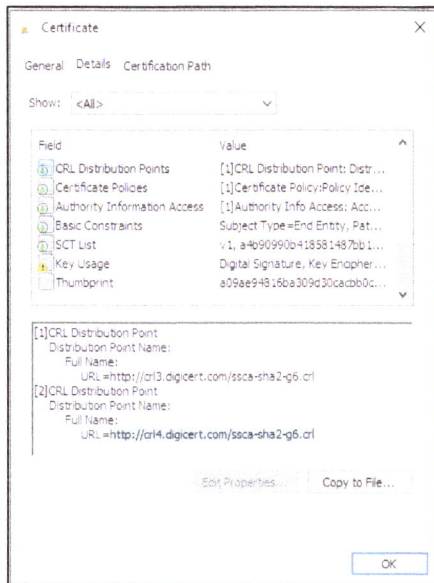

Figure 10.12: Sample CRL Distribution Point URL

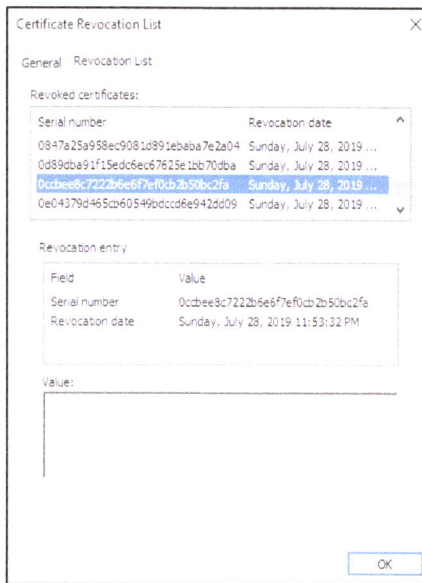

Figure 10.13: Sample Revocation List Content

The Online Certificate Status Protocol (OCSP) helps to resolve the disadvantages of CRLs. It allows the client devices to crosscheck with the CA to find out if a public key credential is still valid. The client device will send an OCSP request to the CA OCSP responder, which will reply with the status of the certificate: Good, Revoked or Unknown.

IEEE 1609.2 Certificate

The PKI-based systems have historically relied on ITU-T X.509 certificate. However, nowadays, there is a new emerging IEEE 1609.2 standard that is specifically designed for the Industrial Internet of Things (IIoT) use cases. Many current IIoT implementations still rely on X.509 certificates as they can offer robust identity and access control. The device manufacturers use the X.509 PKI to protect their devices. However, the X.509 certificates require endpoints with sufficient storage and computational power which might not be readily available in small IoT devices. That's why a new IEEE 1609.2 certificate format has been proposed, and it has approximately half the size of a typical X.509 certificate. It still uses strong elliptic curve cryptographic algorithms like ECDSA and ECDH.

The IEEE 1609.2 standard defines secure message formats and trust infrastructure for wireless access in vehicular environments (WAVE) devices. This standard, however, can possibly be extended well beyond the connected transportation industry to include other use cases across other industries. The 1609.2 certificates stores "devices permission" not "devices identities" in the certificates. Exact details about IEEE 1609.2 certificate format are outside of the scope of CWSA; however, it was briefly mentioned here to explain that X.509 certificates are not the only certificates used in PKI.

Hashing for Integrity

Sometimes our requirement is not concerned with the secrecy of the data being sent; rather we are only concerned with the intact delivery of the data from the sender to the receiver without any modification. Therefore, to ensure data integrity, hashing algorithms can be used. A hashing algorithm takes a variable-length input and generates a one-way fixed-length output that acts as the "digital fingerprint" for the input. The output is generally called a "hash" or "message digest." No key is involved in the process. You cannot take the hashed output and try to recover the original input or even know the length of the message.

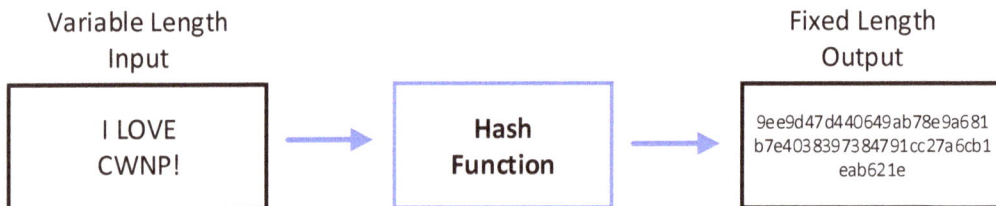

Figure 10.14: Hash Function

Some of the key properties of hashing functions are one-way, irreversible, and collision-free.

One-way means that every input results in single fixed output. Given the input and the output, it is very hard to find another input that yields the same output.

Irreversible means that given the output, it should be very hard to find the original input.

Collision-Free means it is very hard to find two inputs that will generate the same output. A change of a single bit results in a completely different output. For example, we changed one character in the message (Removed "!"), and we got a completely different hash output as shown in Figure 8.

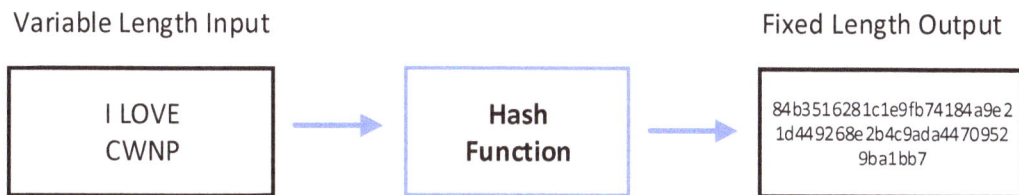

Variable Length Input   Fixed Length Output

| I LOVE CWNP | → | **Hash Function** | → | 84b3516281c1e9fb74184a9e2 1d449268e2b4c9ada4470952 9ba1bb7 |

Figure 10.15: Hash Function (Example 2 – Change 1 Character in Input – Different Output)

Note that we have used this website https://www.browserling.com/tools/sha2-hash to generate the hash for the examples above. You can try it yourself too.

The most commonly used hashing algorithms are listed below. NIST now recommends using SHA-256 or higher from the SHA2 family or any algorithm from SHA3 family[4].

Hashing Algorithms	Message Digest Size (bits)	Description
Message Digest 5	128	Considered Weak – Still widely used
Secure Hashing Algorithm 1 (SHA-1)	160	Considered Weak
SHA-224	224	Part of SHA2 Family
SHA-256	256	Part of SHA2 Family - NIST Approved - Most Popular
SHA-384	384	Part of SHA2 Family - NIST Approved
SHA-512	512	Part of SHA2 Family - NIST Approved
SHA3-224	224	Part of SHA3 Family - NIST Approved

SHA3-256	256	Part of SHA3 Family - NIST Approved
SHA3-384	384	Part of SHA3 Family - NIST Approved
SHA3-512	512	Part of SHA3 Family - NIST Approved
RIPE Message Digest (RIPEMD-160)	160	No known weakness - Less commonly used though
Whirlpool	512	No known weakness - Less commonly used though

Table 10.3: Commonly Used Unkeyed Hashing Algorithms

It is key to understand that hashing, unlike encryption, doesn't guarantee confidentiality. The message will be sent in cleartext. The goal of hashing is to ensure integrity. The sender sends the message in cleartext along with the hash of the message. At the receiver end, the receiver calculates the hash of the received message and compares it with the received hash to confirm that the message wasn't changed. If any bit in the data is changed, the hashed value will be different, and the receiver can determine that the message was altered. It is very critical to understand the importance of data integrity in an IoT environment. An intentional or unintentional modification to the data generated by an IoT sensor or a modification to command sent to a PLC can have direct consequences on system reliability, and possibly human safety.

Unkeyed hashing functions, discussed above, don't provide any way for the receiver to authenticate the message, meaning that it came from the original sender. An attacker can completely change the message and its hash value and send it to the receiver. the receiver will accept it without knowing that the message was altered since the hashed value will be correct. This is where the Message Authentication Code (MAC) and Digital Signatures explained in the following sections can help.

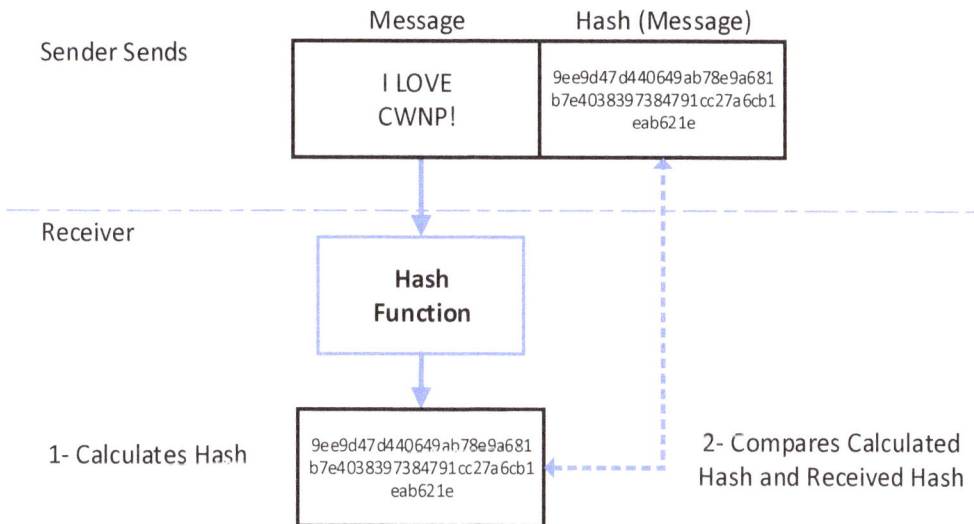

Figure 16: Message Hashing

Finally, hashing alone can be very useful if the hash is transmitted over a different protected channel. This is commonly used with file downloads. Would you ever want to load a firmware for your devices that was manipulated while you were downloading it? For sure, your answer is no. Many websites include a checksum or signature along with the file to be downloaded, as shown in Figure 10.17. As such, you can use tools like QuickHash, Hashtools, HashCompare, Hash Calculator 2…etc. on your PC to compare the hash value of the downloaded file versus the hash value listed on the site. You can thus confirm that the downloaded file is not corrupted nor altered and it matches the file that was hosted at the website.

Lightweight Cryptography (Hashing) for Constrained Devices

Similar to the challenge of applying conventional encryption techniques to constrained devices, it is as well challenging to apply conventional hashing standards to small IoT devices. Actually, none of these approved hash functions by NIST, mainly SHA2 Family and SHA3 Family hash function, are suitable for use in very constrained environments, mainly due their large internal state size requirements [2]. This had led to the development of hashing functions that are optimized for these environments.

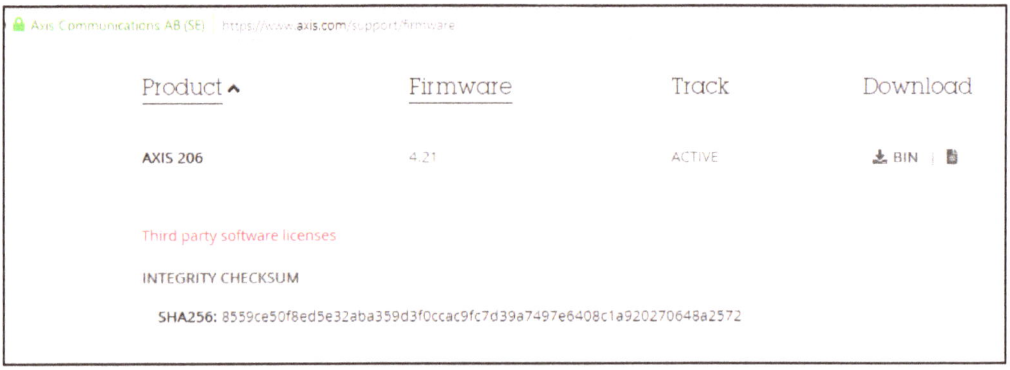

Figure 10.17: Sample Integrity Checksum for File Download

Below is a list of some the lightweight cryptographic hashing algorithms [2]

- PHOTON
- SPONGENT
- Lesamnta-LW
- QUARK

Message Authentication Code for Integrity and Authentication

Message Authentication Code (MAC), also known as a keyed hash, ensures message integrity and protects against message forgery by anyone who doesn't know the secret key. This key is only shared between the sender and the receiver. MAC algorithms can be constructed in different ways like cryptographic hash functions, block cipher algorithms, universal hashing, and more. Following is a list of the most commonly used algorithms where the ones highlighted in bold are the currently NIST approved algorithms[5].

- BLAKE2
- HMAC (Based on Approved Hash Functions)
- KMAC (Based on KECCAK)
- MD6
- OMAC
- CMAC (Based on AES Block Cipher)
- PMAC
- Poly1305-AES
- UMAC

- VMAC

To explain how MAC work, we will use the example in Figures 10.18 and 10.19. The sender calculates the hash value using both the message and a secret key and not the message alone.

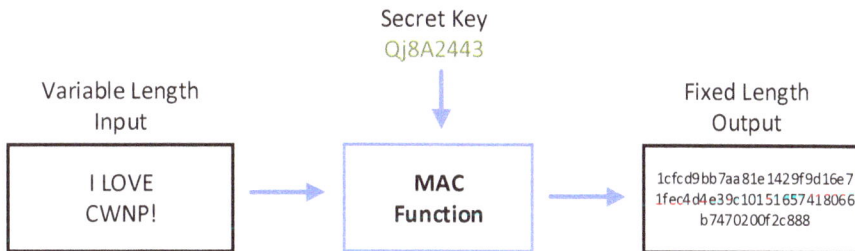

Secret Key
Qj8A2443

Variable Length
Input

Fixed Length
Output

| I LOVE CWNP! | → | MAC Function | → | 1cfcd9bb7aa81e1429f9d16e7 1fec4d4e39c10151657418066 b7470200f2c888 |

Figure 10.18: HMAC Example

The sender then sends the message with the calculated hash of the message and the secret key. At the receiver side, the receiver calculates the hash using the message and the known secret key. If the calculated hash value matches the one sent by the sender, then the receiver is sure that the message was not altered by the third party and it was sent by the sender since it is the one that knows the secret key; thus we have achieved both integrity and authentication.

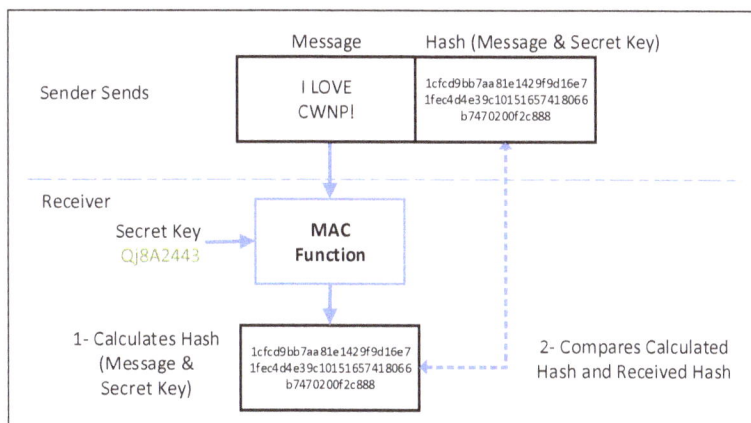

Message   Hash (Message & Secret Key)

Sender Sends

| I LOVE CWNP! | 1cfcd9bb7aa81e1429f9d16e7 1fec4d4e39c10151657418066 b7470200f2c888 |

Receiver

Secret Key
Qj8A2443

MAC Function

1- Calculates Hash (Message & Secret Key)

1cfcd9bb7aa81e1429f9d16e7 1fec4d4e39c10151657418066 b7470200f2c888

2- Compares Calculated Hash and Received Hash

Figure 10.19: HMAC Verification - Receiver Side

An attacker will not be able to change the content of the message and generate a valid hash since the attacker doesn't know the secret key. This is how MAC provides authentication which wasn't available if hashing function was used alone without a secret key.

One limitation for MAC is that it doesn't provide non-repudiation. The receiver can forge any message from the sender and claim that it was sent by the sender since both entities share the same key. Also, in MAC, only the receiver can verify the message since it is the only one that knows the shared secret key. Digital Signatures can be used to overcome these limitations of MAC.

Note that we have used this website https://www.freeformatter.com/hmac-generator.html for the example above with the key "Qj8A2443" and the hashing algorithm SHA-256. You can try it yourself too.

Digital Signatures for Integrity, Authentication & Non-Repudiation

Digital signatures can be used to provide message integrity, authentication, and non-repudiation. A digital signature is signed with a private key and verified with the corresponding public key. Only the holder of the private key can create this signature, since it is the only entity knowing this private key, and anyone knowing the public key can verify it. The main idea of a digital signature is that one entity can sign a message whereas any other entity can verify the correctness of the signature. A digital signature does not provide confidentiality.

To understand how digital signatures work, we will use Figures 10.20 and 10.21 [7]. A sender calculates the hash of the message that he wants to send. The sender then encrypts the hash with the sender's private key. The sender can then send the message along with the signed hash.

At the receiver side, the receiver calculates the hash of the message. In parallel, the receiver decrypts the signed hash using the sender's public key. The receiver compares both hashes. If the hashes are equal, the receiver can be sure that the message was sent by the sender since it is the only one possessing the private key. The receiver is also sure that no one altered the message since the hash values are correct. As such, a digital signature can provide authentication, integrity, and non-repudiation.

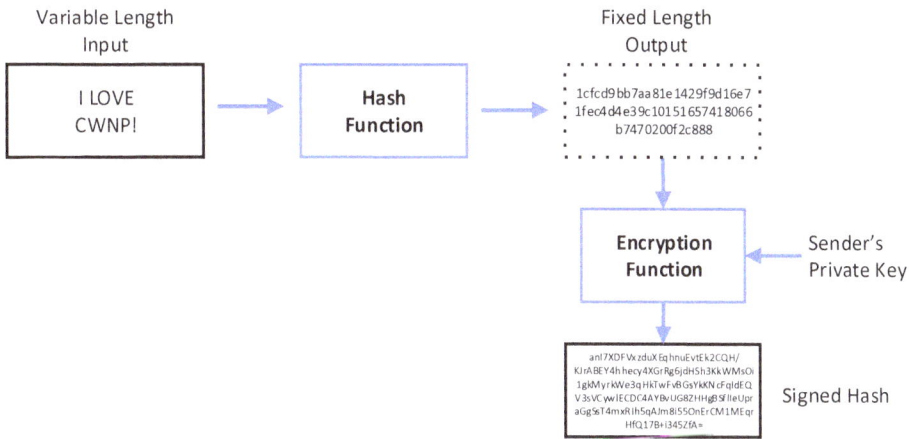

Variable Length Input

I LOVE CWNP!

Hash Function

Fixed Length Output

1cfcd9bb7aa81e1429f9d16e7
1fec4d4e39c10151657418066
b7470200f2c888

Encryption Function

Sender's Private Key

anI7XDFVxzduXEqhnuEvtEk2CQH/
KJrABEY4hhecy4XGrRg6jdHSh3KkWMsOi
1gkMyrkWe3qHkTwFvBGsYkKNcFqIdEQ
V3sVCywlECDC4AYBvUG8ZHHgBSflleUpr
aGgSsT4mxRIh5qAJm8i55OnErCM1MEqr
HfQ17B+i345ZfA=

Signed Hash

Figure 10.20: Digital Signature Signing (Part 1)

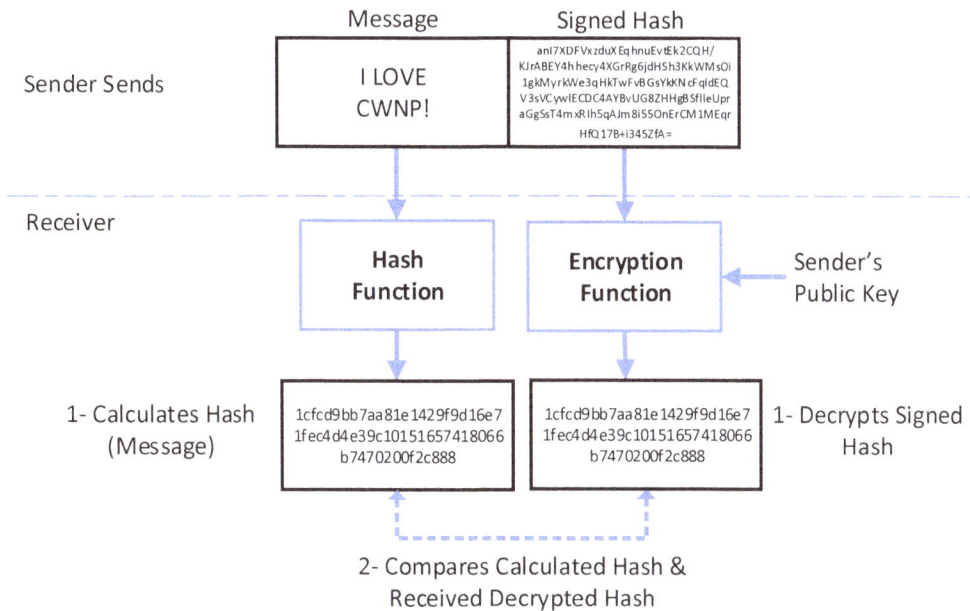

Message

Signed Hash

Sender Sends

I LOVE CWNP!

anI7XDFVxzduXEqhnuEvtEk2CQH/
KJrABEY4hhecy4XGrRg6jdHSh3KkWMsOi
1gkMyrkWe3qHkTwFvBGsYkKNcFqIdEQ
V3sVCywlECDC4AYBvUG8ZHHgBSflleUpr
aGgSsT4mxRIh5qAJm8i55OnErCM1MEqr
HfQ17B+i345ZfA=

Receiver

Hash Function

Encryption Function

Sender's Public Key

1- Calculates Hash (Message)

1cfcd9bb7aa81e1429f9d16e7
1fec4d4e39c10151657418066
b7470200f2c888

1cfcd9bb7aa81e1429f9d16e7
1fec4d4e39c10151657418066
b7470200f2c888

1- Decrypts Signed Hash

2- Compares Calculated Hash & Received Decrypted Hash

Figure 10.21: Digital Signature Verification (Part 2)

As mentioned earlier in Table 3, many asymmetric algorithms can be used to generate digital signatures. Some of these algorithms include:

401

- Rivest–Shamir–Adleman (RSA)
- Elliptic Curve Digital Signature Algorithm
- El Gamal
- Digital Signature Algorithm (DSA)
- Encryption & MAC for Confidentiality and Authentication

The above-mentioned methods are not exclusive. There might be a requirement to provide authentication of the source and confidentiality for the message. Therefore, a combination of the above methods can be used like:

Encrypting the message then calculating the MAC based on the encrypted result. This is commonly known as Encrypt-then-MAC

Creating a MAC for the message then encrypting the message and the MAC. This is commonly known as MAC-then Encrypt

Encrypting the message and calculating the MAC of the message and sending both. This is commonly known as Encrypt-and-MAC

Nonce

To complete our discussion of the key cryptographic technologies, it is important to mention the concept of a nonce. The nonce is very essential to prevent replaying old messages and ensuring message freshness. A nonce, "a number used once," is an arbitrary value that can be used only once in a cryptographic communication. It is often a random or pseudo-random number issued in an authentication protocol to ensure that old communications cannot be reused to perform a replay attack. They can also be used as initialization vectors and in cryptographic hash functions. For example, if the same message is encrypted by the same key, the same ciphered text will be obtained. However, if a nonce is used as an input to the encryption message along with the message, the ciphered text will be different even if the same message is being encrypted since the nonce is different. As such, an attacker will not be able to know that the same message is being sent. In addition, timestamps might also be used to help prevent replay attacks.

Cryptographic Technologies Summary

This section covered in brief the key cryptographic technologies that can be used to secure a network, whether it is wired or wireless. Table 10.4 shows a summary of these techniques and how they can help achieve the associated security goal. It is important to note that cryptographic techniques alone are not able to cover all security goals. We should design the network with multiple security layers that address the different requirements.

	Cryptographic Technique			
Security Goal	Encryption (Ciphering)	MAC	HASH	Digital Signature
Confidentiality	Yes	No	No	No
Integrity	No	Yes	Yes	Yes
Authentication	No	No	Yes	Yes
Non-repudiation	No	No	No	Yes
Kind of Keys	Symmetric or Asymmetric Keys	NA	Symmetric Keys	Asymmetric Keys

Table 10.4: Cryptographic Technique to Security Goal Mapping Summary

Authentication Methods

As explained before, the authentication step acts as the gatekeeper for the other security tasks. With authentication, we can ensure:

- Confidentiality by restricting access to authorized entities only
- Integrity by ensuring modification by authorized entities only
- Non-repudiation by tracing the activity to a particular entity

Authentication can happen in different methods based on

- Something a user knows
- Something a user has
- Something a user is

A combination of the above aka multifactor authentication

Below sections cover the most commonly used authentication methods. Some of these methods are better suited for human to machine types of authentication while others can work with both machine-to-machine or human to machine use cases.

Password-Based Authentication Methods

Password-based authentication is the predominantly used method for a user to authenticate to a system. It falls under the category of "something a user knows." Every entity has an account/username and associated password. The user needs to login by providing the username and password. If the provided username / password combination is correct, the user is given access to the system. This is very similar to what we commonly use to login to any website that requires a username/password for authentication like Facebook, Gmail…etc. This type of authentication is better suited for human to machine type of setup and is not ideal for IoT M2M type of step since:

- It is hard to provision and manage usernames and passwords for a large number of devices
- It is hard to handle the complete account lifecycle process, including the initial deployment and periodic password updates. How easy is it to update the password for thousands of devices?
- It is hard to securely store the passwords, and it might become a backdoor for intrusion. What if the device got stolen?
- It is common to see many IoT attacks happening because the default passwords were kept unchanged
- If the underlying transport/messaging protocol is insecure (like Telnet, HTTP, FTP, MQTT…etc), the username and password might be sent in cleartext

Key-Based Authentication

Key-based authentication falls under the category of "Something a user has". This can be in the form of:

- Shared Symmetric Key [18]: This is the easiest form to produce at scale since the same shared key will be installed on all the devices. However, the risk of such solution outweighs the benefits it brings so this solution should never be practically used.
- Symmetric Key[18]: A symmetric key will be installed on the device and its associated backend platform. This will allow device-to-backend communication using this symmetric key. The challenge of this approach is how to ensure that the symmetric key is adequately secured on the device and on the backend platform.

404

- Trusted Platform Module (TPM) [18]: TPM can be used to securely store keys or even X.509 certificates on the devices. This can offer a more secure authentication method as compared to Symmetric Keys.

Certificate-Based authentication

Certificate-based authentication falls under the category of "Something a user has." It uses the PKI infrastructure where the public key is signed by a trusted CA, as explained previously. The entities can thus securely authenticate each other since they have certificates that are signed by a trusted CA. Certificate-based authentication greatly simplifies device identification and offers a very scalable solution.

Public-key-based digital certificates seem to be the best solution for the majority of IoT authentication use cases that don't involve resource-constrained devices. This setup acts as the foundation to secure the communication between IoT devices, and IoT devices and their cloud platforms. As explained previously, work is done to improve the PKI infrastructure to handle constrained devices like the enhancements being done in lightweight cryptography and in IEEE 1609.2.

Biometric-Based Authentication Methods

Biometric-based authentication methods are being used more often nowadays, especially in areas where security is a top priority like border controls. This authentication method falls under the category of "Something a user is." These methods are as well used as a second-factor authentication in a multifactor authentication setup. Biometric-based setups include fingerprint, retina or iris scan, voice analysis, facial geometry, hand geometry...etc. These setups work well in human-to-machine cases and are not intended for machine-to-machine authentication.

For example, a biometric-based authentication setup can be used to authenticate a technician trying to access the control room. Similarly, we are now seeing more biometric-authentication used in consumer IoT devices like smart biometric locks. For example, Figure 10.22 shows some smart locks with biometric capabilities like voice activation and integration capabilities with Amazon Alexa, Apple Homekit, Google Assitant and Nest.

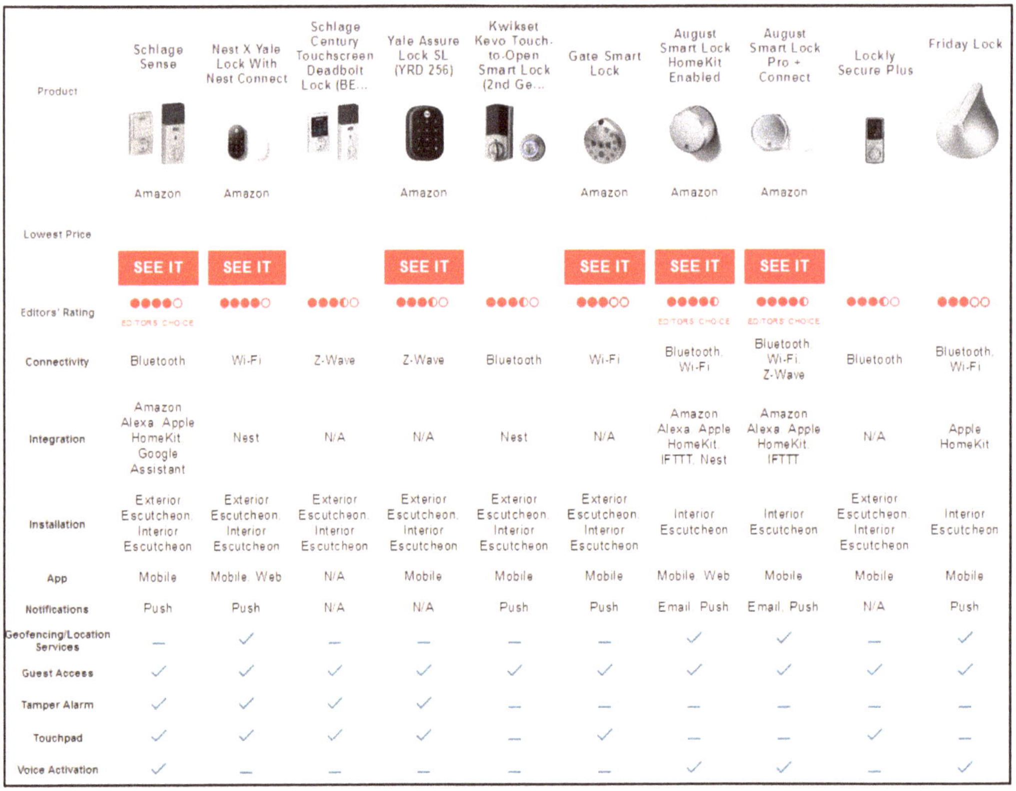

Product	Schlage Sense	Nest X Yale Lock With Nest Connect	Schlage Century Touchscreen Deadbolt Lock (BE...	Yale Assure Lock SL (YRD 256)	Kwikset Kevo Touch-to-Open Smart Lock (2nd Ge...	Gate Smart Lock	August Smart Lock HomeKit Enabled	August Smart Lock Pro + Connect	Lockly Secure Plus	Friday Lock
	Amazon	Amazon		Amazon		Amazon	Amazon	Amazon		
Lowest Price	SEE IT	SEE IT		SEE IT		SEE IT	SEE IT	SEE IT		
Editors' Rating	●●●●○ (Editors' Choice)	●●●●○	●●●○○	●●●●○	●●●●○	●●●○○	●●●●● (Editors' Choice)	●●●●● (Editors' Choice)	●●●●○	●●●○○
Connectivity	Bluetooth	Wi-Fi	Z-Wave	Z-Wave	Bluetooth	Wi-Fi	Bluetooth, Wi-Fi	Bluetooth, Wi-Fi, Z-Wave	Bluetooth	Bluetooth, Wi-Fi
Integration	Amazon Alexa, Apple HomeKit, Google Assistant	Nest	N/A	N/A	Nest	N/A	Amazon Alexa, Apple HomeKit, IFTTT, Nest	Amazon Alexa, Apple HomeKit, IFTTT	N/A	Apple HomeKit
Installation	Exterior Escutcheon, Interior Escutcheon	Exterior Escutcheon, Interior Escutcheon	Exterior Escutcheon, Interior Escutcheon	Exterior Escutcheon, Interior Escutcheon	Exterior Escutcheon, Interior Escutcheon	Exterior Escutcheon, Interior Escutcheon	Interior Escutcheon	Interior Escutcheon	Exterior Escutcheon, Interior Escutcheon	Interior Escutcheon
App	Mobile	Mobile, Web	N/A	Mobile	Mobile	Mobile	Mobile, Web	Mobile	Mobile	Mobile
Notifications	Push	Push	N/A	N/A	Push	Push	Email, Push	Email, Push	N/A	Push
Geofencing/Location Services	—	✓	—	—	—	—	✓	✓	—	✓
Guest Access	✓	—	✓	✓	—	—	—	—	✓	✓
Tamper Alarm	✓	✓	✓	✓	—	—	—	—	—	—
Touchpad	✓	—	✓	✓	—	✓	—	—	✓	—
Voice Activation	✓	—	—	—	—	—	✓	✓	—	✓

Figure 10.22: Consumer Smart Locks with Biometric Capabilities like Voice Activation [17]

Smart Cards

Smart Card is a credit card-sized ID that has an integrated circuit chip embedded in it. It falls under the category of "Something a user has." Most smartcards include a microprocessor and one or more certificates. The certificates are used for asymmetric cryptography, including encryption or digital signatures. Smart Cards are better suited for human to machine type of authentication and not machine to machine.

One-time Password (Tokens)

One-time Password (OTP) is a dynamic password that is only valid for a single session. It falls under the category of "Something a user has." For example, a user can have an OTP sent to his phone. Another option is to have software token, like Google Authenticator, or hardware token, like RSA SecurID to get the OTP used to login to the system. Usually, OTP passwords are used as a second-factor authentication in a multi-factor authentication method. OTPs are better suited for human to machine type of authentication and not machine to machine.

Multi-factor Authentication Methods

Multi-factor authentication (MFA) is an authentication method that uses two or more factoring techniques; something a user knows, something a user has or something a user is. For example, a user needs to provide his password and an OTP password that is sent out of band. Or a user needs to swipe a card and enter a PIN. A user might need to enter a PIN first then complete a fingerprint scan. Multi-factor authentication helps make it more difficult for the attacker to access the target. MFA is better suited for human to machine type of authentication and not machine to machine.

Authorization

Authorization is the second step in the AAA process, and it relies on authentication. So, if authentication can be spoofed or impersonated, then the authorization step is almost useless. Authorization uses access control mechanisms to authorize access to resources. Therefore, it is important to discuss various access control concepts that are relevant to authorization.

Key Access Control Concepts

Implicit Deny

Most access control systems use the "Implicit Deny" principle where access to resources is blocked by default unless it is explicitly allowed for a particular entity. For example, by default, no one is allowed to access your Dropbox files. You can, however, decide to share some files with specific people. Only those specific people will thus have access to these files. Other people will still be blocked from accessing the files. Similarly, a firewall by default blocks all traffic. However, traffic that you need to be permitted can be allowed. All other traffic type is blocked. Therefore, when we are deploying a wireless system, it is key to understand the components involved, their interaction in terms traffic flows and make sure that only needed services are exposed and everything else is protected.

Access Control Matrix

An access control matrix is a table that includes a list of:

- Subjects: meaning entities trying to access the resources
- Objects: meaning the resources that will be access

- Assigned Privilege

For example, in Table 6, the subjects are User 1, User 2, and User 3. They are trying to access the resources or objects: Camera, Door Bell, Door Lock, and Thermostat. Each user will have a different privilege as indicated in the access control matrix. User 1, for example, will have full control over all the devices while user 2 has view-only privilege. User 3 has different privileges depending on the object he/she is trying to access. Please note that this example is simplified to explain the concept. Exact user privileges can be more or less granularly controlled based on the application. Also, instead of having the access matrix based on a particular user, it can be done based on a particular role, making it more scalable.

	Camera	Door Bell	Door Lock	Thermostat
User 1	Full control	Full control	Full control	Full control
User 2	View Only	View Only	View Only	View Only
User 3	Full Control except Network Settings	View Only	No access	Full control

Table 10.5: Access Control Matrix (Simplified Example)

Therefore, when we deploy a wireless network, it is critical to identify the different resources that are available and make sure we sign the right privileges for the entities trying to access these resources.

Constrained Interface

Building upon the example above, User 3 shouldn't have access to the Door Lock. Therefore, the application should have a "constrained interface" where all the settings related to the door lock are removed from User 3 account or at least disabled or dimmed. This is the concept of constrained interface where, depending on the privilege of the user, certain features will be available or unavailable. This, for example, matches with issue number 9 "Insecure Default Settings" in OWASP IoT Top 10 highest priority issues in IoT deployments [10] where many IoT applications will not have the options to restrict access in a granular manner.

Content-Dependent Control

Content-Dependent control checks the access of users to resources based on the content of the resource. For example, an email filter might allow emails to be sent. However, if an email contains a virus, the email filter will block the email even though the user has permissions to send an email. Another common use case is in subscription-based services. If the subscription expires, the user might still be able to login to the portal but the services offered might stop working until the subscription is renewed. The latter example doesn't only impact authorization, but it also affects the overall availability of the system.

Context-Dependent Control

Context-dependent controls check the context of the request before granting access. For example, if we need to add time of day restrictions, we can use context-dependent controls. The user might only have access to the system during working hours. After working hours, even if the user uses the right credentials, he/she might not be able to access the system. Another common example is restricting access based on restricted management IP addresses. Even if the user manages to get access to the management interface and tries to login with the correct credentials, the request will be denied if the connection is not coming from a whitelisted IP. This concept can be used for instance to protect the management network that will be used to manage and monitor the wireless network.

Need to Know

This concept states that subjects should be given access only to what they need to know to complete their job. For example, a technician is installing a new wireless network or a new camera. If he can check that the system is operational after install without the need to login to the backend management interface, then there is no need for the technician to have access to the backend management interface. By limiting access to the system to users who really need access to the system, chances of data leakage or unintentional human errors will be minimized.

Least Privilege

Least privilege concept states that once an entity is authorized, it should be given the lowest privilege needed to complete its task. For example, the security guards checking the cameras should have access to view the camera feeds. However, they shouldn't necessarily be given access full access to the DVR where they can delete some recordings. If the system needs internet access to function, then internet access can be provided. However, if it can work without Internet access, internet access can be blocked. As such, the potential attack surface will be reduced.

Principle of Segregation of Duties

Principle of segregation of duties ensures that sensitive functions are divided across multiple employees each doing a subset of the tasks. For example, in deploying a new PKI environment, securing the private key of the root CA is a very critical function. If this key is compromised, the whole PKI infrastructure will be useless, so we can't rely on a single employee to complete this function. Usually, n-of-m controls are implemented where n employees out of m need to collaborate to access the key.

Authorization Maps Permissions to Entities

Regardless of how authorization is performed, the end goal of authorization is to map the right permissions to entities. The permissions can be based on individual user/device or based on groups of users/devices. Commonly, role-based access control (RBAC) mechanisms are used where each role has a defined set of permissions. A user is thus assigned a role based on its group memberships, and accordingly it gets the permissions linked to those roles. In general, having the authorization done at group/role level helps minimize errors and ensure consistency as there will be fewer roles to manage rather than applying a policy at a specific user/device level. Some of the ways authorization ways to map devices/users connecting to the network to role includes using authentication servers like RADIUS, TACACS+ or DIAMETER.

Nowadays, almost all devices and systems being built have sort of HTTP API to facilitate integrations with third-party systems. Therefore, it is key to not only authorize physical connections but also, it is important to authorize API connections. This is where OAuth 2.0 framework helps.

OAuth 2.0 Authorization Framework

OAuth 2.0 is an authorization framework that enables third-party applications to obtain limited access to user accounts on an HTTP service. It works by delegating user authentication to the service that hosts the user account and authorizing third-party applications to access the user account. OAuth 2.0 provides authorization flows for web and desktop applications, and mobile devices.

OAuth defines four roles:

- Resource Owner: The resource owner is the user who authorizes an application to access his/her account.
- Client: The client is the applications or websites that want to access the user's account. However, the application must be authorized by the user before being given access, and the authorization must be validated by the API.
- Resource Server: The resource server hosts the protected user accounts
- Authorization Server: The authorization server verifies the identity of the user then issues access tokens to the application

Let's take an example from https://developers.nest.com/guides/api/how-to-auth. The resource owner is the user on the left in this case. The client is the application "Your Product." The resource server and authorization servers are shown as "Nest Cloud." At the end of these exchanges, the application "Your Product" will have a token that can be used to call Nest APIs. The application will have permissions as granted by the user.

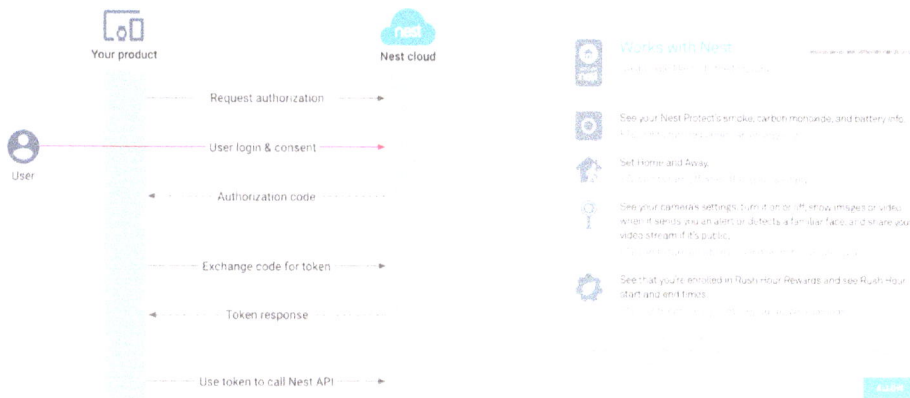

Figure 10.23: Sample Auth 2.0 Workflow - Nest [20]

Figure 10.24: Sample Auth 2.0 - User Prompt to Grant Access - Nest [21]

Many IoT protocol extensions like CoAP (Tschofenig-ACE)[22] [23] and SASL (SASL-OAUTH)[24] are also being modified to fit into the OAuth authentication and authorization framework. This can be very useful in resource-constrained devices.

Monitoring

Like any system or network deployment, it is critical to deploy proper monitoring tools to monitor the system or network post-deployment and raise alerts in case of an anomaly or issue. The monitoring tools should monitor all the components of the solution to ensure availability, security, and proper functionality. Exact criteria to be monitored depends on the deployed system, but in general as a minimum the following information needs to be monitored where applicable.

- Basic Network Reachability: This can be for example tested by a scheduled ping
- Basic Service Reachability: This can be for example tested by accessing reachability to a particular port
- Basic Performance Related Parameters like CPU, Memory, Disk Utilization
- Basic Environmental Related Parameters like Temperature, FAN, Power Status
- Utilization Related Parameters like System Load, Interface Utilization, Top Talkers
- Error Conditions like Packet Drops, Retries, Down Interfaces
- Internet Link Utilization and delay where applicable
- Radio Interface Statistics
- Security Event Logs
- Error Logs

Wherever certificates are used, proper monitoring tools that check and monitor certificates validity should be used to proactively detect certificates that will expire

The time invested in properly deploying a monitoring solution and tuning it will be greatly valued when an issue occurs. A properly configured monitoring solution will help pinpoint issues faster and thus minimize the time needed to resolve them. As such, this can help achieve higher system availability. It is, therefore, crucial to spend a considerable amount of time in this process trying to understand the dependencies and make sure all the dependencies are monitored.

Moreover, integration with third-party monitoring tools, logging systems, helpdesk ticketing, and notification systems, like SMS, should be completed. It is critical to make sure that time is synchronized on the deployed system, preferably via NTP, to be able to properly trace and correlate events when needed. It is very important as well to make sure to forward security logs from the deployed systems to centralized logging servers or security information and event management (SIEM) solutions where applicable.

One very important aspect of monitoring is accounting for both end-user or device access and for admin access. Accounting is the last step in the AAA process. Once access is granted, the account associated with the user or device needs to be regularly monitored to check activity logs. This is key to understand resource usage by individual users or devices, and also to detect any malicious or suspicious activity that can impact the system. Nowadays, there are tools, like User and Entity Behavior Analytics tools, that use machine learning (ML) techniques and artificial intelligence to try to detect such malicious activity. These tools will be more common once we have more IoT devices communicating in a connected M2M setup as it will be almost impossible for human beings to detect malicious activity in the plethora of logs generated.

Accounting for admin access is usually done via RADIUS, TACACS+, or DIAMETER. Audit trail logs as well can be used to determine the activities done by an admin user.

Chapter Summary

In this chapter, we covered the key security concepts that should be addressed to secure wireless networks. We then discussed the core security goals that any security design should address. Afterward, we explained some key cryptographic technologies like encryption, hashing, MAC, digital signatures, PKI…etc. that can be leveraged to achieve some of the security goals. Finally, we explained various authentication techniques, authorization concepts and the importance of deploying a proper monitoring solution.

References

Barker, Elaine, and Allen Roginsky. Transitioning the use of cryptographic algorithms and key lengths. No. NIST Special Publication (SP) 800-131A Rev. 2 (Draft). National Institute of Standards and Technology, 2018.

McKay, Kerry A., et al. "NISTIR 8114 report on lightweight cryptography." National Institute of Standards and Technology (NIST), Gaithersburg (2017).

Buchanan, William J., Shancang Li, and Rameez Asif. "Lightweight cryptography methods." Journal of Cyber Security Technology 1.3-4 (2017): 187-201.

Hash Functions – Source: https://csrc.nist.gov/Projects/Hash-Functions/NIST-Policy-on-Hash-Functions

Message Authentication Codes – Source: https://csrc.nist.gov/Projects/Message-Authentication-Codes

Gallagher, Patrick. "Digital signature standard (dss)." Federal Information Processing Standards Publications, volume FIPS (2013): 186-3.

Liyanage, Madhusanka, et al., eds. A Comprehensive Guide to 5G Security. John Wiley & Sons, 2018.

Padgette, John, Karen Scarfone, and Lily Chen. "NIST Special Publication 800-121 Revision 2, Guide to Bluetooth Security." 2012-6-30) [2015-4-20]. http://csrc.nist.gov/publications/nistpubs/800-121-revl/sp800-121revl. pdf (2017).

PKI: The Security Solution for The Internet Of Things – Source: https://resources.digicert.com/internet-of-things/pki-the-security-solution-for-the-internet-of-things

OWASP Internet of Things Project- Source: https://www.owasp.org/index.php/OWASP_Internet_of_Things_Project

Ballad, Bill, Tricia Ballad, and Erin Banks. Access control, authentication, and public key infrastructure. Jones & Bartlett Publishers, 2010.

TLS Authentication using IEEE 1609.2 certificates – Source: https://tools.ietf.org/pdf/draft-msahli-ise-ieee1609-00.pdf

Shanks, Wylie. "Building and Managing a PKI Solution for Small and Medium Size Business" STI Graduate Student Research - December 23, 2013

Public Key Infrastructure Explained- Source: https://www.kyrio.com/blog/internet-of-things-security/public-key-infrastructure-explained

Bhattacharjee, Sravani "Practical Industrial Internet of Things Security", Packt Publishing
July 2018

Brecht, B., Therriault, D., Weimerskirch, A., et al. "A Security Credential Management System for V2X Communications", Feb 2018

https://uk.pcmag.com/smart-locks/77460/the-best-smart-locks

IoT device authentication options -Source: https://azure.microsoft.com/en-us/blog/iot-device-authentication-options/

Differentiating Between Access Control Terms – Source:
http://www.windowsecurity.com/uplarticle/2/Access_Control_WP.pdf

OAuth 2.0 Authentication and Authorization – Source:
https://developers.nest.com/guides/api/how-to-auth

Authorization Overview – Source:
https://developers.nest.com/guides/api/authorization-overview

Authentication and Authorization for Constrained Environments (ACE): Overview of Existing Security Protocols draft-tschofenig-ace-overview-00.txt – Source:
https://tools.ietf.org/html/draft-tschofenig-ace-overview-00

Authentication and Authorization for Constrained Environments (ACE) using the OAuth 2.0 Framework (ACE-OAuth) draft-ietf-ace-oauth-authz-17 – Source:
https://tools.ietf.org/html/draft-ietf-ace-oauth-authz-17

A SASL Mechanism for OAuth draft-mills-kitten-sasl-oauth-02.txt – Source:
https://tools.ietf.org/id/draft-mills-kitten-sasl-oauth-02.html

Review Questions

1. Which one of the following is not part of the CIA security concept?

    a. Integrity

    b. Availability

    c. Authentication

    d. Confidentiality

2. What security process can provide integrity?

    a. Authentication

    b. Hashing

    c. Encryption

    d. Accounting

3. What security concept prevents an individual from denying they performed an action?

    a. Authentication

    b. Authorization

    c. Non-Repudiation

    d. Ill-Reputation

4. What kind of encryption uses the same key to encrypt and decrypt?

    a. Symmetric

    b. Asymmetric

    c. SHA1

    d. MD5

5. What is the current encryption algorithm most recommend for use in place of 3DES?

    a. RC4

    b. TDES

    c. AES

    d. PKI

6. What is an example of a symmetric key algorithm?
    a. Blowfish
    b. RSA
    c. ECC
    d. DSA

7. What is an example of an asymmetric key algorithm?
    a. AES
    b. RC4
    c. Diffie-Hellman
    d. Twofish

8. What is the most common standard for digital certificates used for authentication and encryption?
    a. SHA1
    b. 802.1X
    c. EAP
    d. X.509

9. What digital certificate standard was specifically designed for IIoT?
    a. X.509
    b. 802.1X
    c. IEEE 1609.2
    d. MD5

10. Which one of the following is a lightweight cryptographic hashing function?
    a. QUARK
    b. SHA1
    c. SHA2
    d. SHA3

Review Answers

1. The correct answer is **C**. Confidentiality, Availability, and Integrity make up the CIA concept.

2. The correct answer is **B**. Hashing algorithms provide integrity. If a hash results in a mismatch from a previous result, the data has changed.

3. The correct answer is **C**. Non-repudiation prevents a person or entity from denying having performed an action.

4. The correct answer is **A**. Symmetric encryption encrypts and decrypts the data with the same key. SHA1 and MD5 are hashing algorithms.

5. The correct answer is **C**. The Advanced Encryption Standard (AES) is the replacement for DES, 3DES, and TDES.

6. The correct answer is **A**. Blowfish as a symmetric algorithm, the others are asymmetric.

7. The correct answer is **C**. Diffie-Hellman is an asymmetric algorithm, the others are symmetric.

8. The correct answer is **D**. The X.509 standard is the most common for digital certificates and is supported by all major PKI solutions.

9. The correct answer is **C**. The 1609.2 standard is specifically designed for IIoT. It is smaller than the X.509 certificates and easier to store in low-power, low-storage IoT devices.

10. The correct answer is **A**. QUARK, PHOTON, SPONGENT, and Lesamnta-LW are all lightweight cryptographic hashing functions that are useful in low-processing-capable IoT devices.

Chapter 11: Troubleshooting Wireless Solutions

Objectives Covered:

4.2   Validate wireless solution implementations including RF communications and application functionality

4.3   Understand and implement basic installation procedures

5.1   Troubleshoot common problems in wireless solutions

Technology has evolved as a tool to bridge humans into a faster and more interconnected web of networks. Wireless technologies and solutions have played a big role in helping untether end-user devices and bridge connectivity where it was previously hard or costly to do so with wired communication technologies. Some wireless technologies allow for automated setup, as well as automatic configuration for wireless links, whether they are simple point-to-point links or more complicated mesh links.

No matter what wireless technology is employed, any solution could be susceptible to errors, faults, and malfunctioning. Issues could arise in the wireless setup itself or could originate from any related system or link in the whole communication channel. Our objective in this chapter is to avoid turning issues into problems. A problematic wireless setup is counter-intuitive since we are utilizing wireless to decrease costs, save time, and save effort while issues and problems will eventually lead to increased costs and defeat the purpose of implementing wireless solutions.

In order to do that, a proper troubleshooting process with the right diagnosis of the possible issues that might arise is important to cover. Moreover, and before delving into the troubleshooting process itself, we want to see how implementing best practices in setting up and configuring wireless implementations can avoid leading to issues and problems in the first place. In the end, a system with all of its components could only be as good as it is designed.

In this chapter, we will cover best practices ranging from the technology itself as well as logistical and prerequisite processes that are related to such setups. We will also discuss how to troubleshoot different issues that might arise in wireless communication technologies while having a proper troubleshooting approach.

Proper Solutions Design

A proper solution design requires technical experience, as well as having the right equipment and solution components to build the solution. A well-designed solution should also take into consideration the resource and solution limitations. Therefore, since we can never usually exploit the maximum expenses for any underlying solution, a well-designed solution is usually an optimal solution. There are so many prerequisites for good design, and as mentioned before, a wireless solution is only as good as it is designed. An optimal solution should always follow the proper process. This section builds on and adds more considerations to that presented in Chapter 3.

To help breakdown a design, we can talk about the simple Deming cycle for quality assurance. Both a solution designer and a solution troubleshooter would need to follow proper planning and take the right action, control, and checking, and again take the necessary action to get back into the cycle of optimizing a good wireless design.

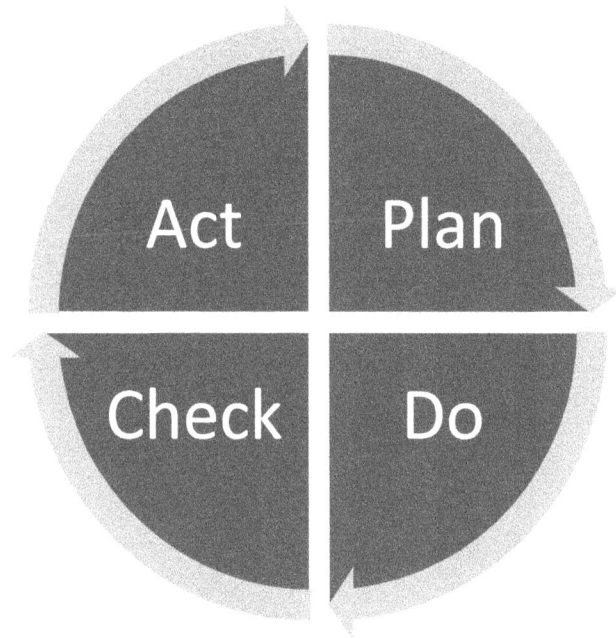

Figure 11.1: The Deming Cycle

The first part of the Deming cycle is planning. Planning can involve the actual design of any solution. Along with planning comes the consideration of the solution requirements, solution limitations, logistics and overhead, timelines, tools, and other factors.

Executing the installation and configuration of the planned solution comes in the next phase.

Checking, or studying the results of the deployed solution is what takes places in the third phase of the cycle. Any testing, post-installation surveys, and analysis take place in this phase.

Finally, any required troubleshooting and tweaking should take place in the fourth phase as a response to any findings from the third phase.

This cycle can be taken into continuity to shift the cycle from a one-time project endeavor into a continuous operation optimization of any deployed solution. Implementors should also adopt this cycle to enhance their approach to different projects, so they create their own template out of best practices that suit different projects and designs based on their own experience.

Some of the common wireless solution planning and implementation tasks will be discussed next.

Designing and Implementing Wireless Solutions

The choice of a specific wireless technology has to be done based on the basis of the customer requirements. Gathering and analyzing the requirements of any solution is the majority of the planning phase. Matching the solution with the requirements means that we need to be ready to design for the proper use-cases, devices, and applications.

Understand and Implement Basic Installation Procedures

At the same time, a clear understanding of each different wireless solution and its installation procedures is a must, or we will fail to deliver based on the design planning and customer requirements.

Having taken the resource limitations into consideration, no shortcuts should be taken in order to deliver a solution faster which, as a result, could skip an important step in the installation procedure. Basic installation procedures can include:

- Selecting Technology & Equipment
- Technology and Vendor-Specific Training
- Executing Site Surveys
- Solution-Specific Considerations
- Complying with Safety Requirements and other Constraints

Selecting Technology & Equipment

This should be done based on business and technical requirements. For example, a long-range communication protocol can be selected for a customer with multiple branch deployment requirements, whereas a shorter-range protocol can be utilized within the same premises of a home or enterprise campus. A proprietary standard could be implemented, but based on cost limitations, the wireless solution designer might be forced to select an alternative technology or connectivity method. Customer or specific use-case requirements might force the selection of the underlying technology and equipment based on customer preference for cost and overall matching of operating conditions such as performance and throughput, power consumption, environmental constraints, size, and licensing (all vendor, industry-specific, and regulatory) among others.

Technology and Vendor-Specific Training

Once the technology is decided on, the right equipment vendor should be selected and prepared for the implementation.

First, understanding the selected technology is mandatory for implementing the wireless solution. All other factors and constraints of any project will depend on this. This is why it is crucial to have certified professionals working on the projects within the area of their expertise.

Second, every vendor has their own guidelines for handling their equipment. Only trained individuals on the selected vendor's equipment should handle that equipment. Otherwise, mishandling and misconfiguration of the equipment might disrupt the implementation by simply utilizing a wrong setup of the equipment but can be as severe as damaging the equipment or leading to the misuse and breach of industry or regulatory constraints.

While trying to reduce the cost or time to implement a solution, some people might resort to non-compliant ways of installing the product. For example, a different mounting method, such as purchasing uncertified mounting kits or even worse, improvising mounting with basic tools, would render an implementation faulty or unsafe. Whether it's a small IoT Bluetooth or Zigbee sensor or a bigger outdoor Wi-Fi access point, or even bigger with GSM antennas, the right mounting equipment should be selected for any specific vendor, and no shortcuts should be taken when installing those. Proper installation procedures should be carried out by the certified individual.

Executing Site Surveys

Different technologies require different equipment and their own configuration, but a constant in a proper design approach is the design and implementation surveys.

Different technologies have different requirements and tools to carry out their own planning. In the same way, surveys are expected to be done for every solution. Link-budgets are a constant variable that is found in all solutions. They are key indicators of solution performance and communication throughput.

Depending on the technology, different tools for carrying out off-site predictive as well as passive and active site surveys can be implemented. Some vendors might offer their own surveying and planning tools, while others would require the purchase of standalone applications/hardware to carry out those surveys. Standalone survey tools can be used to design for multiple wireless technologies such as Wi-Fi and Bluetooth at the same time. Other tools can similarly provide spectrum analysis across different frequency bands to test for the allocation and link-budget calculation.

At the same time, specific wireless technologies such as 4G and LTE might require an altogether different approach. Engineers or technicians from the service provider, telecommunication company, or the vendor itself could be contracted in order to carry our such surveys using their own proprietary tools.

Site access and permission is also a possibility that must be taken into consideration.

Solution-Specific Considerations
Specific solution considerations need to be covered in order for a deployment to be functional.

License to install in the first place should be tackled in required cases. Installing antennas or digging up a pathway for a wired backhaul across the public property, requires the permission of different stakeholders. Planning around this consideration should also be factored into the implementation process flow, cost, time, and effort.

Different wireless technologies can utilize different wireless frequency bands. While many technologies rely on ISM bands, others require a license to operate in a specific frequency band. Even technologies that usually operate in ISM bands, such as Wi-Fi, do require licensing, where a specific country or region's regulatory domain does not allow for an open utilization of those same bands that could be utilized for free in other regions and regulatory domains. Obtaining the license from the telecommunication authorities and regulatory bodies in different countries is required.

Complying with Safety Requirements and other Constraints

We must also consider design variances for different vertical markets as well. Wireless implementations should comply with different applicable standards and constraints required of those markets. Many standards that govern the operation of different industries pertain to health and safety regulations of those industries, mainly workplaces, as well as other environments.

Occupational health and safety bodies and their standards address the safety and well-being of people, including employees and many others who could be affected at those places of employment. Aligned with the same objective, these regulations could be mandated by different health and safety regulation codes, depending on the concerned country, region, or nature of the industry/business.

For example, the Occupational Safety and Health Administration (OSHA) in the USA regulates private employers across 50 states with the objective of providing employees with "employment and a place of employment which are free from recognized hazards that are causing or are likely to cause death or serious physical harm to his employees." At the same time, and within the same country, different state-mandated regulations for health and safety also apply, covering the same workplaces and others within the same scope.

The most important considerations a wireless solution implementor should consider are:

- Having the proper OSHA training to maintain health and safety standard compliance
- Studying any risks incorporated with carrying out specific physical, mechanical and electrical tasks related to the implementation of the wireless solution, such as receiving and storing equipment, mounting wireless gear and antennas, cabling and extra-low voltage systems, etc.
- Abiding by the standards while carrying-out surveys and installations to prevent injuries during and after these procedures that may result in any physical, electrical or any other hazards these acts can impose on the safety of the people in the specific workplace
- Documenting the installation and keeping track of any job-related injury that happens or could happen as a result of the wireless solution implementation

These core requirements, along with others, should usually be considered as part of the planning, designing/doing, and checking of each project to maintain an optimal solution with minimal controlled risks. The proper tools (used by certified people) should also be used for carrying out installation while maintaining compliance.

Different regulatory bodies corresponding to the OSHA in the USA exist in different countries and regions. Please refer to the following website for a comprehensive list:

en.wikipedia.org/wiki/Occupational_safety_and_health#National_legislation_and_public_organizations

OSHA and the corresponding health and safety regulations are not the only constraints that should be followed. In addition, industry-specific regulations and compliance codes, from different bodies like BICSI, ISO, PMI, and JCI among others, apply to different environments and related objectives (quality control/assurance, business continuity, risk management) should also be considered.

General Configuration Considerations

Deploying a wireless solution is not just about implementing wireless technology alone. A solution is usually centered around the technology but needs the hardware that supports that technology. In addition, all systems that need to make the solution work are also required, ranging from the physical, power, and data distribution system that the wireless solution is going to be operated/powered from and augmented on, all the way to the application layer where the business use-cases are achieved for the end-user.

Assuming that the operational requirements, such as having a proper power source and complying with regulatory standards in order to operate, are met, any wireless solution implementation has to address the upper layers to guarantee the optimal solution.

Configure Connectivity (Wireless and Wired)

A wireless system without a wired backhaul, whether to connect client devices, people, or to terminate a point-to-point wireless setup, is useless unless the backhaul itself is a wireless link. Eventually, any wireless link has to be connected somewhere with a data wire.

A wired network can be part of a private (ethernet, locally switched) or a public network (MPLS, municipal fiber optics, ISP-provided). Based on the solution and technology requirements, different backhaul types can be supported, some with seamless integration, having the wired network terminated directly in the wireless equipment, or one or multiple devices could be sitting between the wireless equipment and the main wired network in order to translate the connectivity protocols and do the data routing and switching. The selection of the backhaul and its configuration can be profoundly impacted by the project constraints from a cost perspective as well as technical requirements for the whole system performance.

Connectivity over wireless must be done with the correct equipment and configuration while complying with the solution requirements and regulatory restrictions. Some solutions can be intuitive to configure out of the box, while others would need advanced staging and testing before they are deployed at the production site.

Although out of the box intuitive configuration can sound like a benefit, it might end up causing more issues later on, if some configuration parameters are missed. Being experienced in a specific technology would be more beneficial in configuring any system and even more helpful for any required troubleshooting later on. Understanding the capabilities of the connected endpoints and clients is an important factor.

Most wireless technologies support automatic configuration on the wireless clients by default. In some cases, an intermediate device, like a hub or router, needs to be set up on the distribution network before the stations can start communicating properly. The function of such hubs is to configure local and short-range wireless connectivity while tapping into the main distribution network that is usually using different communication standards. One example could be a Wi-Fi router that is effectively terminating wireless clients while tapping into the wired distribution network for the backhaul where the frames are transformed from wireless to wired frames.

When a custom configuration is required, it is usually for more advanced scenarios such as a mid-range wireless link. Provided that the proper mounting and alignment has been done, especially in the case of a line-of-site (LoS) technology, configuration should be done to get the system up and running and customized to fit the regulatory domain (free or licensed) requirements on the wireless operating frequency and gain to guarantee proper operation of the system. Advanced features that could be configured on wireless gear could be beamforming, data path selection, mesh routing, dynamic frequency selection (DFS), power-saving mode/mechanisms, and other vendor-proprietary features and settings.

Configuration of longer-range wireless communication such as those utilizing the GSM network purely rely on the infrastructure setup done by the telecommunication provider, mobile network operator (MNO) or the ISP. Configuration of the endpoints and client devices require a minimum set of features to be deployed. For example, with NB-IoT operating over cellular radio, the GSMA provides the design and configuration guidelines that help MNOs deploy NB-IoT globally to ensure interoperability and global configuration standards. This should help achieve reliable deployments for a better setup and prevent issues such as:

- Mismatched timers that would affect the client behavior by impacting responsiveness and battery life
- Negative implications for end-to-end security if some (optional) NB-IoT features are not enabled
- No connectivity configuration for one of the roaming visitor's network core servers (Service Capability Exposure Function Gateway – SCEF GW) will lead to loss of communication to the visitor's application server and requiring extra configuration to use alternative communication methods
- Roaming devices might need to be certified for all bands, which will have potentially significant cost implications

Configure the Network Infrastructure

When it comes to the network infrastructure, the wireless solutions administrator can take a service-minded approach. By this, I mean that the administrator can focus on what is needed instead of how it will be accomplished. For example, the following infrastructure services are often required for wireless solutions to work properly:

- **Authentication:** The wireless solution may require specific authentication services, such as 802.1X/EAP authentication, Kerberos authentication, LDAP directory access, certificate provisioning, and standardized Internet-based authentication methods. At times, it is a simple task of opening ports in the Internet firewall. At other times, it requires the complete installation of a multi-server architecture to support the wireless solution.

- **Authorization:** The wireless devices must have access to the required resources. In some cases, the wireless nodes act as identities communicating on the network, and each node must be given access to resources (servers, databases, files, services, etc.). In other cases, an intermediate device acts as a proxy for the wireless nodes and only that device is required to have access.
- **Accounting/Logging:** Organizational policies may require accounting of all network actions. In such cases, the wireless nodes must be considered, and they must be identified in some way such that their actions can be logged and monitored.
- **Name Resolution:** If wireless nodes communicate using IPv4 or IPv6, name resolution will become essential. The Domain Name System (DNS) is the most common method for such name resolution today. Wireless nodes use DNS to resolve the IP address of servers as well as the location of controllers that provide firmware, updates, and control to the wireless nodes. If appropriate, the proper hostnames must be created in the DNS zone to respond to such requests from the wireless nodes. Some end-devices will use the DNS server to locate the controller or configuration server during initial startup. When IPv4 host records are used in DNS, they are called A records. When IPv6 records are used, they are called AAAA records.
- **IP Addressing:** Wireless devices may use IPv4 or IPv6, and some method should be in place to provide them with the proper IP addresses. DHCP is often used for this purpose.
- **Time Synchronization:** In many cases, sensitivity to time variances among devices can either prevent a wireless network from operating or cause devices on the network to provide erroneous information that negatively impacts business decisions. For example, if a wireless sensor is off in time by more than a few milliseconds, it may be impossible to provide proper sensor fusion with other sensors for full-scale analytics. A time server/service, such as the Network Time Protocol (NTP) may be required.
- **File Access:** Wireless nodes may require access to file stores to write logs to the location or to access firmware and software updates.
- **Custom Service Access:** Many wireless solutions have their own custom services that run on the network. These services may run on dedicated servers, in virtual machines, or on network appliances. In any case, the devices will not function if the service is not there. Sometimes, vendors offer the use of local

servers or their cloud service. If the organization chooses to use local servers, the servers must be configured on the network.

The CWSA may be required to install and configure these infrastructure services; however, in many larger organizations, another individual or group will have that responsibility. The CWSA should document the requirements of the wireless solution for the responsible individuals and request appropriate configuration. In smaller organizations, the CWSA can simply configure the DHCP server, DNS server, authentication server, or any other system required.

Configure Cloud Connectivity

As previously mentions, many wireless networks use cloud management today. This is true for Wi-Fi, ZigBee, Z-Wave, LoRa, Bluetooth, 802.15.4, and proprietary protocols, among others. In some cases, vendors offer local management options, and in others, cloud management is the only functional solution. With cloud management, three things must be considered (at a minimum):

- **Licensing of the cloud service:** The first task is to create an account with the cloud service provider so that wireless devices can be registered, authorized, or provisioned into the account. This may require paying an annual licensing fee for utilization of the cloud service. Carefully evaluate the number of devices supported in your contract to ensure you have the proper configuration for deployment. You don't want to find yourself in a scenario where you are half-way through the deployment, and the cloud service is no longer allowing the registration of additional devices.
- **Connecting to the cloud service:** For cloud connectivity, the wireless devices will require access to the Internet or the gateway used by the devices may be the only system that contacts the cloud. Either way, you must ensure the proper ports are open in the firewall to allow for this communication. Check with the vendor literature to verify the proper configuration.
- **Providing sufficient bandwidth:** Finally, it is vital to ensure proper bandwidth allocation is in place. Many cloud-managed wireless services have minimal requirements on Internet bandwidth, but whatever the requirements are, they must be accounted for in the implementation. Figure 11.2 shows the Cellular Data Calculator for Monnit sensors. This is used with sensors that work through a cellular gateway to reach the Monnit cloud. As you can see, with

1,273 sensors communicating with the cellular gateway every 10 minutes and the gateway communicating with the iMonnit cloud every 5 minutes, the total monthly bandwidth is still only 91.547 megabytes per month.

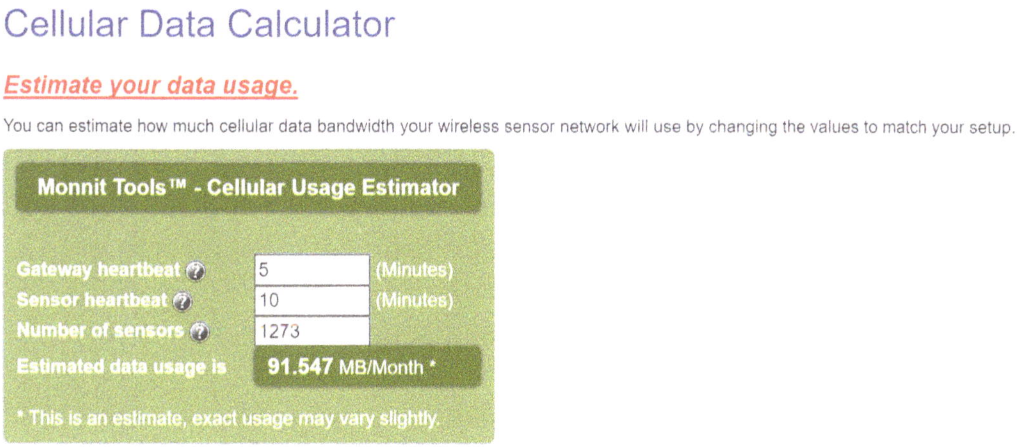

Cellular Data Calculator

Estimate your data usage.

You can estimate how much cellular data bandwidth your wireless sensor network will use by changing the values to match your setup.

Figure 11.2: Monnit Cellular Usage Calculator

Configure Enhanced Features

Some wireless solutions will require extra configuration for advanced features that may be optional or required. These include:

- **Video:** solutions utilizing video, including wireless surveillance systems and end-user video streaming devices, will place extra strain on the network. Proper Quality of Service (QoS) solutions should be implemented, and careful throughput planning should be performed as well.
- **Voice:** voice communications require very low latency (delay in the network) and jitter (variance in network delay). Implementing separate networks for voice can be beneficial, but in less strained networks, proper QoS configuration may be sufficient.
- **Captive Portals:** used mostly in Wi-Fi access, captive portals may be used on guest networks to authorize user access, provide advertising opportunity, or require acceptance of use policies. The captive portal will force the client device to navigate to a specified web page and not allow the device to communicate with the rest of the Internet until it is unlocked through a back-end authorization solution or simply requiring the user to click a particular button.

In some cases, the user is forced to a captive portal web page initially, but no action is required, and the user can immediately start using the Internet as usual.

- **Location Services:** if location services are used, such as BLE beacons, Wi-Fi locationing, RFID locationing, sensor-based locationing, or any other system, it may require more careful planning. The fact that locationing is being used indicates that mobility is assumed. Therefore, network coverage must exist everywhere that you desire to locate tracked devices.

- **Telemetry:** tracking telemetry data, often performed in industrial deployments, transportation, and other networks, will simply require that the appropriate communications be allowed through the network. Check with the vendor literature to verify that any required protocols are supported and that communications can occur.

- **Mobile Device Management (MDM):** MDM solutions allow you to control mobile devices and often provide a centralized management dashboard. The mobile devices must be reachable from the centralized system. Once again, this is about ensuring full communications through the network as required.

- **Network Function Virtualization (NFV):** a network abstraction solution that controls network forwarding and other communications separate from the hardware that performs the operation. NFV is usually abstracted, itself, in configuration interfaces so that the administrator is simply indicating what should happen and is not concerned about the way in which NFV accomplishes the task. NFV is focused on the granular details of network operations. NFV is used heavily in 5G network deployments.

- **Software Defined Networking (SDN):** a network abstraction solution that controls either networking hardware or NFV functions for management of network control and forwarding. SDN is usually abstracted, itself, in configuration interfaces so that the administrator is simply indicating what should happen and is not concerned about the way in which SDN accomplishes the task. SDN is focused on the big picture of the network.

- **Contain-Based Apps:** Application containers are self-contained solutions containing all dependencies of an application in a single bundle (container). You must ensure that you have access to the proper cloud service provider that supports the container type you will be using or that you have the internal server infrastructure to implement the containers. The "contained" applications effectively run in a sandbox on the target system. Common container types include Docker, rkt, LXD, Windows Containers, and Hyper-V containers.

Troubleshooting and Remediation

Different wireless technologies have been discussed throughout this book (more than 10 chapters worth so far). It would be challenging to cover each and every wireless technology with the potential issues that might arise in implementing them. However, it is essential for every wireless solution administrator to understand the OSI layers stack that will enable them to diagnose the problem more efficiently by following the right process, as discussed, all the way back in Chapter 1.

Troubleshooting begins by identifying the problem and then working to gather the facts to help identify its possible causes. There are multiple troubleshooting processes that different vendors follow, or that your company or you individually might have developed.

Referring to the Deming Cycle, troubleshooting has to do most with checking (CHECK) the performance and configuration of a wireless system, especially if there are any issues or problems, and then implementing (ACT-DO) a planned fix (PLAN) to the setup and configuration with the required fix and then keep monitoring the solution for any issues (CHECK).

Following a troubleshooting methodology congruent to the Deming Cycle is key to make sure operation and problem resolution are done in an optimal and well-documented approach, saving time and making it easy for any resources handling the system to continue operations. The CWNP methodology for troubleshooting comes in handy here to help wireless administrators identify, plan, resolve and document the troubleshooting of wireless systems.

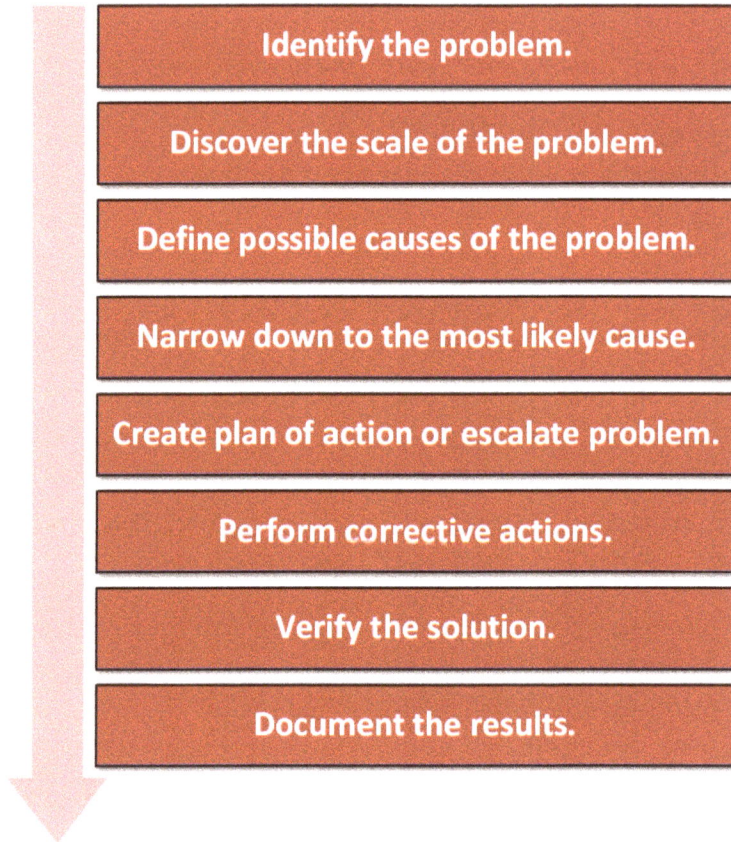

Figure 11.3 The CWNP Troubleshooting Methodology (steps ordered from top to bottom as 1-8)

Troubleshoot common problems in wireless solutions

Being able to identify the problem and then narrow down to the root cause of that problem are the two most important steps of the troubleshooting methodology. Otherwise, a lot of time, effort, and operational cost will be lost incurring more financial losses. A bad experience for a home user of a wireless solution will likely cause frustration and the home user to post a bad review of their experience and drop the product altogether, which would negatively impact the vendor. Disruption of services for commercial users of a long-range wireless communications solution/product would cause them to shift to an infrastructure or product of a different ISP or MNO incurring business losses on their original provider.

Timely identification of the problem and its root-cause is highly tied with a troubleshooter's understanding of the underlying wireless technology, as well as the whole system that provides the overall solution setup. While someone might be able to identify the cause of a problem from previous experience, others might need to test and troubleshoot different components of the solution to be able to come up with such identification. Basic OSI layer understanding comes into play here to help in identifying the right layer where the problem is happening and, as a result, identify the root cause.

Starting from the lowest layer, we can work our way up as we try to identify the cause(s) and build on our own expertise for similar troubleshooting in the future.

At the Physical and Data Link Layers, common issues related to hardware and the physical medium of the technologies deployed in the solution might be causing problems.

Malfunctioning Hardware

Damaged devices can sometimes be the most straightforward cause for a wireless connectivity issue. Failure to abide by a vendor's recommended practices for installation and operation could be one cause of faulty hardware. Hardware can also fail due to end-of-life of the product where products have simply crossed their expected lifetime until they fail. With outdoor deployments being a large part of wireless solutions, external factors could be affecting the function of the hardware. These could include:

- Fluctuating power source/input
- Environmental factors like exceptional weather conditions that damaged or affected the hardware in some way like the alignment of an antenna due to strong winds
- Change of operating conditions such as the construction or presence of new obstacles that are blocking wireless signals. This could be as simple as some birds who have created their own nests on top of a direct LoS solution, or a tree growing within a coverage Fresnel zone, or even a new tower constructed in the pathway of the wireless signal.
- Intentional damage of the hardware by manhandling them or vandalism of installed systems

As you can see, any of these different reasons might lead to a hardware malfunction. Once identified, the root-cause can be resolved by implementing a plan, applying the plan, and documenting it to resolve the problem. Following vendors' troubleshooting recommendations can help fast-track the resolution, while following the recommended installation and operation requirements would help give any installation the expected longevity. Scheduled maintenance for hardware equipment can help identify and eliminate many causes for hardware malfunctioning so that troubleshooting can be done proactively.

Interference

The medium which any wireless technology utilizes lies on the lower layer levels of the OSI stack and should be tackled next when trying to identify the problem while troubleshooting.

As mentioned earlier, different technologies utilize different frequency bands. Each technology might have its own mechanism of frequency selection, hopping, spread spectrum, and interference detection and/or mitigation. However, when a solution faces issues, wireless interference should always be evaluated by the troubleshooter.

Interference could be caused by the combination of different signal sources operating in the same space and on specific gain levels that could lead to change in the noise levels causing an increase in communication errors to the extent of total communication failure, or what can be considered a denial of service (DoS).

Different kinds of interference types can be considered:

- **Narrow-band:** can affect a single or a few channels of communication frequencies, which causes errors and disrupts communications on those frequencies. A high-gain signal operating on the same wireless frequency as another co-existent technology is an example of narrow-band interference.
- **Wide-band:** can affect a whole frequency band leading to total failure of the wireless solution. A frequency generator or jammer can be one example of a wide-band interferer.
- **All-band:** can affect a whole frequency band because of the nature of the technology utilizing all channels leading to increased errors and disruption of services. For example, wireless technologies that utilize a spread spectrum mechanism that utilizes the full band can disrupt other co-existent wireless

solutions that need to jump between a few channels of the same frequency band.

The leading cause of interference is the operation of multiple wireless solutions utilizing the same or different technologies that could be utilizing the same frequency space. Other wireless interference issues can happen because of improper frequency planning and operation or because of incidental radiation from non-wireless devices, such as motors and some lighting equipment. This could be due to many reasons including:

- misconfiguration, such as full reliance on automated vendor solutions for frequency selection/interference mitigation mechanism or therefore lack-off such mechanisms, whether automated or manual or selection of wrong operation mode and regulatory domains
- selection of hardware, which utilizes different frequency ranges, that was originally built for a different region
- failure to obtain the proper licenses to operate in the allocated frequencies which could sometimes guarantee exclusivity of frequency bands/channels

Wireless spectrum analyzers are critical tools that must be utilized during wireless communication planning and troubleshooting in order to detect any source of interference and to identify its nature and source. If we are concerned about a specific technology, the proper spectrum analyzers with the capabilities to scan the operating frequencies of that technology must be utilized. These could be in the form of a standalone spectrum analyzer provided by the same vendor of the wireless solution or a third-party vendor. Some wireless vendors, depending on the technology, can offer an integrated spectrum analyzer that is built-into the wireless product itself and has its own software and/or application to run the spectrum analysis features.

If we are looking for a tool that could cover all wireless technologies, that would drive us to look for more advanced spectrum analyzers, with more substantial capabilities to identify interference across all and different known wireless operation frequency bands, but those would typically cost more to acquire and to operate. For example, a spectrum analyzer that can analyze bands from 100 MHz through to 60 GHz and is portable in nature would cost several thousand dollars in US currency.

Assuming all reasons for misconfiguration, hardware selection, operating license are covered, interference problems are solved in most cases by removing the main cause/source of such interference or changing the channels on the systems experiencing interference. Once the source of interference is identified, the proper procedure should be followed to tackle the removal of the interference source. If removal is not an option, then the proper mechanism must be configured for the interference to be avoided.

Signal strength

Every wireless technology employs a recommended basic set of link budget, fade margin, and error capacity to match the transmission capabilities and throughput speed based on the modulation of the technology. What is common for all technologies is the expected signal strength and noise level so that demodulation on the receiver's end can take place to turn symbols into bits and then send those bits to upper layers so the proper information can be processed.

If a technology fails to match its minimum requirements for the total link budget, errors will occur when trying to demodulate a received signal and disruption to the communication will take place. One significant factor potentially violating this requirement could be the signal strength. In order to troubleshoot, we can consider different reasons for signal strength change or weakness by breaking down the communication model into the three main components including:

- Transmitter: If the transmitted signal has a lower power (or gain) than usual, this could result in the link budget not being met with other factors that make the signal weaker as it travels to the receiver that makes it non-readable or prone to errors.
- Receiver: If the receiver doesn't have enough power to listen to the received communication, this can also lead to communication errors. Receiver sensitivity is also an important factor.
- Wireless Medium: Changes to the wireless medium such as a change in distances or any conditions that affect free-space path loss (FSPL) can cause the link budget and fade margins to change.

Transmitter/receiver issues:

- Using the wrong antenna model with the required gain to match the link budget
- Misconfiguration of gain levels or utilization of automated calibration methods that might allow for a lower than optimal signal level gain and sensitivity
- A lower power source that might lead to lower gain and sensitivity
- Hardware misconfiguration with the wrong cabling and wiring for power and antennas
- Hardware malfunction which causes the loss of radio/antenna capabilities
- Hardware misalignment that leads to utilizing a lower power signal

Wireless medium issues:

- Introduction of an interference source that changes the SNR or base noise levels
- Change of weather conditions for outdoor wireless links that might change the FSPL and overall attenuation
- Node failure in a mesh setup that leads to switching to an alternate node could be a suitable resiliency mechanism but can change the hop distance which affects the link budget
- Introduction of an obstructing body, such as a newly constructed project or natural tree growth that blocks or attenuates the signal

Drivers

At the Network and Transport layers, connectivity issues related to routing and interconnecting different systems together might be causing communication disruptions or drops altogether. This behavior can be due to faulty drivers. If the device requires driver installation for operation, always check the vendor website to see if updated drivers have been released that can resolve the issue. Tools that help resolve driver errors include:

- Operating system logs
- Operating system analyzers (like Microsoft Message Analyzer)
- Debug tools
- Network protocol analyzers

Network Errors

Upper layer issues can be identified all the way up to the application layer, where software issues or application misconfiguration can also lead to service disruptions.

Logs in the application server may be helpful in troubleshooting such scenarios as well as logs in the end devices. If IP communications are used, a protocol analyzer may prove useful in evaluating the transmissions to ensure that what should be communicated is being communicated. Tools that help resolve network errors include:

- Log analyzers
- Protocol analyzers
- Dashboards
- Handheld network testers

Software/Firmware Issues

Software and firmware issues are also common in wireless networks today. The complexity of wireless networks "behind the radio" has increased significantly in recent years. Today, cloud solutions drive the networks, on-premises solutions drive the networks, and multi-tiered applications drive the networks. With this extra complexity, we must consider software on the devices and software supporting the devices.

The software on the devices is often firmware. Depending on the device, a quick view of a vendor's website may reveal one or two firmware updates since device release, or it may reveal dozens of updates.

BEYOND THE EXAM: It's the Firmware Stupid

My name is Tom Carpenter, and I am a firmware failure. While I did not write this chapter for the CWSA Study and Reference Guide (though I did write several other chapters), during my role as general editor I thought it useful to share my non-wireless experience with firmware problems.

About a year before the time of writing, I built a new tower computer with a super powerful motherboard, super-powerful processor, 64 gigabytes of RAM, tens of terabytes of drive space, and a beastly powerful video card (I needed it for work, honey… just in case my wife reads this). In the time since the build, I have reloaded the operating system at least six or seven times. The second or third rebuild required that I stop using the M.2 socket drive on the motherboard

– it just quit working. So, I used 2.5-inch SSD drives instead. But stability continued to be a major problem.

Several months ago, it finally dawned on me that I should check to see if the motherboard had any firmware updates. Oh my! There were dozens of updates beyond my version. More importantly, eight of them, that's right, eight of them, were focused specifically on resolving problems with the M.2 socket.

Needless to say, the rest is history. I updated the firmware on the motherboard to, in this case, the last update that fixed an M.2 socket problem and guess what? The M.2 socket is working fine. I am typing on that computer right now (see, it really is for work) and I have had no stability problems with, at times, more than 60 windows open at the same time while running several virtual machines.

The moral of the story is simple for wireless solutions. Check the vendor website for firmware updates anytime you're having problems across multiple instances of the same devices running the same firmware. Don't be a Tom Carpenter, be a firmware updater.

Resolving software problems does not end with the device itself. The supporting software on the network must be appropriately configured and bug-free as well. Given that the vendor provides the software, any bugs will need to be reported to them for resolution, but the configuration part is in your control. We will address this more in the later section titled *Improper Configuration*.

Faulty Custom Software Code – Faulty installation

Many wireless solutions support APIs for customization of the software or access to gathered information. Custom software code is often more prone to bugs because of the fact that less testing is often performed before releasing to production. If such problems are detected, report them to the software developers quickly for resolution.

Faulty installation typically comes down to the improper location of wireless devices or improper configuration, which we will talk about next.

Improper Configuration

When it comes to configuration, both the devices and the supporting services and network must be considered. Proper configuration starts with proper planning. Be sure to take the following steps to prevent problems:

- Consult vendor literature for best practices in configuration.
- Define a standard configuration baseline to use for devices.
- Document configuration settings for supporting services and the network.
- Document any changes made to the configuration settings.
- Share all documentation with all relevant personnel.

Sometimes configuration problems occur because of upgrades. The configuration set in use worked before the upgrade, but due to some change in system functionality, it does not work after the upgrade. Always check vendor literature before upgrades to see if any existing configuration parameters may become problematic after the upgrade.

Tools that can be useful in troubleshooting configuration problems include:

- Log analyzers
- Protocol analyzers
- Knowledge systems (documentation)
- Vendor websites

Security Misconfiguration

A specific area of improper configuration is security. If the end devices are not properly configured for security to match the network requirements, they will be unable to connect. Additionally, if Network Access Control (NAC) solutions are in use, they may prevent the end devices from reaching portions of the network if they do not match policy-based health parameters.

Always verify that the security parameters are configured appropriately if devices are failing to connect. This step is particularly important when the signal strength in the area is sufficiently strong, and no interferers have been detected. Before checking other configuration parameters, check the security settings.

Tools that can be useful in troubleshooting security configuration problems include:

- Log analyzers

- Protocol analyzers
- Knowledge systems (documentation)

Chapter Summary

In this chapter, you explore implementation through troubleshooting of wireless networks. The good news is that experience with any wireless technology helps master any other technology. Whether you are troubleshooting Wi-Fi networks or wireless sensor networks, many of the same skills are required. In the next and final chapter, you will explore integration and automation options through the power of APIs and scripting/programming languages.

Review Questions

1. What is the first step in the Deming cycle that can be applied to wireless design and troubleshooting?
    a. Do
    b. Check
    c. Act
    d. Plan

2. What organization, in the United States, is representative of an organization that provides regulations for worker safety?
    a. IEEE
    b. OSHA
    c. IRS
    d. FCC

3. For what might and end-device use DNS during initial startup?
    a. Locating the controller or configuration server
    b. Acquiring an IP address
    c. Synchronizing time
    d. None of these

4. Which one of these is not one of the three minimum considerations when configuring cloud access for wireless networks?
    a. Providing sufficient bandwidth
    b. Licensing of the cloud service
    c. Localizing the cloud service database
    d. Connecting to the cloud service

5. Which networking virtualization solution is more focused on the granular operations of the network: NFV or SDN?
    a. NFV
    b. SDN

6. According to the CWNP troubleshooting methodology, what should be performed after narrowing down to the most likely cause?
    a. Perform corrective actions
    b. Discover the scale of the problem
    c. Create plan of action or escalate the problem
    d. Document the results

7. What kind of interference exists only on small portions of the frequency band in use?
    a. All-band
    b. Narrow-band
    c. Wide-band
    d. None of these

8. What is a wireless medium (the space between transceiver antennas) issue that can impact signal strength?
    a. Output power
    b. Receive sensitivity
    c. Weather
    d. Antenna alignment

9. What tool can be helpful in troubleshooting device driver issues for wireless devices?
    a. Spectrum analyzer
    b. Dashboards
    c. Operating system logs
    d. Wi-Fi scanner

10. What might require adjustment after a software upgrade to a wireless solution?
    a. Configuration parameters
    b. Antenna alignment

Review Answers

1. The correct answer is **D**. The four steps or phases are Plan > Do > Check > Act, in that order.
2. The correct answer is **B**. The Occupational Safety and Health Administration (OSHA) is the regulatory agency for worker safety in the United States.
3. The correct answer is **A**. Some devices will use DNS to locate other configuration devices.
4. The correct answer is **C**. The primary benefit of a cloud service is the lack of requirement for locally hosted applications or data.
5. The correct answer is **A**. SDN is more focused on the big picture, and NFV is more focused on the granular operations.
6. The correct answer is **C**. The next step is to create a plan of action or escalate the problem to someone who can.
7. The correct answer is **B**. Narrow-band interference exists only on a portion of the frequency band in use.
8. The correct answer is **C**. Weather, other interference sources, new construction, and other items can cause issues between the two transceivers. The other items are transmitter/receiver issues.
9. The correct answer is **C**. When troubleshooting device driver issues, consider using operating system logs, operating system analyzers, debug tools, and protocol analyzers.
10. The correct answer is **A**. It is not uncommon for a system to implement changes during an upgrade that will require configuration changes after the upgrade.

Chapter 12: Programming, Scripting, & Automation

Objectives Covered:

3.1   Identify and document the wireless system requirements

3.5   Understand the wireless solution and consider key issues related to automation, integration, monitoring, and management

3.7   Plan for the technical requirements of the wireless solution

5.2   Understand and determine the best use of scripting and programming solutions for wireless implementations

5.3   Understand application architectures and their impact on wireless solutions

Historically, wireless vendors (and some 3rd party developers) have built standalone systems that included all the pieces and parts to accomplish a fixed set of features and functionality. Over time this led to these systems becoming increasingly large and complex as they are modified to bring additional functionality to the product. This feature sprawl caused them to grow into massive, resource-hungry, and often closed systems that provided almost no flexibility to meet the needs of evolving organizations.

Changes to modern applications have brought about a departure from these large, single-tier systems. Points of integration between applications are easily accessible through REST APIs, and standard methods of configuring the same features across multiple manufacturers enable integrators and developers the opportunity to quickly and easily build integrations or enhancements to existing applications. Due to this drastic shift in knowledge requirements, the traditional silos are dying, and programming no longer belongs solely to the application developers. Modern network integrators should have a basic understanding of not only the traffic flowing across the network but also every system attached to it - including the knowledge to retrieve information from all corners of the network and use it to increase operational efficiencies while reducing management overhead.

In this chapter, we will start by defining what an API is, common communication methods, and language selection when interacting with APIs. Then we'll cover types of data that may be important and why you may want to build an API integration. Finally, a brief overview of the various application and integration architectures used to connect disparate systems. This chapter will not go in-depth on any specific method, protocol, or language details as that is beyond the scope of the CWSA exam.

What is an API?

While you've likely seen or heard the initialism **API** before, you may not know exactly what it means or even what the letters stand for. An ***Application Programming Interface***, API for short, when used in the context of computing, is defined on Wikipedia as *"a set of subroutine definitions, protocols, and tools for building application software. In general terms, it is a set of clearly defined methods of communication between various software components"*. In the world of networking, a more refined definition could be *"a well-defined methodology for communication between multiple systems for the purposes of easily sharing information useful for the management, monitoring, and overall* *health of a network.*" Or even more simply, an API is an interface for sending or receiving network configuration or health information between systems or applications. Before we jump straight into the deep end building integrations, let's expand the definition of an API and how it works, starting with the categories of API.

Categories of APIs

When interacting with (or even building your own) APIs, one of the very first things that will need to be determined is what category, or type, of API you'll be interacting with. There are three types of API in general, with a fourth combined type. While this high-level categorization won't necessarily affect the overall build of the integration, it will directly impact the type of authentication used, security requirements, and locations that your integration services need access too (e.g. internal network, VPN, Internet).

Open

An open API is exactly what it sounds like - completely open to the public with no access restrictions. While rare, there are some data sources available, such as weather information, you may need to pull data from that are of this type.

Partner

A partner API is on that is only available to business partners or customers. These APIs have access restrictions that require some sort of authentication or verification before interacting. Partner APIs are typically available across the Internet, but the owner will provide some sort of credentials, a token or another authentication mechanism, before granting access. Examples of this type of API are services such as Twitter, Facebook, AWS, Cisco Meraki, or Mist Systems.

Internal

Internal APIs are systems that only expose data to systems housed in the same organization's network. With internal APIs, you'll most likely be dealing with a software appliance or application that resides in your datacenter. Systems such as network access control (NAC) or location-based systems (LBS) will often have internal APIs.

Composite

A composite API is simply a mixture of any of the three methods above. Aggregation of data from public, partner, and internal APIs into a single usable data source is a composite API.

Common API Communication Methods

When building an integration of systems, a standard communication method will be defined in the documentation for the remote system. In a large majority of cases, this is going to be a web-based technology utilizing encrypted Hypertext Transfer Protocol (HTTP). The encryption method used will be Transport Layer Security (TLS) although it may still be referred to as Secure Sockets Layer (SSL) - the now-deprecated predecessor to TLS. Both non-encrypted and TLS-encrypted HTTP are the common connection method of the Internet. Utilizing an established HTTP connection, clients and servers will interact using REST, Webhooks, RESTCONF, OpenConfig, or many others.

With some APIs, HTTP is not utilized, and the remote service may be using a WebSocket. While similar to HTTP, WebSockets provide a full-duplex - two-way - communication channel enabling low-overhead, two-way communication between the server and the client over the same IP ports and in some cases, the exact same web servers. In addition to basic two-way communication, WebSockets provide a mechanism for streaming messages which can be necessary for data flows providing system telemetry.

Finally, non-HTTP based methods exist, such as Message Queuing Telemetry Transport (MQTT) and Network Configuration Protocol (NETCONF). MQTT is a publish-subscribe messaging protocol designed for Machine-to-Machine (M2M) connectivity. Primarily used for Internet of Things (IoT) sensor-type devices, MQTT requires minimal resources and provides efficient distribution of information in a one-to-one or one-to-many model. NETCONF is an XML-encoded standard developed to provide mechanisms for installation, updates, and removal of configuration items specifically in network devices using the Remote Procedure Call (RPC) layer.

Choosing a Language

To start building any automation or integration, you must first choose the language that you will be using. Often times this is a trivial decision based solely on your familiarity and comfort level, but depending on the platforms, data conversion needs, or tasks that need to be completed, you may be forced outside of your comfort zone. In order to quickly adapt, you should be aware of a few high-level differentiators between types of languages, styles of programming, and the strengths and weaknesses each possess so you can effectively build the solution.

Managed (Interpreted) vs. Unmanaged (Compiled) Languages

Yes, all scripting languages are programming languages while the reverse is not necessarily true. There will be syntactical and idiomatic differences between all languages, but the most significant difference for our purposes is that scripting is done with interpreted languages versus compiled languages. A compiled language is one that is written and before being run is translated - compiled - into another target language and run directly by the host operating system. Interpreted languages, on the other hand, are essentially read by a service or runtime engine installed on the host operating system and interpreted into intermediate languages live - a process called runtime compilation. There is a bit more intricacy to the processes, but for this book that information is out of scope. Another name for these language types is **Unmanaged** for compiled and **Managed** for interpreted languages.

Common Managed Languages	Common Unmanaged Languages
1. Python	6. C
2. Java	7. C++
3. Go	8. Pascal
4. PHP	9. Delphi
5. Javascript	10. Visual Basic

At the surface, managed languages may be more appealing since they can be easily run, don't require direct interaction with system resources, and have quite a bit more flexibility but that comes at a cost. Runtime compilation allows for quick modification of code and much easier testing, but they are much slower. Unmanaged languages, typically being compiled directly down to machine code, not only run much faster but also allow (or require depending on your point of view) more in-depth system functions like resource management and hardware access. This is due to the unmanaged approach to underlying functions - hence the name.

Functional vs. Object-Oriented Programming

Besides being managed, scripting is a familiar method for most network admins. If you're at all familiar with network devices or Linux operating systems, you are already scripting in a manner of speaking. One common scripting language is Bash, the default shell and command language found on most Linux variants. Using Bash, you can write a set of commands to automate operating system interactions into a simple script that will run as if you were entering each command manually. Similarly, in network operating systems such as Cisco's IOS or Aruba's ArubaOS, the startup and running-config can be considered scripts. The same commands you would manually enter in the shell are stored and executed in order. While a rudimentary version, these examples are a style of programming that is often called *Functional* or *Procedural*. Functional programming is a style that separates data and behaviors. Any data created is immutable, or unchangeable, and the data is run through functions which will return a new data.

The alternative to functional programming is *Object-Oriented Programming (OOP)* and rather than basic data containers, utilizes *objects*. These objects are structures based on and inherit properties from *classes*, frameworks for objects, that contain information (known as attributes) and code functions (known as methods) for data manipulation. An object is changeable and able to be stored and shared throughout the program. Some languages will handle each style slightly differently so your style must be taken into account before deciding which to use. While there is much debate about which is better or more powerful, OOP or functional, this is a choice that depends almost solely on the situation.

Structured vs. Unstructured Data

When reviewing the data that will be passed between the remote system and your integration, the makeup of the data is important. There are two universal types of data when programming - Structured and Unstructured. Structured data is made up of information with clearly defined layout and format. Conversely, unstructured data is information that has no defined models or schema. Structured data makes searching, manipulation, and storage easy to deal with while unstructured data generally has to be interpreted and parsed for meaning before anything can be done with it. An example of structured data for those unfamiliar with programming would be an address as printed on an envelope. The information has a standard format, and each field has meaning, from the street and number to the city, state, and zip code, each piece can be easily extracted and has a defined meaning. Unstructured data could be represented by a simple text document or a posting on a social media site. There is meaning to the information contained within, but it isn't clearly defined and requires some level of interpretation to extract. One additional note with data is that some information can be considered semi-structured. For instance, the message body of an email is plain text and therefore unstructured, but when you evaluate the entire message, headers and all, the data becomes semi-structured. Some languages deal with unstructured data better than others, allowing the programmer to build models for the data and quickly parse it into structured data. This should be taken into account when deciding on a language.

Additionally, you must consider the data modeling languages often supported by the development environment. For example, XAML is most commonly used with .NET. If you already know .NET and XAML is an available return set from the system you plan to automate, it will likely be a good choice for you.

Familiarity & Access to Information

Finally, when deciding on a language (and this is probably the most important factor) experience, comfort, and access to helpful resources are going to be paramount. When developing an integration with a remote system, a specific language may be more capable and better suited, but if you do not have any background with it or the ability to learn it before writing the integration, you'll end up spending a lot more time than necessary getting it right.

Later in this chapter, some details for a few of the most commonly used languages today, the potential need for data conversion between models/structures, and examples on how to use them.

Structured Data	Unstructured Data
1. Date	5. Media (audio & video)
2. Address	6. Text file
3. Phone number	7. Office document
4. Credit card number	8. Tweet

Why Are We Integrating Systems?

There are many types of integrations, but in the world of networking and wireless they will mostly all fall into one of two types: *management* or *automation*. There are many off-the-shelf solutions you may be familiar with exist that combine both of these into a single interface or platform, but for the purposes of this book, we will address them individually.

Management

> "Measurement is the first step that leads to control and eventually to improvement. If you can't measure something, you can't understand it. If you can't understand it, you can't control it. If you can't control it, you can't improve it."
>
> ~ H. James Harrington

There is a point where measurement of how things work is paramount to increasing our understanding. Automation and programmability should first increase our ability to measure how systems perform, providing data that help in our understanding and ultimately gives us the ability to decide what action is necessary. Managing a large or distributed network requires the administrator to be in constant contact with their devices. By gathering a wide range of information and calculating metrics, they can have a clear view of the health and operation of the network. Additionally, storing this information allows for the creation of a baseline (a point-in-time description of a device or devices used for defining state and subsequent changes) and historical information for issue and configuration tracking.

Historically, basic management and monitoring were done using some antiquated polling methods, but in the age of the programmatic network and the DevOps (development operations) engineer, there are better ways. Modern management and monitoring typically rely on a push/pull model of data gathering - where the integration will either periodically poll or receive data sent from the system being managed. The process performing this function will typically be a daemonized (background) application or script running on a central server and storing the information in relational databases (more later in this chapter), configuration files, or even plain text files. Information gathered this way can be simple status (online/offline), device configuration state, health information, or access and authorization control logs. This information can be used to monitor and verify basic details such as the configuration against a baseline or to keep records of management access.

More in-depth monitoring is often categorized differently and is referred to as network *telemetry* or *analytics*. Telemetry is one of the easiest and often overlooked, integrations that can be built. Telemetry (derived from the Greek *tele-*, meaning "at a distance," and *-metry*, meaning "of or related to measuring") is defined on Wikipedia as "an automated communications process by which measurements and other data are collected at remote or inaccessible points and transmitted to receiving equipment for monitoring." In the context of a wireless network, this would include data such as client count, channel utilization, CPU utilization, traffic throughput, location-based information, or one of numerous other key performance indicators (KPIs). Telemetry systems are often where the data for reporting of internal policy or regulatory compliance are stored. These systems will often use a publisher/subscriber model (if available) of data collection and will store the information in a time-series database (more later in this chapter).

Automation

> "The first rule of any technology used in a business is that automation applied to an efficient operation will magnify the efficiency. The second is that automation applied to an inefficient operation will magnify the inefficiency."
>
> ~ Bill Gates

Personally, I call this "*engineering amplification*" because of the way a properly implemented system can give a single administrator the power of many. Automation is the other type of common integration used by network administrators to perform a repetitive task or procedure with minimal or no human interaction. The tasks can range from simple management tasks to updated configuration deployment to proactive operational optimizations. The process can be run as a background service (daemon), a manual administrator-initiated script, or an automatically triggered script (proactive or reactive based on configured inputs). Automation allows numerous advantages in modern networks such as reduced operational costs by reducing management overhead, improved operations by removing human error, and increased performance by building a self-improving network. Automation is often paired in conversations with telemetry and machine learning as a way to detect anomalies better and faster than any human could allowing for precise configuration as well as security enforcement.

Automation systems can come in many flavors; custom in-house, open-source packages, and enterprise software packages. Each one offers benefits, and the decision of which to use will depend entirely on the goals, budget, and the size of the systems that will be integrated into the platform. Several freely-available packages can be deployed in a number of ways including in your own custom scripts.

Application & Integration Architectures

Before beginning a project, an integration developer will need to find out the application architecture used by all systems to be involved as well as the integration architecture in use by the organization and work to understand it in great detail. The application architecture will determine what, if any, hardware or software that may need to be purchased, deployed, or altered as well as the resource and security implications. The integration architecture is a reference model for how multiple systems within an enterprise interact and share information (if at all). These individual systems can be directly related, indirectly reliant on one another, or completely independent from one another. Care needs to be taken when building a greenfield integration where no application or integration architecture exists so as to not introduce issues by choosing the wrong type - the type that doesn't grow with the organization or applications. This decision could have long-lasting repercussions and technical debt for an organization.

Application Architectures

The application architecture can either be dictated by the applications and systems that will be sharing information or through design choices made by the integration designer. Based on the project and integration needs, it should be fairly easy to determine which type to choose.

Monolithic

A monolithic application is the most traditional, familiar, tried-and-true architecture and how most enterprise applications are deployed. By definition, it consists of a single system utilizing a narrow set of technologies and dependencies to provide a service or services for an enterprise using a shared codebase and libraries. Monolithic applications are typically maintained by a focused team with deep institutional knowledge and familiarity with the entire system. While a monolithic application is still quite common and relevant, it is very difficult to scale and very difficult to add new features due to the shared codebase and dependencies on both system libraries and language intertwined throughout the system. This architecture is (very) slowly being phased out when possible for more modular architectures.

Microservices

Microservices are a group of loosely coupled services (each one similar to a monolithic application) separated into smaller, function-driven components. These systems are typically deployed in a containerized model (i.e. Docker, Kubernetes, AWS). Each component in this architecture is lightweight, modular, self-contained, and can be independently scaled based on needs - often dynamically. By making each part of the system independently deployable, a microservices architecture is highly scalable and has the ability to react to sudden spikes in resource needs.

The disadvantages of this architecture can be quick to cause issues and sometimes hard to overcome. There is a measurable increase in the effort needed during the initial design, and without a proper plan, the levels of granularity can become overwhelming. Additionally, the added complexity of a containerization service to manage the underlying systems and testing needed to vet the application as a whole are increased.

Serverless

Serverless applications are event-driven systems that separate the application from the resources removing the "always-on" server components and provide a highly available on-demand application. These applications are run on Function-as-a-Service (FaaS) and Backend-as-a-Service (BaaS) offerings such as AWS Lambda, Google Cloud Functions, Azure Functions. They are easy to deploy, cost less than traditional cloud offerings, and scale automatically as load increases. Due to the cloud provider nature of this architecture though, vendor lock-in is almost guaranteed.

Integration Architectures

As mentioned above, the integration architecture is an important decision to make if it has not been made already.

Point-to-Point

In a **Point-to-Point** integration architecture, every application is tightly coupled to its partners, meaning it has hard-coded connections to every other application it needs to communicate and share data with. In this architecture, all messaging and data transformation are designed and handled to specifically work between the specific member systems. Managing each integration individually, including member application revisions will eventually cause complexity to become overwhelming. The benefit of a point-to-point architecture is that a reference design will likely exist for gathering data from one system and sharing it directly to another. If a reference design does not exist, it is fairly simple to build a proof-of-concept integration that can be translated directly into a production environment.

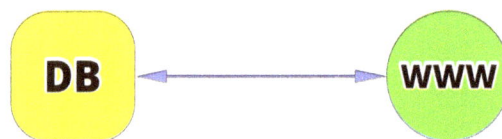

Figure 12.1: Point-to-point integration showing two applications

The disadvantage with this model, however, though it can and does work quite well at a small scale, is that as systems grow and applications are added to the enterprise, the scalability becomes highly problematic.

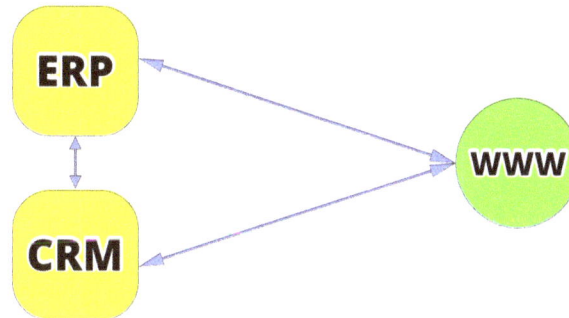

Figure 12.2: Point-to-point integration showing multiple applications

Service Bus

Much more flexible than a point-to-point integration, a **Service Bus** or **Enterprise Service Bus (ESB)** architecture provides flexibility through a central system that brokers communication between individual applications. In this architecture, all messaging and data transformation are handled by the broker. Each application will have a connector to the ESB that acts as a client and/or server depending on the function it serves in the enterprise. In the event a single application has revisions that change the way data is presented or consumed, the only update needed would be at the broker, and it would remain transparent to all other applications. This architecture is highly scalable and can often offer increased application performance decreased development cycles because, in order to launch a new integration, no other systems will be impacted. While fewer than a point-to-point architecture, the disadvantages of a service bus are that it requires custom data translation models to be built for each application sharing or requesting information through it and it is not as easily deployed in a proof-of-concept environment that can replicate the production systems.

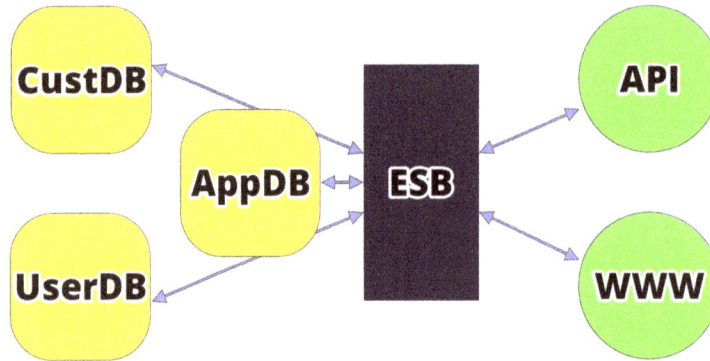

Figure 12.3: Enterprise Service Bus integration showing multiple applications

It may seem like the service bus architecture is the only choice based on the information here, but that is not the case. There are a variety of reasons to choose one over the other, including regulated data controls, API category, or project budget.

Additional Application Considerations

Integrations and applications based on any of the above architectures can be designed in a few different ways that separate the individual functions into separate tiers (or a single-tier).

Application Layers

An application can be divided into logical layers based on a distinct function within the architecture. The commonly accepted layers are *Presentation*, *Business Logic*, and *Data Access*.

Presentation Layer

The presentation layer is where the user or remote service interacts with the application. Typically running at this layer is a web server that relays requests from the client to the business logic layer and back. It provides any transport-dependent translation to the data being received or returned and packages it accordingly. Examples of common web servers deployed at this layer are Apache and Nginx.

Business Logic Layer

The business logic layer is where rules are applied that determine how data is created, stored, and changed. It enforces the methods and data routing used to provide information to the presentation layer. This is where software development occurs and interacts with the upper and lower application layers.

Data Access Layer

The data access layer is where logging, traffic routing, and any other services required to support the business logic layer occur. This is where database and file system access is handled. The specific file system will be operating system and hardware dependent but will operate the same way for all; a request from the business logic layer to retrieve a file will be handled, and the file data returned. As for the databases that may reside here, quite a few different types exist; relational, non-relational, and time-series are the primary ones we're concerned with, in the context of network systems integrations.

SQL (Relational) Database

Relational databases are collections represented by a schema that defines the structure of data types and attributes contained in tables consisting of rows and columns of information logically similar to a spreadsheet. These databases can be accessed and updated using a Structured Query Language (SQL). The name "relational" comes from the use of *keys* which are used to reference data in other tables, rows, or columns. This allows the data to stay relatively compact and quickly accessible while maintaining some level of structured information. Examples of SQL databases include: Microsoft SQL, MySQL, PostgreSQL

NoSQL (Non-relational) Database

Opposite to the strict schemas of relational databases, non-relational databases are schema-agnostic. There are several subtypes of NoSQL databases that store and access data in slightly different ways but the common feature is that they all allow unstructured and semi-structured data to be stored and manipulated. NoSQL architectures allow for greater horizontal scaling and performance but often sacrifices immediate consistency between the nodes for eventual consistency of data. Examples include Redis, MongoDB, Cassandra, Riak, and Berkley DB.

Time-series Database (TSDB)

A time-series database is a system specifically designed to efficiently store and index data that has a timestamp associated with each entry. This is especially useful for telemetry/analytics data where a given KPI changes over time and you may want to create baseline information for automation systems to be able to react to drastic changes or to create visualizations for management dashboards. Examples of TSDB include InfluxDB, Prometheus, RRDTool and Graphite.

Application Architecture Tiers

Single-Tier

A single-tier application is one where all application layers and data reside locally (or on shared storage), and everything is self-contained. This application can have multiple services running performing as different application layers (e.g. web server, database, application logic) all sharing resources.

Multi-Tier

A multi-tier application is one where the layers of an application are divided into separate systems. This allows for systems that are more easily deployed and scaled. While multi-tier doesn't necessarily describe how the architecture is laid out in a precise manner, its type can be subdivided into more descript types ; *2-tier* (client-server), *3-tier* (web-based), or *N-tier* (distributed) applications.

Operating System Services

In addition to the application layers and their associated services, the operating system runs additional services to provide basic functionality. While not a part of whatever application configured to run on the system, these services provide basic and required functionality for an application to run. This list includes some of the common services that are important to know but is in no way comprehensive.

The operating system manages network communications, ensuring that the resource is correctly shared between all applications requesting access and properly routing packets up and down the stack. Underneath this service are several other services including the Dynamic Host Configuration Protocol manager (DHCP), Domain Name System service (DNS), Network Time Protocol service (NTP), and the system firewall.

DHCP service

The DHCP service (if configured) manages the request for dynamically assigned network addresses. If configured, this service will request an address from a remote DHCP server and use the response to configure the local network interface with the appropriate network, host, DNS, and time server addresses, as well as any other optional services, configured.

DNS service

The operating system runs a DNS lookup service that relays queries for the domain name to IP address translation from the applications to the configured remote (or local) servers.

NTP service

A service that maintains operating system time synchronization with internal or Internet-based time servers. Time-specific applications will require time to be synchronized across all participating systems.

Firewall

A network security service design to monitor and police incoming and outgoing traffic based on a set of predetermined rules. This service may be able to provide both network and application-layer filtering to permit or deny traffic based on a series of rules. A network-based rule will permit or deny traffic based on the source or destination IP address of the traffic and can be configured as widely as a large network segment or as granularly as an individual host. An application-based rule will permit or deny traffic based on the source or destination TCP/IP port of the traffic.

Data Structure and Types

When accessing an API endpoint or passing data between applications, a known data type and structure must be used in order for the information to be usable by either side of the exchange. The defined data types and the structures that contain them dictate the programmatic understanding and available operations that can be performed on each piece of information. In simpler terms, a data type is a classification of a piece of information while a data structure is a well-defined way of organizing a grouping of information.

While an entire book could be written (and likely has been) about the various data types and structures and how they translate to a particular programming language, for the purposes of this book we will only take a high-level look at a few data types common to popular programming and scripting languages.

Data Types

For an application or script to be able to understand and perform operations on a particular piece of data that data must conform to a specific set of rules. These rules are defined by the type of data. At a high-level view of the most common scripting languages, data will fall into one of the following types:

- **Number**: An item that refers to a number. This type is often divided into integers or floating-point numbers depending on the language, with each having a unique set of operations and limitations.
- **String**: A character or characters.
- **Boolean**: A representation of true or false and depending on the language may be a subtype of a number (true = 1, false = 0).
- **Array**: A sequence of items stored as a group. Items in an array are typically accessed by specific location within the array.
- **Dictionary**: A key/value paired group of data where a key (most commonly a string) is provided as a pointer to the item (value) being stored. Items stored in dictionaries are accessed by calling their assigned key.

Data Structures

A data structure is simply a way for data to be organized and stored for easy access to the contained information. By having a clearly defined structure, this information can be shared across processes or applications without the need for custom decoding, parsing, or mapping functions since that can all be handled internal to the language and transparent to the developer.

JSON

One of the most common data exchange formats in use today is JavaScript Object Notation or JSON. While the name may elicit language dependence, JSON is completely language-independent. It is a lightweight format that is easy for humans as well as computers to both read and write. Due to the small size and ease of use, JSON is ideal for exchanging data between disparate systems or processes and has quickly become the de-facto standard for web-based communications. As a common internet data exchange format, JSON's internet media type is defined as application/json and the most common file extension used is .json.

Starting in the early 2000s, JSON has gone through multiple iterations and, as of 2017, is defined by IETF RFC 8259. Using data types and structures common to nearly all modern languages used by developers today, including (but not limited to) Go, JavaScript, Java, Perl, and Python, JSON is able to provide flexibility to both API producers and consumers by removing the need for either side to know or care what the other is using to write their applications.

The complete specification for JSON as well as a listing of language-specific libraries and implementations can be found at https://json.org.

JSON's data structure is comprised of 2 types of containers; objects and arrays. The values in both structural elements can be one of the supported data types or either one of the core data structures (object or array).

object

- An unordered collection of key/value pairs
- Language-specific equivalents: record, struct, or dictionary
- Starts and ends with curly braces
- Keys are strings
- Values can be any supported data type or structure

{"key": "value"}

array

- An ordered collection of values
- Language specific equivalents: array, list, or sequence
- Starts and ends with square brackets
- Values can be any supported data type or structure

["value_1", "value_2", "value_3", "value_4"]

JSON supports the following base data types: number, string, boolean, and null in addition to allowing nesting of the data structure.

number

- A numeric value
- Language specific equivalents: integer, floating point

string

- A sequence of characters
- Starts and ends with double-quotes
- Language-specific equivalents: character, string

boolean

- A representation of truth
- Valid values are true or false

null

- A representation of no value
- Language specific equivalents: None, nil

Example

```
{
  "String": "This is a string",
  "Number": 100,
  "Boolean": true,
  "Null value": null,
  "Nesting": {
    "Array": ["This", "is", "an", "array", "of", "strings."],
    "Object": {
      "String": "This is a string inside a nested object."
      ...
    }
  }
}
```

```
["This", "is", "an", "array", "of", 8, "objects", {"key": "with mixed data types"}]
```

XML

A document-based, open standard, markup language developed by the World Wide Web Consortium that is both human and machine-readable, XML (Extensible Markup Language) is designed for both data storage and transport. While numerous variations exist today such as RSS, SOAP, SVG, and XHTML we will simply be focusing on the core data structure and types within XML. As a common internet data exchange and data storage format, XML's internet media type is defined as application/xml or text/xml and the most common file extension used is .xml. With version 1.0 being released in 2008, XML has gone through several changes and updates over the years, and as of 2014, the latest revision (1.1) is defined by IETF RFC 7303.

The complete specification for XML can be found at the W3C Site and at the time of this writing is version 1.1 (second edition).

At its core, XML is a hierarchical structure consisting of a string of characters and is often referred to as a tree. For our purposes, an XML tree consists of elements, tags, attributes, and content. If you're familiar with HTML, XML will have a very similar feel due to the use of opening and closing tags around each element. An element is comprised of a start () and end () tag enclosed within angle brackets, and any information contained within those tags is the element's content. All documents start with an XML element declaring the version and encoding type used in the document.

```
<? xml version="1.1" encoding="UTF-8" ?>
```

An example of a logical tree structure and its XML representation shown here:

Logical Structure

```
AccessPoint
├──Name
└──Radio_0
   └──Band
   └──Channel
   └──Power
   └──SSID
      └──VLAN
      └──Security
```

XML Equivalent

```
<AccessPoint>
  <Name>AP-01</Name>
  <Radio_0>
    <Band>2.4</Band>
    <Channel>3</Channel>
    <Power>20</Power>
    <SSID>CWNP-Open</SSID>
      <VLAN>10</VLAN>
    <SSID>CWNP-Secure</SSID>
      <VLAN>20</VLAN>
      <Security>WPA2-PSK</Security>
  </Radio_0>
</AccessPoint>
```

In addition to the elements and content, XML supports element attributes as a way to add name/value pairs of information to an element. This can reduce the length of the document as well as add more descriptive capabilities to your XML as shown here using the same information as above but resulting in a much more compact piece of data:

XML Equivalent with Attributes

```
<AccessPoint name="AP-01">
  <Radio_0 band="2.4" channel="3" power="20">
    <SSID vlan="10">CWNP-Open</SSID>
    <SSID vlan="20" security="WPA2-PSK">CWNP-Secure</SSID>
  </Radio_0>
</AccessPoint>
```

Attributes must be unique per element, and if multiple are required in your data you can use a comma, semi-colon, or space-delimited list (depending on the data provided). Below you can see a space-delimited list for the SSID attribute security which tells us this SSID utilizes 2 types of security - WPA-PSK and WPA2-PSK.

XML Equivalent with Multiple Attributes

```
<AccessPoint name="AP-01">
  <Radio_0 band="2.4" channel="3" power="20">
    <SSID vlan="10">CWNP-Open</SSID>
    <SSID vlan="20" security="WPA-PSK WPA2-PSK">CWNP-Secure</SSID>
  </Radio_0>
</AccessPoint>
```

Tags can also be set as empty by adding a forward slash before the final angle bracket. This can be used to simplify and compact your data even further. Below we can see the empty SSID element with an added attribute of name that replaces the element content.

XML Equivalent with Multiple Attributes

```
<AccessPoint name="AP-01">
  <Radio_0 band="2.4" channel="3" power="20">
    <SSID name="CWNP-Open" vlan="10" />
    <SSID name="CWNP-Secure" vlan="20" security="WPA-PSK WPA2-PSK" />
  </Radio_0>
</AccessPoint>
```

YAML

A powerful data serialization language, YAML (*YAML ain't markup language*) is very similar in function to JSON but primarily serves a different purpose. YAML is a document-based data storage structure that uses Python-like indentation and supports strings, integers, floats, lists, and associative arrays and is a common configuration file type. It is used for numerous open source automation systems such as SaltStack, Ansible, and Nornir and can be quite useful to manage configurations and state when building your own automation or telemetry gathering systems. Similar to JSON, YAML is built using common data types and structures in order to maintain language agnosticism and usability in your language of choice. The file extensions most commonly associated with YAML are .yml and .yaml although the latter is considered best practice.

The complete specification for YAML can be found at https://yaml.org and at the time of this writing is version 1.2.

YAML's data structure is comprised of 2 types of containers; associative arrays and lists. The values in both structural elements can be one of the supported data types or either one of the core data structures (associative arrays or lists).

associative array

- An unordered collection of key/value pairs

- Language-specific equivalents: record, struct, or dictionary

- Colon and space-separated key/value pair

- Keys are strings

- Values can be any supported data type or structure

```
# Associative array
key_1: value_1
key_2: "value_2"
key_3: 'value_3'
```



```
# Inline associative array
{ key_1: value_1, key_2: 'value_2', key_3: "value_3"}
```

list

- An ordered collection of values

- Language-specific equivalents: array, list, or sequence

- Individual lines starting with a hyphen or a comma-separated sequence enclosed within square brackets

- Values can be any supported data type or structure

```
# List of items
- value_1
- "value_2"
- 'value_3'
- value_4
```

```
# YAML also assumes a list based on indentation and a starting hyphen
- "value_1"
value_2
'value_3'
value_4
```

```
# Inline list of items
["value_1", value_2, value_3, 'value_4']
```

YAML supports the following base data types: number, string, boolean, and null in addition to allowing nesting of the data structure.

string

- A sequence of characters

- Does not require quotes but can be enclosed in either single or double quotes

integer

- A numeric value representing a positive or negative whole number

float

- A numeric value representing a positive or negative number that may or may not contain decimal places

Example

```
--- # This is an example YAML document
name: "YAML ain't markup language"
supported_data_types: # List of items
  - "strings"
    integers
    "floats"
    'lists'
    dictionaries
file_extensions: [".yml", ".yaml"] # Also a list of items
--- # Multiple documents supported in a single file
nested_items: # Associative array of items
  key_1: value   # String
  key_2: 2.0     # Float
  key_3: 3       # Integer
```

CSV

Dating back to the early days of computing and preceding personal computer by more than a decade (likely beginning with IBM's Fortran in the early 1970s) information being transferred between systems of varying architectures needed to have a better, less error-prone, plain-text format that was easy to use for both humans and machines - mostly because of the cumbersome nature of punch-card systems and the fixed-width columnar data schemes that were widely used. For the new data format, information was separated into records that contained fields to store information. Records were most commonly determined by a line-break (CRLF - carriage return + line feed, a leftover from the days of mechanical typewriters) but in order to separate the fields within each record, something needed to delimit the information, and for that task the comma was chosen, giving the data type the name Comma-Separated Value or CSV.)

After decades of expanding use, the IETF built a standard, defined in RFC 4180, in 2005, which clearly outlined the CSV format and its MIME content type.

If you've ever dealt with a spreadsheet, you'll be familiar with the CSV format. CSV is mostly used when storing fairly simple data (numbers and text) in a file with each line, or record, containing a series of values, fields, that are traditionally delimited by commas - hence the name - although commas are not necessarily required (other delimiters can be used depending on the implementation and systems utilizing the information). Some other commonly used delimiters are semicolons or tabs, and even though they are not utilizing commas, still retain the CSV name and file extension, .csv.

While typically used to refer to raw spreadsheets and data files, a CSV can also be used when transferring data between systems through APIs or other means. Typically in these cases, HTTP/S is used, and the media type will be defined in the request headers as the MIME type text/csv to inform the receiving system of what is to be expected.

The CSV format is fairly simple and consists of records, or rows, of information. Since it is plain-text, there are no data types or containers used - it is all just text. If other data types need to be utilized, such as integers, that is up to the system receiving the data to interpret and parse.

Referring to the RFC, there are 7 basic requirements when composing a CSV file shown in the list here (with examples). In these examples, the line-break is called out as CRLF, but this is no longer needed in practice due to computers handling that part for us now.

1. Each record must be on a new line

```
record_1_field_1,record_1_field_2,record_1_field_3 CRLF
record_2_field_1,record_2_field_2,record_2_field_3
```

2. The last line may or may not end with a line-break

 record_1_field_1,record_1_field_2,record_1_field_3 CRLF
 record_2_field_1,record_2_field_2,record_2_field_3 CRLF

3. A header row can be added to provide the field names

 header_1,header_2,header_3 CRLF
 record_1_field_1,record_1_field_2,record_1_field_3 CRLF
 record_2_field_1,record_2_field_2,record_2_field_3

4. Each record contains one or more fields, all records within a file must have the same number of fields, spaces should be interpreted as a part of the field, and the last field of a record must not be followed by a comma

 header_1,header_2,header_3 CRLF
 record 1 field 1,record 1 field 2,record 1 field 3 CRLF
 record_2_field_1,record_2_field_2,record_2_field_3

5. Fields can be double-quoted but it is not required

 "record_1_field_1","record_1_field_2","record_1_field_3" CRLF
 record_2_field_1,record_2_field_2,record_2_field_3

6. Any field that contains a line-break, double-quotes, or commas must be enclosed in double-quotes

 "record 1,CRLF
 field 1","record 1, field 2",record_1_field_3

7. Any double-quoted field that contains double-quotes must be escaped by adding an additional double-quote

 "record 1""field 1""",record_1_field_2,record_1_field_3

YANG

YANG is a data modeling language used to model configuration and state data manipulated by the Network Configuration Protocol (NETCONF), NETCONF remote procedure calls, and NETCONF notifications. (IETF RFC 6020)

The RFC goes on to state:

> YANG is a language used to model data for the NETCONF protocol. A YANG module defines a hierarchy of data that can be used for NETCONF-based operations, including configuration, state data, Remote Procedure Calls (RPCs), and notifications. This allows a complete description of all data sent between a NETCONF client and server.

> YANG models the hierarchical organization of data as a tree in which each node has a name, and either a value or a set of child nodes. YANG provides clear and concise descriptions of the nodes, as well as the interaction between those nodes.

> YANG structures data models into modules and submodules. A module can import data from other external modules, and include data from submodules. The hierarchy can be augmented, allowing one module to add data nodes to the hierarchy defined in another module. This augmentation can be conditional, with new nodes appearing only if certain conditions are met.

> YANG models can describe constraints to be enforced on the data, restricting the appearance or value of nodes based on the presence or value of other nodes in the hierarchy. These constraints are enforceable by either the client or the server, and valid content MUST abide by them.

YANG defines a set of built-in types, and has a type mechanism through which additional types may be defined. Derived types can restrict their base type's set of valid values using mechanisms like range or pattern restrictions that can be enforced by clients or servers. They can also define usage conventions for use of the derived type, such as a string-based type that contains a host name.

YANG permits the definition of reusable groupings of nodes. The instantiation of these groupings can refine or augment the nodes, allowing it to tailor the nodes to its particular needs. Derived types and groupings can be defined in one module or submodule and used in either that location or in another module or submodule that imports or includes it.

YANG data hierarchy constructs include defining lists where list entries are identified by keys that distinguish them from each other. Such lists may be defined as either sorted by user or automatically sorted by the system. For user-sorted lists, operations are defined for manipulating the order of the list entries.

YANG strikes a balance between high-level data modeling and low-level bits-on-the-wire encoding. The reader of a YANG module can see the high-level view of the data model while understanding how the data will be encoded in NETCONF operations.

YANG is an extensible language, allowing extension statements to be defined by standards bodies, vendors, and individuals. The statement syntax allows these extensions to coexist with standard YANG statements in a natural way, while extensions in a YANG module stand out sufficiently for the reader to notice them.

YANG resists the tendency to solve all possible problems, limiting the problem space to allow expression of NETCONF data models, not arbitrary XML documents or arbitrary data models. The data models described by YANG are designed to be easily operated upon by NETCONF operations.

It is important to note that OpenConfig has standardized on YANG as the data modeling language it uses.

Proprietary Data Structures

While becoming less common, proprietary data structures have been used for as long as companies have been protecting intellectual property, processes, and methods, so they still exist and over the course of a career developing integrations between systems you are almost guaranteed to run into them in one form or another at some point. A proprietary data structure is fairly straightforward in definition - a data structure that wasn't developed by a standards body and most likely its definitions are not publicly available. There can be numerous reasons for this such as intellectual property and trade secret protection (as mentioned above), data protection in-transit or at rest, or even performance requirements not met by any currently available open standards. Some vendors choose to implement open protocols with slight variations that can be figured out with a little bit of work while others may build a protocol that cannot be reasonably decoded by anyone without the specification and details (such as a custom binary protocol or encryption suite). It may be challenging to find information to work with a proprietary protocol and not all vendors are willing to share the structure and definitions. Make sure you have a clear understanding of what is available to you and what kind of support you can expect before embarking on a project utilizing proprietary data structures.

Some well-known examples of proprietary data structures in the wireless industry are:

- Cisco's Network Mobility Services Protocol (NMSP) which provides wireless LAN controller (WLC) telemetry and client location information to external systems

- HPE Aruba's Advanced Monitoring Protocol (AMON) which is used to provide statistics and details of the wireless network as well as client information

Due to the closed nature of proprietary data, there are no examples to provide.

API Types

Now that you understand the languages available and the types of data you may receive, it's time to review the common API types you'll be interacting with. The most common type of API you will deal with is REST. REST stands for REpresentational State Transfer. In REST, you are only concerned with the transfer of the current state. This means that we don't really care how the underlying system works.

RESTful APIs have risen in popularity due to their simple nature. Unlike previous protocols like SOAP (Simple Object Access Protocol), there isn't a need to understand the underlying architecture of the system you are connecting to. Also, REST doesn't dictate the format of the data being transferred. Whereas SOAP required the data to be in XML, REST does not require a specific format, meaning JSON, YAML, XML, CSV and others can all be valid RESTful data payloads.

RESTful APIs

REST APIs are modeled around HTTP, specifically as a set of rules about how to use the HTTP protocol to exchange data. A typical REST request looks much like an HTTP request and has very similar behavior.

REST uses what is called "resources" to describe data structures within a system. When interacting with wireless systems, you might see resources like clients, WLANs, access points, etc. To access a resource, you make a call to the URI or Uniform Resource Identifier. This looks a lot like an HTTP URL and is essentially an address referencing a resource.

An example of a URI might be https://192.0.2.1/api/device/001122334455 and the resource you would be a device with the MAC address of '00:11:22:33:44:55.

REST Verbs

REST also uses verbs to describe what it is you are attempting to do. If in the previous example you were trying to get information about the device, you would use a GET verb to get the details of that device.

GET:

Retrieves the state information for the specified object. This verb requires a URI and a header. Parameters are optional and can be inserted into the URI of the header. Payload is not a required field for this verb.

POST:

Creates an object with the specified payload. This verb requires the payload in addition to the URI and header.

PUT:

This method is used to update an existing entry. If you wanted to update a WLAN with a new parameter, you would use a PUT. PUTs can also be used to create a new object, usually with an identifier that is specified in the PUT.

PATCH:

This method is also used to update an existing entry; however generally this method only needs the information that is to be changed. PUT may require you to pass all parameters, while PATCH should only require the parameters to be changed.

DELETE:

This method is used to delete a resource.

REST Status Codes

Just like an HTTP request, a REST request has a status code. These give you a high-level indication of the status of the request made. In HTTP, many of us are familiar with the status code 404 (Page Not Found).

1xx Status Codes:

These series of status codes are meant to be information codes about the HTTP protocol.

2xx Status Codes:

Success messages typically reside in the 2xx status codes. For example, the response to our GET regarding a specific MAC would likely be returned with a 200 (OK) status code. For creating a new object, we would often see 201 (Created), and for updates we would likely see a 202 (Accepted).

3xx Status Codes:

300 level status codes generally discuss some sort of redirection. If the URL moved, or the API resource changed, a status 300 might be telling you that the resource is now somewhere else.

4xx Status Codes:

Generally, this means something the client did was wrong. The very generic 400 (Bad Request) could mean that you did something that made your request invalid. Or 401

(Unauthorized) is common as well, indicating you may have failed to authenticate properly to the API, or that you are attempting to do something you are not authorized for. Another common response might be 404 (Not Found) perhaps if you tried a resource that didn't exist. In our previous example of getting device details for a MAC address, it's possible the system doesn't have that client in the database.

5xx Status Codes:

Generally, this means that the server had an issue performing the request. While the authors wish this was always the case, it's often possible that 5xx status codes can be caused by the client passing data to the API that it did not know how to handle. It's still the client's fault for sending bad data, but the server did not handle the error properly.

An easy way to remember status codes is this:

- 100 FYI
- 200 OK
- 300 Go somewhere else
- 400 Client did something wrong
- 500 Server did something wrong

While the HTTP status codes for APIs are generally a high-level indicator whether the request was successful or not, often the API will include information about the success of the action in addition to the overall success. For example, you might successfully create the WLAN, but it failed to set the value for a field because a prerequisite is not met.

Guiding Principles of REST

REST is governed by six guiding principles. We won't go into detail on all of them, as some of them are intended for those writing APIs instead of consuming them.

Client-Server:

RESTful APIs are meant to be client-server, meaning that a client makes a request of the server to perform some action. There are use-cases for Server-Client and Server-Server architectures, but these are RESTful architectures.

Stateless:

All the information needed to complete a request must be contained within the request. REST doesn't have the ability to prompt for additional information; missing info will result in a failed request.

Layered System:

REST is layered in such a way that you only have access to the layer of the system you are currently accessing. This abstracts a lot of the system interworking, meaning you might be accessing data on a database server, but you don't know which one or even what type of database it is.

The remaining REST principles are:

- Cacheable
- Uniform Interface
- Code on Demand

You can read more about the REST guiding principles at https://restfulapi.net/

Webhook APIs

Where RESTful APIs are meant to be a Client-Server architecture, Webhook is often considered to be a Server-Server architecture. In this scenario, one system will send a POST to another system. This is highly useful for event-driven behavior when you need to notify another system of an event happening. For example, if you wanted to be notified when a device leaves a geofenced area, the system can send a POST to a webhook to wipe the device.

While it is possible to achieve this with a traditional REST API, it would require the client to repeatedly poll the REST API to check the status. In these situations, Webhooks are much more efficient, as it operates on demand.

It should be noted that Webhooks require IP and port reachability between the servers.

Websockets APIs

So far we have talked about REST (Client-Server) and Websocket (Server-Server). Websockets are a method for Server-Client communication. In Websocket, the client opens a socket (connection) to the server, and this is left open. When an event happens, the Server can send a web-response back down the TCP socket to the client.

This is useful for clients to receive real-time information from servers, but when the server does not have direct IP reachability to the client. It's also advantageous for clients where they do not have a local webserver running to receive a POST from the Server.

Standard HTTP GET/POST Processing

In some limited capability systems, data can be retrieved through standard HTTP requests or HTTPS requests. The process does not really implement a RESTful API, but if the application console is accessed through a web browser, it may be possible to retrieve web pages that contain the desired data and then parse that data yourself in your script or application. The work is much more tedious than when you have proper APIs, but it can suffice.

> The major problem with hacking a web application like this to get what you want is that, if the vendor updates the web application and changes the structure of pages, it is likely to break your application. Whereas APIs are more likely to remain constant.

Chapter Summary

In this chapter, we were introduced to the integration of systems using programming/scripting and APIs. We discussed the different categories and types of APIs, common API communication methods, and the high-level language differences between numerous classifications of programming languages. We also covered the two common types of network systems integrations, *management* and *automation*, and how an integrator can reduce the management overhead of a network while increasing operational efficiencies using them. Finally, we covered the many architectures related to applications and system integrations and all the additional considerations that need to be taken into account when designing a network system integration.

It's important to remember when building an integrated system that there are many, many decisions to make but the most important place to start is determining the *Integration Architecture* as well as ensuring the chosen *Application Architecture* and programming language match the API category, data needs, and desired integration type.

Now that we know what actions need to be taken, coordinating all the pieces and parts that need to participate in the action is where network systems integrations come into their own. As Linus Torvalds said, "Talk is cheap, show me the code."

Review Questions

1. What category provides restricted, verified access to an API?

    A. Open

    B. Composite

    C. Internal

    D. Partner

2. What type of encryption is used to secure HTTP traffic?

    A. SSL

    B. TLS

    C. MQTT

    D. REST

3. Unmanaged programming languages do not require resource management due to being a runtime compiled language.

    A. True

    B. False

4. An email message, including the headers, is considered what kind of data?

    A. Structured data

    B. Unstructured data

5. What is a baseline?

    A. A push/pull model of data gathering

    B. Key performance indicators (KPIs)

    C. A point-in-time description of a device or devices used for defining state and subsequent changes

    D. Engineering amplification

6. How can are automation system integration typically run? (Select all that apply)

    A. Background service (daemon)

    B. Time-series database

    C. Manual administrator-initiated script

    D. Automatically triggered script

7. Monolithic applications are event-driven systems that separate the application from the resources removing the "always-on" server components and provide highly available on-demand application.

    A. True

    B. False

8. In an Enterprise Service Bus (ESB) integration architecture, what actions are handled by the system acting as a centralized broker? (Choose all that apply)

    A. Data storage

    B. Messaging

    C. Data transformation

    D. Business logic

9. In the Data Access application layer, which type of databases allows for greater horizontal scaling and performance but often sacrifices immediate consistency between the nodes for eventual consistency of data?

    A. TSDB

    B. SQL databases

    C. Non-relational databases

    D. InfluxDB

10. In a distributed multi-tier application architecture, what is the maximum number of application layers that can be used?

    A. 3

    B. 2

    C. 7

    D. No maximum

Review Answers

1. The correct answer is **D.** Partner APIs are typically available across the Internet, but the owner will provide some sort of credentials, a token or another authentication mechanism, before granting access.

2. The correct answer is **B.** The encryption method used to encrypt HTTP traffic is Transport Layer Security (TLS) although it may still be referred to as Secure Sockets Layer (SSL) - the now-deprecated predecessor to TLS.

3. The correct answer is **B.** Unmanaged languages do not have an intermediate language and are compiled directly into machine code; this requires the programmer to handle all low-level system functions such as resource management.

4. The correct answer is **A.** The message body of an email is plain text and therefore unstructured, but when you evaluate the entire message, the data is considered semi-structured.

5. The correct answer is **C.** A baseline is defined as "a point-in-time description of a device or devices used for defining state and subsequent changes" and is used to determine the normal operation of a network.

6. The correct answers are **A**, **C**, and **D.** The automation system process can be run as a background service, a manual administrator-initiated script, or an automatically triggered script. A time-series database is a system specifically designed to efficiently store and index data that has a timestamp associated with each entry.

7. The correct answer is **B.** Monolithic applications are the most traditional, familiar, tried-and-true architecture and how most enterprise applications are deployed while serverless applications are event-driven systems that separate the application from the resources.

8. The correct answers are **B**, and **C.** The ESB acts as a broker and handles messaging between integrated systems. Business logic and data storage are application layers.

9. The correct answer is **C.** Non-relational or NoSQL databases are schema-agnostic and allow unstructured and semi-structured data to be stored and manipulated.

10. The correct answer is **D.** A distributed multi-tier application architecture or N-tier architecture, is not limited to any specific number of application layers.

Glossary: A CWNP Universal Glossary

40 MHz Intolerant: A bit potentially set in the 802.11 frame allowing STAs to indicate that 40 MHz channels should not be used in their BSS or in surrounding networks. The bit is processed only in the 2.4 GHz band.

4-Way Handshake: The process used to generate encryption keys for unicast frames (Pairwise Transient Key (PTK)) and transmit encryption keys for group (broadcast, multicast) (Group Temporal Key (GTK)) frames using material from the 802.1X/EAP authentication or the pre-shared key (PSK). The PTK and GTK are derived from the Pairwise Master Key (PMK) and Group Master Key (GMK) respectively.

802.11: A standard maintained by the IEEE for implementing and communicating with wireless local area networks (WLANs). Regularly amended, the standard continues to evolve to meet new demands. Several Physical Layer (PHY) methods are specified and the Medium Access Control (MAC) sublayer is also specified.

802.11a: An 802.11 amendment that operates in the 5GHz band. It uses OFDM modulation and is called the OFDM PHY. It can support data rates of up to 54 Mbps.

802.11aa: An 802.11 amendment that added support for robust audio and video streaming through MAC enhancements. It specifies a new category of station called a Stream Classification Service (SCS) station. The SCS implementation is optional for a WMM QoS station.

802.11ac: An 802.11 amendment that operates in the 5GHz band. It uses MU-MIMO, beamforming, and 256 QAM technology, up to 8 spatial streams and OFDM modulation. Support is included for data rates up to 6933.3 Mbps.

802.11ae: An 802.11 amendment that provides prioritization of management frames. It defines a new Quality of Service Management Frame (QMF). When the QMF service is used, some management frames may be transmitted using an access category other than the one used for voice (AC_VO). When communicating with stations that do not support the QMF service, the station uses access category AC_VO to transmit management frames. When QMF is supported, the beacon frame includes a QMF Policy element.

802.11ah: An 802.11 draft that specifies operations in the sub-1 GHz range. Frequencies used vary by regulatory domain. The draft supports 1, 2, 4, 8 and 16 MHz channels with OFDM modulation.

802.11ax: An 802.11 draft that will support bi-directional MU-MIMO, higher modulation rates and sub-channelization. It is too early to know the final details of this amendment at the time of writing; however, it is planned to operate in the 2.4 GHz and 5 GHz band.

802.11b: An IEEE 802.11 amendment that operates in the 2.4GHz ISM band. It uses HR/DSSS and earlier technology. It can support data rates of up to 11Mbps.

802.11e: An 802.11 amendment, now incorporated into the most recent rollup, that provided quality of service extensions to the wireless link through probabilistic prioritization based on the contention window. The Wi-Fi Multimedia (WMM) certification is based on this amendment.

802.11g: An IEEE 802.11 amendment that operates in the 2.4GHz ISM band. It uses ERP-OFDM and earlier technology. It can support data rates of up to 54Mbps.

802.11i: An 802.11 amendment, now incorporated into the most recent rollup, which provided security enhancements to the standard and resolved weaknesses in the original WEP encryption solution. It provided for TKIP/RC4 (now deprecated) and CCMP/AES cipher suites and encryption algorithms.

802.11n: An IEEE 802.11 amendment that operates in the 2.4 ISM and 5GHz UNII/ISM bands. It uses MIMO, HT-OFDM and earlier technology. It can support data rates of up to 600Mbps.

802.11k: An IEEE 802.11 amendment that specifies and defines WLAN characteristics and mechanisms.

802.11r: An IEEE 802.11 amendment that enables roaming between access points.

802.11u: An IEEE 802.11 amendment that adds features for mobile communication devices such as phones and tablets.

802.11w: An IEEE 802.11 amendment to increase security for the management frames.

802.11y: An IEEE 802.11 amendment that allows registered stations to operate at a higher power output in the 3650-3700 MHz band.

802.1X: 802.1X is an IEEE standard that uses the Extensible Authentication Protocol (EAP) framework to authenticate devices attempting to connect to the LAN or WLAN. The process involves the use of a supplicant to be authenticated, authenticator, and authentication server.

802.11 State Machine: The 802.11 state machine defines the condition of the connection of a client STA to another STA and can be in one of three states: Unauthenticated/Unassociated, Authenticated/Unassociated, or Authenticated/Associated.

802.3: A set of standards maintained by the IEEE for implementing and communicating with wired Ethernet networks and including Power over Ethernet (PoE) specifications.

AAA Framework: Authentication, Authorization, and Accounting is a framework for monitoring usage, enforcing policies, controlling access to computer resources, and providing the correct billing amount for services.

AAA Server Credential: The AAA server credential is the validation materials used for the server. When mutual authentication is required, a server certificate is typically used as the AAA server credential.

Absorption: Occurs when an obstacle absorbs some or all of a radio wave's energy.

Access Category (AC): An access category is a priority class. 802.11 specifies four different priority classes – voice (AC_VO), video (AC_VI), best effort (AC_BE), and background (AC_BK).

Access Layer Forwarding: Data forwarding that occurs at the access layer, also called *distributed data forwarding*. The data is distributed from the access layer directly to the destination without passing through a centralized controller.

Access Point: An access point (AP) is a device containing a radio that is used to create an access network, bridge network or mesh network. The AP contains the Distribution System Service.

Access Port: An AP used for mesh networks and that connects to the wired or wireless network at the edge of the mesh.

Acknowledgement Frame: A frame sent by the receiving 802.11 station confirming the received data.

Access Control List (ACL): ACLs are lists that inform a STA or user what permissions are available to access files and other resources. ACLs are also used in routers and switches to control packets allowed through to other networks.

Active Mode: A power-save mode in which the station never turns the radio off.

Active Scanning: A scanning (network location) method in which the client broadcasts probe requests and records the probe responses in order to determine the network with which it will establish an association.

Active Survey: A wireless survey conducted on location that involves measuring throughput rates, round trip time, and packet loss by connecting devices to an AP and transmitting data during the survey.

Ad-Hoc Mode: The colloquial name for an Independent Basic Service Set (IBSS). STAs connect directly with each other and an AP is not used.

Adjacent Overlapping Channels: Adjacent overlapping channels are channels whose bands interfere with their neighboring channels on the primary carrier frequencies. Non-overlapping channels are channels whose bands do not interfere with neighboring channels on the primary carrier frequencies.

Adjacent Channel Interference (ACI): ACI occurs when channels near each other (in the frequency domain) interfere with one another due to either partial frequency overlap on primary carrier frequencies or excessive output power.

AES (Advanced Encryption Standard): The encryption cipher used with CCMP and WPA2 providing improved security over WEP/RC4 or TKIP/RC4.

AID: Association ID (AID) is an identification assigned by a wireless STA (AP) to another STA (client) in order to transmit the correct data to that device in an Infrastructure Basic Service Set.

AirTime Fairness: Transmits more frames to client STAs with higher data rates than those with lower data rates so that the STAs get fair access to the air (medium) instead of having to wait for slower data rate STAs.

Aggregated MAC Protocol Data Units (A-MPDU): A-MPDU transmissions are created by transmitting multiple MPDUs as one PHY frame as opposed to A-MSDU transmissions, which are created by passing multiple MSDUs down to the PHY layer as a single MPDU.

Aggregated MAC Service Data Unit (A-MSDU): See *Aggregated MAC Protocol Data Unit.*

Amplification: The process of increase a signal's power level.

Amplifier: A device intended to increase the power level of a signal.

Amplitude: The power level of a signal.

Antenna: A device that converts electric power into radio waves and radio waves into electric power.

Association: The condition wherein a client STA is linked with an AP for frame transmission through the AP to the network.

Announcement Traffic Indication Message (ATIM): A traffic indication map (sent in a management frame) in an Ad-Hoc (IBSS) network to notify other clients of pending data transfers for power saving purposes.

Attenuation: The loss of signal strength as an RF wave passes through a medium.

Attenuator: A device that intentionally reduces the strength of an RF signal.

Authentication: The process of user or device identity validation.

Authentication and Key Management (AKM): The protocols used to authenticate a client STA on a WLAN and generate encryption key for use in frame encryption.

Authentication Server: The authentication server validates the client before allowing access to the network. In an 802.1X/EAP implementation for WLANs, the authentication server is often a RADIUS server.

Authenticator: The device that provides access to authentication services in order to allow connected devices to access network resources. In an 802.1X/EAP implementation for WLANs, the authenticator is typically the AP or controller.

Automatic Power Save Delivery (APSD): APSD is a power saving method which uses both scheduled (S-APSD) and unscheduled (U-APSD) frame delivery methods. S-APSD sends frames to a power save STA from the AP at a planned time. U-APSD sends frames to a power save STA from the AP when the STA sends a frame to the AP. The frame from the STA is considered a trigger frame.

Autonomous AP: An AP that can perform security functions, RF management, and configuration without the need for a centralized WLAN controller or any other control platform.

Azimuth Chart: A chart showing the radiation pattern of an antenna as viewed from the top of the antenna. Also called an H-Plane Chart or H-Chart.

Backoff timer: The timer used during CSMA/CA to wait for access to the medium, which is selected from the contention window.

Band Steering: A method used by vendors to encourage STAs to connect to the 5 GHz band instead of the 2.4 GHz band, which is more congested. Typically implemented by ignoring probe requests for some period of time before allowing connection to the 2.4 GHz radio by clients known to have a 5 GHz radio based on previous connections to the AP or controller.

Bandwidth: The frequencies used for transmission of data. For example, a 20 MHz wide channel has 20 MHz of bandwidth.

Basic Service Area (BSA): The coverage area provided by an AP wherein client STAs may connect to the AP to transmit data on the WLAN or through the AP to the network.

Basic Service Set (BSS): An AP and its associated STAs. Identified by the BSSID.

Basic Service Set Identification (BSSID): The ID for the BSS. Often the MAC address of the AP STA. When multiple SSIDs are used, another MAC address-like BSSID is generated.

Beacon Frame: A frame transmitted periodically from an AP that indicates the presence of a BSS network and contains capabilities and requirements of the BSS. Also colloquially called a beacon instead of the full phrase, beacon frame.

Beamforming: Directing radio waves to a specific area or device by manipulating the RF waveforms within the different radio chains.

Beamwidth: The width of the radiated signal lobe from the antenna in the intended direction of propagation. It is usually measured at the point where 3 dB of loss is experienced.

Bill of materials (BOM): A list of the materials and licenses required to assemble a system, in the case of WLANs, including APs, controllers, PoE injectors, licenses, etc.

Bit: A basic unit of information for computer systems. A bit can have a value of 1 or 0. Used in binary math.

Block Acknowledgement: An acknowledgement frame that groups together multiple ACKs instead of transmitting each individual ACK when a block transmission has been received.

Bridge: A device used to connect two networks. Wireless bridges create the connection across the wireless medium.

BSS Transition: Roaming that occurs between two BSSs that are part of the same ESS.

Byte: A basic unit of information that typically consists of 8 bits. Also called an octet.

Capacity: The number of clients and applications a network or AP can handle.

Captive Portal: Authentication technique that re-routes a user to a special webpage to verify their credentials before allowing access to the network. Commonly used in hotel and guest networks.

Guest Networks: A segregated network that is designed for use by temporary visitors.

CardBus: A PCMCIA PC Card standard interface that supports 32-bits and operates at speeds of up to 33 MHz. It is primarily used in laptops.

Carrier Frequencies: The frequency of a carrier signal or the frequencies used to modulate information.

Carrier Sense Multiple Access (CSMA): CSMA is a protocol that allows a node to detect the presence of traffic before sending data on a shared network. Used in CSMA/CA.

Carrier Sense Multiple Access with Collision Avoidance (CSMA/CA): CSMA/CA is the method in 802.11 networks in which a node only sends data if the shared network is idle in order to avoid collisions.

CCMP: Counter Cipher Mode with Block Chaining Message Authentication Code Protocol (CCMP) is an key management solution that provides for improved security over WEP.

CCMP/AES: CCMP used with AES, as it is in 802.11 networks, is a key management and encryption protocol that provides more security than WEP. It is based on the AES standard and uses a 128 bit key and 128 bit block size.

Centralized Forwarding: Every forwarding decision is made by a centralized forwarding engine, such as the WLAN controller.

Certificate Authority (CA): A server that validates the authenticity of a certificate used in authentication and encryption systems. The CA may issues certificates or it may authorize other servers to do the same.

CompactFlash (CF): Originally produced in 1994 by SanDisk, CF is a flash memory mass storage device format that can support up to 256 GB. CF devices can also function as 802.11 WLAN adapters.

Channel: A specified range of frequencies used in the 802.11 standard used by devices to communicate on the network.   Channels are commonly 20, 40, 80 and 160 MHz in width in WLANs. Newer standards will support 1, 2, 4, 8 and 16 MHz channels in sub-1 GHz networks.

Channel Width: The range of frequencies a single channel encompasses.

Clear Channel Assessment (CCA): CCA is a feature defined in the IEEE 802.11 standard that allows a client to determine idle or busy state of the medium based on energy levels of a frame or raw energy levels as specified in each PHY.

Client Utilities: Software installed on devices that allows the device to connect to, authenticate with and participate in a WLAN.

Co-Channel Interference (CCI): Congestion cause by the normal operations of CSMA/CA when multiple BSSs exist on the same channel. Commonly called co-channel congestion (CCC) today as well.

Collision Avoidance (CA): A method in which devices attempt to avoid simultaneous data transmissions in order to prevent frame collisions. Used in CSMA/CA.

Coding: A process used to encode bits to be transmitted on the wireless medium such that error recovery can be achieved. Part of forward error correction (FEC) and defined in the modulation and coding schemes (MCSs) from 802.11n forward.

Containment: A process used against a detected rogue AP to prevent any connected clients from accessing the network.

Contention Window: A number range defined in the 802.11 standard and varying by QoS category from which a number is selected at random for the backoff timer in the CSMA/CA process.

Control Frame: An 802.11 frame that is used to control the communications process on the wireless medium. Control frames include, RTS frames, CTS frames, PS-Poll frames and ACK frames.

Controlled Port: In an 802.1X authentication system, the virtual port that allows all frames through to the network, but only after authentication is completed.

Controller-Based AP: An AP managed by a centralized controller device. Also called a lightweight AP or thin AP.

Coverage: 1) The colloquial term used for the BSA of an AP. 2) The requirement of available WLAN connectivity throughout a facility, campus or area. Often specified in minimum signal strength as dBm; for example, -67 dBm.

Clear-to-Send (CTS) Frame: A CTS frame sent from one STA to another to indicate that the other STA can transmit on the medium. The duration value in the CTS frame is used to silence all other STAs by setting their NAV timers.

Data Frame: An 802.11 frame specified for use in carrying data based on the general frame format. Also used for some signaling purposes as null data frames.

Data Rate: The rate at which data is sent across the wireless medium. Typically represented as megabits per second (Mbps) or gigabits per second (Gbps). The data rate should not be confused with throughput rate, which is a measurement of Layer 4 throughput or useful user data.

dBd (decibel to dipole): A relative measurement of antenna gain compared to a dipole antenna. Calculated as 2.14 dB greater than dBi as a dipole antenna already has 2.14 dBi gain.

dBi (decibel to isotropic): A relative measurement of antenna gain compared to a theoretical isotropic radiator. When necessary, calculated as 2.14 dB less than dBd.

dBm (decibel to milliwatt): An absolute measurement of the power of an RF signal based on the definition of 0 dBm = 1 milliwatt (mW).

Distributed Coordination Function (DCF): A protocol defined in 802.11 that uses carrier sensing, backoff timers, interframe spaces and frame duration values to diminish collisions on the wireless medium.

Elevation Chart: A chart showing the radiation pattern of an antenna as viewed from the side antenna. Also called an E-Plane Chart or E-Chart.

Deauthentication Frame: A notification frame sent from an 802.11 STA to another STA in order to terminate a connection between them.

Decibel (dB): A logarithmic, relative unit used when measuring antenna gain, signal attenuation, and signal-to-noise ratios. Strictly defined as 1/10 of a bel.

Delay: The time it takes for a bit of data to travel from one node to another. Also called latency.

Delivery Traffic Indication Message (DTIM): A message sent from an AP to clients in the Beacon frame indicating that it has data to transmit to the clients specified by the AIDs.

Differentiated Services Code Point (DSCP): A Layer 3 QoS marking system. IP packets can include DSCP markings in the headers. Eight precedence levels, 0-7, are defined.

Diffraction: The bending of waves around a very large object in relation to the wave.

Direct-Sequence Spread Spectrum (DSSS): A modulation technique where data is coupled with coding that spreads the data across a wide frequency range. Provides 1 or 2 Mbps data rates in 802.11 networks.

Disassociation Frame: A frame sent from one STA to another in order to terminate the association.

Distributed Forwarding: See *Access Layer Forwarding*. Also called, *distributed data forwarding*.

Distribution System (DS): The system that connects a set of BSSs and LANs such that an ESS is possible.

Distribution System Medium (DSM): The medium used to interconnect APs through the DS such that they can communicate with each other for ESS operations using either wired or wireless for the DS connection.

Domain Name System (DNS): A protocol and service that provides host name resolution (looking up the IP address of a given host name) and recursive IP address lookups (finding the host name of a known IP address). Also, colloquially used to reference the server that provides DNS lookups.

Driver: Software that allows a computer to interact with a hardware device such as a WLAN adapter.

Duty Cycle: A measure of the time a radio is transmitting or a channel is consumed by a transmitting device.

Dynamic Frequency Selection (DFS): A setting on radios that dynamically changes the channel selection based on detected interference from radar systems. Many 5 GHz channels require DFS operations.

Dynamic Rate Switching (DRS): The process of reducing a client's data rate as frame transmission failures occur or signal strength decreases. DRS results in lower data rates but fewer transmissions required to successfully transmit a frame.

Encryption: The process of converting data into a form that unauthorized users cannot understand by encoding the data with an algorithm and a key or keys.

Enhanced Distributed Channel Access (EDCA): An enhancement to DCF introduced in 802.11e that implements priority based queuing for transmissions in 802.11 networks based on access categories.

Equivalent Isotropically Radiated Power (EIRP): The output power required of an isotropic radiator to equal the measured power output from an antenna in the intended direction of propagation.

Extended Rate Physical (ERP): A physical layer technology introduced in 802.11g that uses OFDM (from 802.11a) in the 2.4 GHz band and offers data rates up to 54 Mbps.

Extended Service Set (ESS): A group of one or more BSSs that are interconnected by a DS.

Extensible Authentication Protocol (EAP): An authentication framework that defines message formats for authentication exchanges used by 802.1X WLAN authentication solutions.

Fade Margin: An amount of signal strength, in dB, added to a link budget to ensure proper operations.

Fast Fourier Transform (FFT): A mathematical algorithm that takes in a waveform as represented in the time or space domain and shows it in the frequency domain. Used in spectrum analyzers to show real-time views in the frequency domain (Real-time FFT).

Fragmentation: The process of fragmenting 802.11 frames based on the fragmentation threshold configured. Fragmented frames have a greater likelihood of successful delivery in the presence of sporadic interference.

Frame Aggregation: A feature in the IEEE 802.11n PHY and later PHYs that increases throughput by sending more than one frame in a single transmission. Aggregated MSDUs or aggregated MPDUs may be supported.

Frame: A well-defined, meaningful set of bits used to communicate management and control information on a network or transfer payloads from higher layers. Frames are defined at the MAC and PHY layer.

Free Space Path Loss: The natural loss of amplitude that occurs in an RF signal as it propagates through space and the wave front spreads.

Fresnel Zones: Ellipsoid shaped zones around the visual LoS in a wireless link. The first Freznel zone should be 60% clear and would preferably be 80% clear to allow for environmental changes.

Frequency: The speed at which a waveform cycles in a second.

Full Duplex: A communication system that allows an endpoint to send data to the network at the same time as it receives data from the network.

Gain: The increase in signal strength in a particular direction. Can be accomplished passively by directing energy into a smaller area or actively by increasing the strength of the broadcasted signal before it is sent to the antenna.

Group Key Handshake: Used to transfer the GTK among STAs in an 802.11 network if the GTK requires updating. Initiated by the AP/controller in a BSS.

Group Master Key (GMK): Used to generate the GTK for encryption of broadcast and multicast frames and is unique to each BSS.

Group Temporal Key (GTK): Used to encryption broadcast and multicast frames and is unique to each BSS.

Guard Interval (GI): A period of time between symbols within a frame used to avoid intersymbol interference.

Half Duplex: A communication system that allows only sending or receiving data by an endpoint at any given time.

Hidden Node: The problem that arises when nodes cannot receive each other's frames, which can lead to packet collisions and retransmissions.

High Density: A phrase referencing a WLAN network type that is characterized by large numbers of devices requiring access.

Highly-Directional Antenna: An antenna, such as a parabolic dish or grid antenna, that has a high gain in a specified direction and a low beamwidth measurement as compared to semi-directional and omnidirectional antennas.

High Rate Direct Sequence Spread Spectrum (HR/DSSS): An amendment-based PHY (802.11b) that increase the data rate in 2.4 GHz from the original 1 or 2 Mbps to 5.5 and 11 Mbps while maintaining backward compatibility with 1 and 2 Mbps.

High Throughput (HT): An amendment-based PHY (802.11n) that increased the data rate up to 600 Mbps and added support for transmit beamforming and MIMO.

Hotspot: A term referencing a wireless network connection point that is typically open to the public or to paid subscribers.

Independent Basic Service Set (IBSS): A set of 802.11 devices operating in ad-hoc (peer-to-peer) mode without the use of an AP.

Institute of Electrical and Electronics Engineers (IEEE): A standardization organization that develops standard for multiple industries including the networking industry with standard such as 802.3, 802.11 and 802.16.

Intentional Radiator: Any device that is purposefully sending radio waves. Signal strength of the intentional radiator is measured at the point where energy enters the radiating antennas.

Interference: In WLANs, an RF signal or incidental RF energy that is radiated in the same frequencies as the WLAN and that has sufficient amplitude and duty cycle to prevent 802.11 frames from successful delivery.

Interframe Space (IFS): A time interval that must exist between frames. Varying lengths are used in 802.11 and a references as DIFS, SIFS, EIFS and AIFS in common use.

Internet Engineering Task Force (IETF): An open group of volunteers develops Internetworking standards through request for comments (RFC) documents. Examples include RADIUS, EAP and DNS.

Isotropic Radiator: A theoretical antenna that spreads the radiaton equally in every directon as a sphere. None exist in reality, but the concept is used to measure relative antenna gain in dBi.

Jitter: The variance in delay between packets sent on a network. Excessive jitter can result in poor quality for real-time applications such as voice and video.

Jumbo Frame: An Ethernet frame that contains more than 1500 bytes of payload and up to 9000 to 9216 bytes.

Latency: The time taken data to move between places. Typically synonymous with delay in computer networking.

Layer 1: The physical layer (PHY) that is responsible for framing and transmitting bits on the medium. In 802.3 and 802.11 the entirety of Layer 1 is defined.

Layer 2: The data-link layer that deals with data frames moving within a local area network (LAN). In 802.3 and 802.11, the MAC sublayer of Layer 2 is defined.

Layer 3: The network layer where packets of data are routed between sender and receiver. Most modern networks use Internet Protocol (IP) at Layer 3.

Layer 4: The transport layer where segmentation occurs for upper layer data and TCP (connection oriented) and UDP (connectionless) are the most commonly used protocols.

Lightning Arrestor: A device that can redirect ambient energy from a lightning strike away from attached equipment.

Line of sight (LoS): When existing, the visual path between to ends. RF LoS is different from visual LoS. RF LoS does not require the same clear path for the remote receiver to hear the signal. When creating bridge links, visual LoS is often the starting point.

Link Budget: The measurement of gains and losses through an intentional radiator, antenna and over a transmission medium.

Loss: The reduction in the amplitude of a signal.

MAC filtering: A common setting that only allows specific MAC addresses onto a network. Ineffective against knowledgeable attackers because the MAC address can be spoofed to impersonate authorized devices.

Management Frame: A frame type defined in the 802.11 standard that encompasses frames used to manage access to the network including beacon, probe request, prober response, authentication, association, reassociation, deauthentication and disassociation frames.

Master Session Key (MSK): A key derived between an EAP client and EAP server and exported by the EAP method. Used to derive the PMK, which is used to derive the PTK. The MSK is used in 802.1X/EAP authentication implementations. In personal authentication implementations, the PMK is derived from the pre-shared key.

Maximal Ratio Combining (MRC): A method of increasing the signal-to-noise ratio (SNR) by combining signals received on multiple radio chains (multiple antennas and radios).

Mesh: A network that uses interconnecting devices to form a redundant set of connections offering multiple paths through the network. 802.11s defined mesh for 802.11 networks.

Mesh BSS: A basic service set that forms a self-contained network of mesh stations.

Milliwatt (mW): A unit of electrical energy used in measuring output power of RF signals in WLANs. A mW is equal to 1/1000 of a watt (W).

Mobile User: A user that physically moves while connected to the network. The opposite of a stationary user.

Modulation: The process of changing a wave by changing its amplitude, frequency, and/or phase such that the changes represent data bits.

Modulation and Coding Scheme (MCS): Term used to describe the combination of the radio modulation scheme and the coding scheme used when transmitting data, first introduced in 802.11n.

MPDU: A MAC protocol data unit (MPDU) is a portion of data to be delivered to a MAC layer peer on a network and it is data prepared for the PHY layer by the MAC sublayer. The MAC sublayer receives the MSDU from upper layers on transmission and creates the MPDU. It receives the MPDU from the lower layer on receiving instantiation and removes the MAC header and footer to create the MSDU for the upper layers.

MSDU: A MAC service data unit is a portion of transmitted data to be handled by the MAC sublayer that has yet to be encapsulated into a MAC Layer frame.

Maximum Transmission Unit (MTU): The largest amount of data that can be sent at a particular layer of the OSI model. Typically set at layer 4 for TCP.

Multi-User MIMO (MU-MIMO): An enhancement to MIMO that allows the AP STA to transmit to multiple client STAs simultaneously.

Multipath: The phenomenon that occurs when multiple copies of the same signal reach a receiver based on RF behaviors in the environment.

Multiple Channel Architecture (MCA): A wireless network design using multiple channels strategically designed so that the implemented BSSs have minimal interference with one another.

Multiple Input/Multiple Output (MIMO): A technology used to spread a stream of data bits across multiple radio chains using spatial multiplexing at the transmitter and to recombine these streams at the receiver.

Narrowband Interference: Interference that covers a very narrow band of frequencies and typically not the full with of an 802.11 channel when used in reference to WLAN interferers.

Near-Far: A problem that occurs when a high powered device is closer to the AP in a BSS and a low powered device is farther from the AP. Most near-far problems are addressed with standard CSMA/CA operations in 802.11 networks.

Network Allocation Vector (NAV): The NAV is a virtual carrier sense mechanism used in CSMA/CA to avoid collisions and is a timer set based on the duration values in frames transmitted on the medium.

Network Segmentation: The process used to separate a larger network into smaller networks often utilizing Layer 3 routers or multi-layer switches.

Noise: RF energy in the environment that is not part of the intentional signal of your WLAN.

Noise Floor: The amount of noise that is consistently present in the environment, which is typically measured in dBm.

Network Time Protocol (NTP): A protocol used to synchronize clocks in devices using centralized time servers.

Octet: A group of eight ones and zeros. An 8-but byte. Sometimes simply called a byte.

Orthogonal Frequency Division Multiplexing (OFDM): A modulation technique and a named physical layer in 802.11 that provides data rates up to 54 Mbps and operates in the 5 GHz band. The modulation is used in all bands, but the named PHY operates only in the 5 GHz band.

Omni-Directional Antenna: An antenna that propagates in all directions horizontally. Creates a coverage area similar to a donut shape (toroidal). Also known as a dipole antenna.

Dipole Antenna: An antenna that propagates in all directions horizontally. Creates a coverage area similar to a donut (toroidal) shape. Also known as a omni-directional antenna.

Open System Authentication: A simple frame exchange, providing no real authentication, used to move through the state machine in relation to the connection between two 802.11 STAs.

Opportunistic Key Caching (OKC): A roaming solution for WLANs wherein the keys derived from the 802.1X/EAP authentication are cached on the AP or controller such that only the 4-way handshake is required at the time of roaming.

OSI (Open Systems Interconnection) Model: A theoretical model for communication systems that works by separating the communications process into seven, well-defined layers. The seven layers are Application, Presentation, Session, Transport, Network, Data Link and Physical.

Packet: Data as represented at the network layer (Layer 4) for TCP communications.

Passive Gain: An increase in strength of a signal by focusing the signal's energy rather than increasing the actual energy available, such as with an amplifier.

Passive scanning: A scanning (network location) method wherein a STA waits to receive beacon frames from an AP which contain information about the WLAN.

Passive survey: A survey conducted on location that gathers information about RF interference, signal strength and coverage areas by monitoring RF activity without active communications.

Passphrase Authentication: A type of access control that uses a phrase as the pass key. Also called personal in WPA and WPA2.

Phase: A measurement of the variance in arrival state between to copies of a wave form. Waves are said to be in phase or out of phase by some degree. The phase can be manipulated for modulation.

PHY: A shorthand notation for physical layer which is the physical means of communication on a network to transmit bits.

Physical (PHY) Layer: The physical (PHY) layer refers to the physical means by which a message is communicated. Layer one of the OSI model.

PLCP: Physical Layer Convergence Protocol (PLCP) is the name of the service within the PHY that receives data from the upper layers and sends data to the upper layers. It is the interaction point with the MAC sublayer.

PMD: Physical Medium Dependent (PMD) is the service within the PHY responsible for sending and receiving bits on the RF medium.

PMK Caching: Stores the PMK so a device only has to perform the 4 way handshake when connecting to an AP to which it has already connected.

Pairwise master Key (PMK): The key derived from the MSK, which is generated during 802.1X/EAP authentication. Used to derive the PTK. Used in unidirectional communications with a single peer.

PoE Injector: Any device that adds Power over Ethernet (PoE) to ethernet cables. Come in two variants, endpoint (such as switches) and midspan (such as inline injectors).

Point-to-Multipoint (PtMP): A connection between a single point and multiple other points for wireless bridging or WLAN access.

Point-to-Point (PtP): A connection between two points often used to connect two networks via bridging.

Polarization: The technical term used to reference the orientation of antennas related to the electric field in the electromagnetic wave.

Power over Ethernet (PoE): A method of providing power to certain hardware devices that can be powered across the Ethernet cables. Specified in 802.3 as a standard. Various classes are defined based on power requirements.

PPDU: PLCP Protocol Data Unit (PPDU) is the prepared bits for transmission on the wired or wireless medium. Sometimes also called a PHY Layer frame.

Preauthentication: Authenticating with an AP to which the STA is not intending to immediately connect so that roaming delays are reduced.

Pre-shared Key (PSK): Refers to any security protocol that uses a password or passphrase or string as the key from which encryption materials are derived.

Primary Channel: When implementing channels wider than 20 MHz in 802.11n and 802.11ac, the 20 MHz channel on which management and control frames are sent and the channel used by STAs not supporting the wider channel.

Probe Request: A type of frame sent when a client device wants information about APs in the area or is seeking a specific SSID to which it desires to connect.

Probe Response: A type frame sent in response to a probe request that contains information about the AP and the requirements of BSSs it provides.

Protected Management Frame (PMF): Frames used for managing a wireless network that are protected from spoofing using encryption. Protocol defined in the 802.11w amendment.

Protocol Analyzer: Hardware or software used to capture and analyze networking communications. WLAN protocol analyzers have the ability to capture 802.11 frames from the RF medium and decode them for display and analysis.

Protocol Decodes: The way information in captured packets or frames is interpreted for display and analysis.

PSDU: PLCP Service Data Unit (PSDU) is the name for the contents that are contained within the PPDU, the PLCP Protocol Data Unit. It is the same as the MPDU as perceived and received by the PHY.

PTK (Pairwise Transient Key): A key derived during the 4-way handshake and used for encryption only between two specific endpoints, such as an AP and a single client.

Quality of Service (QoS): Traffic prioritization and other techniques used to improve the end-user experience. IEEE 802.11e includes QoS protocols for wireless networks based on access categories.

QoS BSS: A BSS supporting 802.11e QoS features.

Radio Chains: A reference to the radio and antenna used together to transmit in a given frequency range. Multi-stream devices have multiple radio chains as one radio chain is required for each stream.

Radio Frequency (RF): The electromagnetic wave frequency range used in WLANs and many other wireless communication systems.

Radio Resource Management (RRM): Automatic management of various RF characteristics like channel selection and output power. Known by different terms among the many WLAN vendors, but referencing the same basic capabilities.

RADIUS: Remote Authentication Dial-In User Service (RADIUS) refers to a network protocol that handles AAA management which allows for authentication, authorization and accounting (auditing). Used in 802.11 WLANs as the authentication server in an 802.1X/EAP implementation.

RC4 (Rivest Cipher 4): An encryption cipher used in WEP and with TKIP. A stream cipher.

Real-Time Location Service (RTLS): A function provided by many WLAN infrastructure and overlay solutions allowing for device location based on triangulation and other algorithms.

Reassociation: The process used to associate with another AP in the same ESS. May also be used when a STA desires to reconnect to an AP to which it was formerly connected.

Received Channel Power Indicator (RCPI): Introduced in 802.11k, a power measurement calculated as INT((dBm + 110) * 2). Expected accuracy is +/- 5 dB. Ranges from 0-220 are available with 0 equaling or less than -110 dBm and 220 equaling or greater than 0 dBm. The value is calculated as an average of all received chains during the reception of the data portion of the transmission. All PHYs support RCPI and, though 802.11ac does not explicitly list its formulation, it references the 802.11n specification for calculation procedures.

Received Signal Strength Indicator (RSSI): A relative measure of signal strength for a wireless network. The method to measure RSSI is not standardized though it is constrained to a limited number of values in the 802.11 standard. Many use the term RSSI to reference dBm, and the 802.11 standard uses terms like DataFrameRSSI and BeaconRSSI and defines them as the signal strength in dBm of the specified frames, so the common vernacular is understandable. However, according to the standard, "absolute accuracy of the RSSI reading is not specified" (802.11-2012, Clause 14.3.3.3).

Reflection: An RF behavior that occurs when a wave meets a reflective obstacle large than the wavelength similar to light waves in a mirror.

Refraction: An RF behavior that occurs as an RF wave passes through material causing a bending of the wave and possible redirection of the wave front.

Regulatory Domain: A reference to geographic regions management by organizations like the FCC and ETSI that determine the allowed frequencies, output power levels and systems to be used in RF communications.

Remote AP: An AP designed to be implemented at a remote location and managed across a WAN link using special protocols.

Resolution Bandwidth (RBW): The smallest frequency that can be extracted from a received signal by a spectrum analyzer or the configuration of that frequency. Many spectrum analyzers allow for the adjustment of the RBW within the supported range of the analyzer.

Retry: That which occurs when a frame fails to be delivered successfully. A bit set in the frame to specify that it is a repeated attempt at delivery.

Return Loss: A measure of how much power is lost in delivery from a transmission line to an antenna.

RF Cables: A cable, typically coaxial, that allows for the transmission of electromagnetic waves between a transceiver and an antenna.

RF Calculator: A software application used to perform calculations related to RF signal strength values.

RF Connector: A component used to connect RF cables, antennas and transmitters. RF connectors come in many standardized forms and should match in type and resistance.

RF Coverage: Synonymous with coverage in WLAN vernacular. Reference to the BSA provided by an AP.

RF Link: An established connection between two radios.

RF Line of Sight (LoS): The existence of a path, possibly including reflections, refractions and pass-through of materials, between two RF transceivers.

RF Propagation: The process by which RF waves move throughout an area including reflection, refraction, scattering, diffraction, absorption and free space path loss.

RF Signal Splitter: An RF component that splits the RF signal with a single input and multiple outputs. Historically used with some antenna arrays, but less common today in WLAN implementations.

RF Site Survey: The process of physically measuring the RF signals within an area to determine resulting RF behavior and signal strength. Often performed as a validation procedure after implementation based on a predictive model.

Roaming: That which occurs when a wireless STA moves from one AP to another either because of end user mobility or changes in the RF coverage.

Robust Security Network (RSN): A network that supports CCMP/AES or WPA2 and optionally TKIP/RC4 or WPA. To be an RSN, the network must support only RSN Associations (RSNAs), which are only those associations that use the 4-way handshake. WEP is not supported in an RSN.

Robust Security Network Association (RSNA): An association between a client STA and an AP that was established through authentication resulting in a 4-way handshake to derive unicast keys and transfer group keys. WEP is not supported in an RSNA.

Rogue Access Point: An access point that is connected to a network without permission from a network administrator or other official.

Rogue Containment: Procedures used to prevent clients from associating with a rogue AP or to prevent the rogue AP from communicating with the wired network.

Rogue Detection: Procedures used to identify rogue devices. May include simple identification of unclassified APs or algorithmic processes that identify likely rogues.

Role-Based Access Control (RBAC): An authorization system that assigns permissions and rights based on user roles. Similar to group management of authorization policies.

RSN Information Element: A portion of the beacon frame that specifies the security used on the WLAN.

Request to Send/Clear to Send (RTS/CTS): A frame exchange used to clear the channel before transmitting a frame in order to assist in the reduction of collisions on the medium. Also used as a backward compatible protection mechanism.

RTS Threshold: The minimum size of a frame required to use RTS/CTS exchanges before transmission of the frame.

S-APSD: See *Automatic Power Save Delivery*.

Scattering: An RF behavior that occurs when an RF wave encounters reflective obstacles that are smaller than the wavelength. The result is multiple reflections or scattering of the wave front.

Secondary Channel: When implementing channels wider than 20 MHz in 802.11n and 802.11ac, the second channel used to form a 40 MHz channel for data frame transmissions to and from supporting client STAs.

Semi-Directional Antenna: An antenna such as a yagi or a patch that has a propagation pattern which maximizes gain in a given direction rather than an omni-directional pattern, having a larger beamwidth than highly directional antennas.

Service Set Identifier (SSID): The BSS and ESS name used to identify WLAN. Conventionally made to be readable by humans. Maximum of 32 bytes long.

Signal Strength: A measure of the amount of RF energy being received by a radio. Often specified as the RSSI, but referenced in dBm, which is not the proper definition of RSSI from the 802.11 standard.

Single Channel Architecture (SCA): A WLAN architecture that places all APs on the same channel and uses a centralized controller to determine when each AP can transmit a frame. No control of client transmissions to the network is provided.

Single Input Single Output (SISO): A radio transmitter that supports one radio chain and can send and receive only a single stream of bits.

Signal to Noise Ratio (SNR): A comparison between the received signal strength and the noise floor. Typically presented in dB. For example, given a noise floor of -95 dBm and a signal strength of -70 dBm, the SNR is 25 dB.

Space-Time Block Coding (STBC): The use of multiple streams of the same data across multiple radio chains to improve reliability of data transfer through redundancy.

Spatial Multiplexing (SM): Used with MIMO technology to send multiple spatial streams of data across the channel using multiple radio chains (radios coupled with antennas).

Spatial Multiplexing Power Save (SMPS): A power saving feature from 802.11n that allows a station to use only one radio (or spatial stream).

Spatial Streams: The partitioning of a stream of data bits into multiple streams transmitted simultaneously by multiple radio chains in an AP or client STA.

Spectrum Analysis: The inspection of raw RF energy to determine activity in an area on monitored frequencies. Useful in troubleshooting and design planning.

Spectrum Analyzer: A hardware and software solution that allows the inspection of raw RF energy.

Station (STA): Any device that can use IEEE 802.11 protocol. Includes both APs and clients.

Supplicant: In 802.1X, the device attempting to be authenticated. Also the term used for the client software on a device that is capable of connecting to a WLAN.

Sweep Cycle: The time it takes a spectrum analyzer to sweep across the frequencies monitored. Often a factor of the number of frequencies scanned and the RBW.

System Operating Margin (SOM): The actual positive difference in the required link budget for a bridge link to operate properly and the received signal strength in the link.

Temporal Key Integrity Protocol (TKIP): The authentication and key management protocol supported by WPA systems and implemented as an interim solution between WEP and CCMP.

Transition Security Network (TSN): A network that allows WEP connections during the transition period over to more secure protocols and an eventual RSN. An RSN does not allow WEP connections.

Transmit Beamforming (TxBF): The use of multiple antennas to transmit a signal strategically with varying phases so that the communication arrives at the receiver such that the signal strength is increased.

Transmit Power Control (TPC): A process implemented in WLAN devices allowing for the output power to be adjusted according to local regulations or by an automated management system.

Uncontrolled Port: In an 802.1X authentication system, the virtual port that allows only authentication frames/packets through to the network and, when authentication is successfully completed, provides the 802.1X service with the needed information to open the controlled port.

User Priority (UP): A value (from 0-7) assigned to prioritize traffic that correspond to different access categories for WMM QoS.

Virtual Carrier Sense: The 802.11 standard currently defines the Network Allocation Vector (NAV) for use in virtual carrier sensing. The NAV is set based on the duration value in perceived frames within the channel.

Voltage Standing Wave Ratio (VSWR): The Voltage Standing Wave Ratio is the ratio between the voltage at the maximum and minimum points of a sanding wave.

Watt: A unit of power. Strictly defined as the energy consumption rate of one joule per second such that 1 W is equal to 1 joule per 1 second.

Wavelength: The distance between two repeating points on a wave. Wavelength is a factor of the frequency and the constant of the speed of light.

Wired Equivalent Privacy (WEP): A legacy method of security defined in the original IEEE 802.11 standard in 1997. Used the RC4 cipher like TKIP (WPA), but implemented it poorly. WEP is deprecated and should no longer be used.

Wi-Fi Alliance: An association that certifies WLAN equipment to interoperate based on selected portions of the 802.11 standard and other standards. Certifications include those based on each PHY as well as QoS and security.

Wi-Fi Multimedia (WMM): A QoS certification created and tested by the Wi-Fi Alliance using traffic prioritizing methods defined in the IEEE 802.11e.

Wi-Fi Multimedia Power Save (WMM-PS): A power saving certification designed by the Wi-Fi Alliance and optimized for mobile devices and implementing methods designated in the IEEE 802.11e amendment.

Wireless Intrusion Prevention System (WIPS): A system used to detect and prevent unwanted intrusions in a WLAN by detecting and preventing rogue APs and other WLAN threats.

Wireless Local Area Network (WLAN): A local area network that connects devices using wireless signals based on the 802.11 protocol rather than wires and the common 802.3 protocol.

WPA-Enterprise: A security protocol designed by the Wi-Fi Alliance. Requires an 802.1X authentication server. Uses the TKIP encryption protocol with the RC4 cipher. Implements a portion of 802.11i and the older, no deprecated TKIP/RC4 solution.

WPA-Personal: A security protocol designed by the Wi-Fi Alliance. Does not require an authentication server. Uses the TKIP encryption protocol with the RC4 cipher. Also known as WPA-PSK (Pre-Shared Key).

WPA2-Enterprise: A security protocol designed by the Wi-Fi Alliance. Requires an 802.1X authentication server. Uses the CCMP key management protocol with the AES cipher. Also known as WPA2-802.1X. Implements the non-deprecated portion of 802.11i.

WPA2-Personal: A security protocol designed by the Wi-Fi Alliance. Does not require an authentication server. Uses the CCMP key management protocol with the AES cipher. Also known as WPA2-PSK (Pre-Shared Key).

Wi-Fi Protected Setup (WPS): A standard designed by the Wi-Fi Alliance to secure a network without requiring much user knowledge. Users connect either by entering a PIN associated with the device or by Push-Button which allows users to connect when a real or virtual button is pushed.

www.ingramcontent.com/pod-product-compliance
Lightning Source LLC
Chambersburg PA
CBHW050104220326
41598CB00043B/7377